# PLANETS, STARS, AND GALAXIES

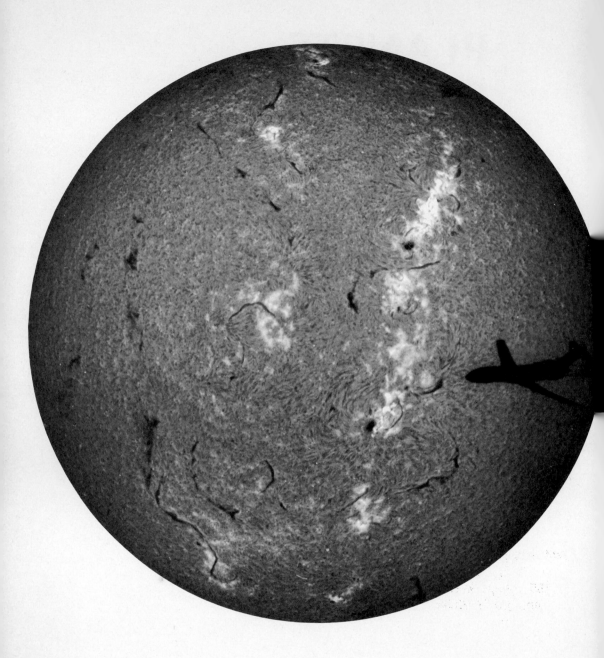

**Frontispiece**
A jet airplane, by chance, silhouetted against the sun. (Astronomy Department, Pennsylvania State University)

# PLANETS, STARS, AND GALAXIES

## AN INTRODUCTION TO ASTRONOMY

### THIRD EDITION

## STUART J. INGLIS
### ROGUE COMMUNITY COLLEGE

## JOHN WILEY & SONS, INC.
### NEW YORK • LONDON • SYDNEY • TORONTO

Star Plates 2–5 copyright © by the California Institute
of Technology and Carnegie Institution of Washington.

Library of Congress Catalog Card Number: 79–169164
ISBN 0–471–42740–3

Printed in the United States of America

10 9 8 7 6 5 4 3 2 1

To Isa

# PREFACE

The purpose of this book has not changed in this, the third edition; there is still a need to present the subject of astronomy to those who, without being well-grounded in mathematics, nevertheless feel that man must have some sense of his place in and his relation to the physical universe. With the ever-increasing ability of man to observe and to travel in the universe it becomes even more important that knowledge of astronomy not be restricted to the scientist.

The fear that nonscience students have of science is not new, it is generations old. However, that fear can be alleviated. There is nothing that arouses fear quite so much as the unknown. If the nonscience student is permitted to graduate from college without having taken a course in science designed for him, and in which he can achieve some degree of success, then he continues to be fearful and even suspicious of science. If he is obliged to take a course in which excessive use of mathematics causes him to lose interest, then his fears either remain or are intensified. Science will then remain an unknown, hiding behind a cloak of mathematics.

I have revised this book with continued interest in the liberal arts student. It is intended to be used as a college text for a one-semester course. It is my experience that the liberal arts student today is better prepared than students of ten years ago. He is better able to grasp concepts and in fact he is more apt to insist that he understand. Because of this, I have eliminated some of the less rigorous parts and introduced some new observations and ideas which demand more of the student. However, mathematics is still used very little in this book.

The chapters devoted to the solar system have been extensively rewritten. New techniques and new observations have stimulated study of the solar system. To live in the years of the inauguration of interplanetary space probes is exciting. This increased activity in astronomy and space science, however, makes it more difficult for the student to learn solely from the textbook, for textbooks cannot be entirely up to date. New advances are apt to occur even while the book is in press. The student relies on the teacher who must keep abreast of the latest developments. The student may also read from periodicals.

It is interest in the latest developments in astronomy which has encouraged me to let the student become more involved in the subject. I feel that he can greatly benefit by writing a short paper on a subject of his interest. With this in mind I have expanded the bibliography at the end of each chapter. These bibliographies are prepared for the student's use and so contain repeated references to *Sky and Telescope* and *Scientific American*.

In preparing this revision a number of professors have made suggestions which have been a great help to me. In particular I would like to thank Professor Donat G. Wentzel and his colleagues at the University of Maryland who graciously pointed out a number of errors in the first printing of the second edition. Mr. Jack Barss, formerly of Phillips Academy, Andover, Massachusetts, and Dr. Billy Smith of Chabot College, Hayward, California, also made very helpful suggestions. I thank Mr. Otto Barrett of Contra Costa College for his lasting contributions. But it is I who accept responsibility for any errors or misstatements which may remain in spite of their comments.

Perhaps it is only the author who understands the debt of gratitude he owes his family. My son, Jeff, and my daughters, Adrienne and Jennifer, have not only helped me directly, but been patient with me. My wife, Isa, has typed, read, and been very understanding. For all their help and consideration, I am truly grateful.

Stuart J. Inglis

Grants Pass, Oregon

# CONTENTS

ix

CHAPTER
5
CHAPTER

# PLANETS IN MOTION            108

CHAPTER
6
CHAPTER

# SATELLITES            130

CHAPTER
7
CHAPTER

# MINOR PLANETS, METEORITES, AND COMETS                              166

CHAPTER
**8**
CHAPTER

# THE AGE AND ORIGIN OF THE SOLAR SYSTEM     202

CHAPTER
**9**
CHAPTER

# THE SUN     220

CHAPTER 10 CHAPTER

# STABLE STARS 256

CHAPTER 11 CHAPTER

# NONSTABLE STARS                                              290

## CHAPTER 17 CHAPTER

# THE UNIVERSE AND RELATIVITY     446

# MAN AND
# THE HEAVENS

# CHAPTER CHAPTER 1

# MAN AND THE HEAVENS

Man is curious about the heavens. Other animals see the stars; apparently some birds navigate by the stars during their long migrations. Yet man alone is curious about the heavens.

Without this curiosity, scientists, for example, would not have devised better microscopes to investigate the tiny microbes they had discovered. Since these microbes are, in themselves, interesting creatures, the early studies were not done with any thought of helping or improving the lot of mankind. Those studies were conducted for their own sake—because of man's inherent curiosity.

Yet the early studies of microbes paid huge dividends when, a century and a half later, it was discovered that many diseases are caused by those tiny microscopic creatures. None of the early scientists could have foreseen the intimate relationship between the objects of their study and the health of mankind.

Astronomers study celestial objects, but not because it will directly benefit mankind. They study the universe because the universe is, of itself, interesting. The studies of the universe have not only had an immense influence on man's concept of himself and his world, but they have added to the understanding of physics, lately even to chemistry, and most recently to the possible understanding of the origin of life on Earth. Organic compounds that may have led to more complex organic molecules are being found in meteorites that were formed in the solar system but beyond the influence of the Earth.

If, the astronomer argues, the sun has planets about it, one with life on it, then why shouldn't other stars have planets and some of them have life? Would that life be like ours? Why not? What observations will help us answer these questions?

Although searching for life on other worlds is not the main occupation of astronomers, the recognition of such life will come only as a result of a more thorough understanding of the celestial objects, many of which we see as stars in the nighttime sky. Mankind wants to understand his environment, and that environment extends beyond our own backyard or the nearest polluted river. It extends to the microscopic beings and to the distant galaxies. If our horizons of interest shrink completely to our

**Chapter Opening Photo**
Stonehenge, an astronomical observatory in southern England, built perhaps 3800 years ago. (From *Stonehenge Decoded*, courtesy Gerald S. Hawkins, British Crown Copyright.)

**Figure 1-1**

The stars of the constellation Orion setting behind buildings of Lick Observatory. This photograph is a time exposure taken with a stationary camera. (Lick Observatory photograph)

mundane lives, then our philosophical outlook of life shrinks accordingly, and man becomes smaller.

In the beginning of astronomy, the celestial objects were thought to be gods who intimately affected the people's lives. It was noticed, for example, that the seasons, which meant so much to the agrarian peoples of 3000 B.C., could be correlated with the motions of the sun, and later with the risings and settings of certain stars (Figure 1-1). Accordingly the sun, the stars, and the moon, too, played an important role in ancient astronomy.

As the sun stays in the sky for a longer period each day the weather becomes warmer. It is time to plant the crops. To the people who lived in ancient Egypt the lengthening day meant something more: the Nile, the source of their livelihood, would overflow its banks and supply water to the ground so that their crops could grow.

To be able to predict when the Nile would overflow became a vital concern of the ancient Egyptians. Their priests noticed that when a very bright star, the one we call Sirius, rose above the horizon concurrently with the sun, they could expect the river to overflow within a matter of days. The importance of the Nile in their lives led the priests to study more carefully the motions of the sun.

They found that the sun seemed to travel not only across the sky, but through the field of stars as well. If the sun rose with Sirius one morning, it would rise a little later than Sirius the next. After about 91 days Sirius would be high in the sky when the sun appeared on the eastern horizon. After about 182 days Sirius would be setting in the west when the sun was just rising above the horizon in the east. After 365 days, however, the sun would again rise with Sirius, the Nile would overflow, and a new year would begin. Thus the calendar was born.

The month was born of the motions of the moon. There is a new moon every $29\frac{1}{2}$ days, which made it difficult to establish a year with an integral number of full months. This could not and cannot be done simply because $29\frac{1}{2}$ does not divide into 365 an integral number of times.

When Sirius and the sun rise concurrently they appear at different points on the eastern horizon, Sirius more to the southeast than the sun. Huge temples were built with long narrow corridors directed to the exact spot where Sirius would appear. These dark corridors eliminated most of the light of the dawn and enabled the priests to see Sirius more clearly. Through prolonged observation, coupled with this improved method, they came to realize that Sirius rose concurrently with the sun not once every 365 days but every $365\frac{1}{4}$ days.

If the priests set up a calendar with only 365 days for every year, in 4 years the sun would rise with Sirius one day later than their calendar predicted; in 8 years it would be 2 days late. After 100 years the Nile would overflow its banks 25 days later than the calendar date which had been set 100 years previously.

The priests realized that in order to correct this error in the length of the year they would have to add one day every fourth year and thus make their predictions more accurate. The calendar itself, however, was not changed until the time of Julius Caesar when the Romans took over this knowledge and officially adopted leap year.

In the course of the ensuing centuries it was found that the simple leap year overcorrected the calendar since the year is actually 365.2422 days long or a trifle less than $365\frac{1}{4}$ days. This inaccuracy of the Julian Calendar, as it is called, was corrected in the sixteenth century by Pope Gregory XIII. The Gregorian Calendar, the one we use today, drops leap year at the close of each century excepting every fourth century. The century year is not a leap year unless it is a multiple of 400. The year 2000 will be a leap year, the year 1900 was not.

The Egyptians had other troubles, too. After many years of observation they found that Sirius could no longer be used for predicting the overflow of the Nile. The year stayed the same length, but something had happened that made Sirius and the other stars appear to move very slowly in the sky. As a result, the priests had to find other stars that rose concurrently with the sun when the Nile was to overflow and new temples were built to observe these stars.

The Egyptians were well aware of the four cardinal points of the compass (North, South, East, and West). They noticed that each night the stars seemed to travel through the sky in circles centered on a common point in the north (Figure 1-2). For many years there was a star near that point, but this star did not stay there through the centuries; it moved just as Sirius seemed to move in the sky. After hundreds of years another star was at that special place in the sky, about which the stars seem to rotate.

If we look out into the sky at night we can locate this spot, the north celestial pole, because there is a fairly bright star near it. This star is now called the North Star, or Polaris. Just as 3,000 years ago a different star marked the north pole of the sky, 3,000 years from now Polaris will not mark the north celestial pole; in fact there will be no star so close to that point. Polaris, however, will again be the North Star 26,000 years from now. Even before the time of Christ the ancient astronomers knew about this apparent motion of the stars, a motion we now call precession, and whose cycle lasts not 24 hours nor $365\frac{1}{4}$ days, but 26,000 years.

**Figure 1-2**

Diurnal (daily) motion of the stars about the North Celestial Pole. (Lick Observatory photograph)

The brighter stars were named by the ancient peoples, chiefly the Arabs and the Greeks. Such names as Aldebaran, Betelgeuse, Deneb, and Vega are Arabic in origin. Others, such as Antares, Canopus, Procyon, and Pollux, were given to us by the Greeks. A few, such as Capella, were derived from the Latin.

These ancient peoples watched the sky closely and told stories about the figures they imagined to be represented by the configurations of stars. These figures, our constellations, divide our sky as counties divide a state. Examples are the Big Dipper (correctly called Ursa Major, the Big Bear); Orion, the hunter; Gemini, the twins (from which we derive our expression "by Jiminy"); Scorpius; and the archer, Sagittarius. Most of the constellations seen from the northern hemisphere are the subject of much myth and folklore.

The myth and folklore, generated by the ancient Babylonians, Greeks, and Arabs telling stories night after night under the clear desert sky, led to the idea that the stars are somehow related to our daily lives. And thus astrology was born. Astrology depends on the motions of the planets and on the ability to judge the time of birth of an individual. Consequently, as astrology became more popular, it stimulated interest in increasing the accuracy of astronomical observations.

Astronomy has become more precise, the fulfillment of its predictions today would be classed as miracles by the ancients. The accuracy and predicting ability of astrology, however, has not become noticeably better than it was 2,000 years ago.

The fainter stars were not named in antiquity but are now named according to the catalog in which they appear. For example, B.D. $+30°3639$ is a star in the catalog compiled in Bonn called the *Bonner Durchmusterung*. Nonstellar objects are also named after the catalog in which they are listed. M31 is in Messier's catalog, and N.G.C. 6523 is listed in the New General Catalog. In addition to catalog numbers some of the more famous nonstellar objects have names that derive from the appearance of the object (for example, the Crab nebula, the Owl nebula) or from the constellation in which they appear (for example, the Orion nebula).

Many people take pride in our particular modern civilization with its many miraculous inventions, but should we not take pride also in the other great civilizations that mankind has produced? Many people in different parts of the world have known a great deal about the very complicated motions of the sun, moon, stars, and planets without the aid of our technology.

In 2254 B.C. the early Chinese civilizations had a calendar that was essentially correct. In A.D. 1279, at the time of Kubla Khan, Chinese astronomers built the first observatory in the world to be equipped with instruments. The Mayas, of what is now Central America, had a calendar that accounted for not only the motions of the sun but also the motions of the moon, Venus, and Mars. This calendar was revised in the year A.D. 1091. The ancient Hindus had developed a calendar that included periods of time up to 4,320,000,000 years. This, interestingly enough, is about the age of the Earth.

Over 2,000 years ago the Phoenicians were using the stars to guide their ships on long voyages across the Mediterranean Sea, through the Pillars of Hercules, and up to the British Isles. The Polynesians made even more astounding sea voyages. Unlike the Phoenicians, they had no coast lines which could be used as landmarks. Sometimes the Polynesians sailed between islands more than a thousand miles apart.

This seems amazing to us who tend to look at any but our modern technical society as primitive. We have newspapers, radios, and printed calendars to tell us the day of the month. As a result the average person today is neither aware of nor concerned with the movements of celestial objects. The sun and stars are not our timepieces; the clock and the calendar serve us better. But both our calendar and the rate of the clock are determined by the motion of the Earth as it turns on its axis once a day and as it makes its complete trip around the sun each year. Modern man, living in large metropolitan areas ablaze with lights and shrouded with the gases and dust of civilization, scarcely sees the stars on the *celestial sphere* over his head. He hardly glances at the beauty of the celestial sphere, the myriads of stars, and the graceful motions of the planets. He is aware only of the sun and occasionally of the moon, but he is scarcely aware that both the sun and the moon move relative to the stars on the celestial sphere—the dome of the heavens.

Astronomy was a practical science in its early conception, when it was needed to navigate and to develop a calendar. In 1609, however, the approach to astronomy changed. It was then that Galileo first looked through his homemade telescope and saw the mountains on the moon, as well as four other moons revolving about Jupiter. He noticed that Mars and Saturn are not like the stars, for when they are magnified they appear as disks, while the stars remain but points of light. He saw that Venus is different from the other planets in that it goes through phases like the moon. Galileo was fascinated by these phenomena, and his interest in them for their own sake helped turn astronomy into a pure science.

Galileo was concerned with things so distant, so vast, and so mysterious that he followed the urge to learn about our universe even though what he learned might have no direct effect upon men's lives. He and the other astronomers who have followed have seen the unseen; they are revealing mysteries that seem almost beyond the comprehension of mankind.

Yet studying the vastness of space and time may have some effect, not on our daily life perhaps, but upon our thinking and concept of life. Little by little we realize that the Earth, once thought to be the very center of the entire universe, is really just a bit of matter revolving around one of billions of stars that form, together with vast volumes of cosmic dust and gas, a huge galaxy. Comprehending this, we must next confront the fact that this huge galaxy of stars, which we call the Milky Way system, is only one of billions of such galaxies, each with its billions of stars. We are lost in the universe so long as we speak in terms of miles and not light years; so long as we look with our eyes and not with huge telescopes; so long as we think of mankind as having control of vast amounts of energy when he controls the nucleus of the atom, and fail to compare this energy with that released by an exploding star.

Perhaps this concept of the universe will, after all, have some effect upon our daily lives: that Earth, with powerful mankind scurrying over its surface, is less than insignificant in size and influence when measured on a universal scale.

Perhaps, as man travels out into space and looks back, he will finally realize that the world's inhabitants are common occupants of the space craft we've called The Earth. We are all neighbors, for to travel to the far ends of the Earth is like going next door when compared with a trip to the moon. Perhaps, when we realize what close neighbors we are, we will learn to get along better one with another. Then space travel will have made a truly magnificent contribution to society.

## QUESTIONS AND PROBLEMS

1. England was late in adopting the Gregorian calendar. What trouble did this cause the British government?
2. Locate Polaris, and by watching the stars during the night verify the statements that each travels in a circle concentric with the North Celestial Pole.
3. With a little effort, a great deal of enjoyment can be gained by learning to recognize the more important constellations.

## FOR FURTHER READING*

Bernhard, H. J., et al., *New Handbook of the Heavens*, McGraw-Hill paperback, New York, 1964.

Hawkins, G. S., *Stonehenge Decoded*, Dell Publishing Co. paperback, New York, 1965.

Menzel, D. H., *A Field Guide to the Stars and Planets*, Houghton Mifflin Co., Boston, 1964.

Ronan, C., *The Astronomers*, Hill and Wang, New York, 1964.

Thiel, R., *And There Was Light*, Mentor, New York, 1960.

Emlen, S. T., "The Celestial Guidance System of a Migrating Bird," *Sky and Telescope*, p. 4 (July 1969).

Graubard, M., "Under the Spell of the Zodiac," *Natural History*, p. 10 (May 1969).

Kals, W. S., "Polynesian Navigation," *Sky and Telescope*, p. 358 (June 1967).

Price, D. J. de S., "Unworldly Mechanics," *Natural History*, p. 9 (March 1962).

Price, D. J. de S., "Astronomy's Past Preserved at Jaipur," *Natural History*, p. 48 (June–July 1964).

"The Sun," *Natural History*, p. 38 (April 1963).

"Visits to Stonehenge and Herstmonceux," *Sky and Telescope*, p. 197 (Oct. 1970).

*A list of books and articles appears at the end of each chapter. This list is arranged for the student's convenience. Books appear first, then full articles with authors, and finally short articles without authors.

**CHAPTER 2 CHAPTER**

# BASIC TOOLS AND METHODS

The astronomer's study of the distant celestial objects is fascinating in part because the distances of the objects are so great. Seen from another standpoint, however, those vast distances act as a barrier that limits the astronomer in his studies. Distant galaxies can be studied only by the light and other forms of radiation that they emit. The only extraterrestrial materials that man has handled and studied in his laboratories are the meteorites (bits of material that move in orbits about the sun much as the Earth does and that occasionally fall onto the Earth), and the rocks brought back from the moon.

Since the astronomer is dependent upon light we must first consider how it is possible to learn so much about the stars, planets, and other astronomical objects from it alone. To understand this we will need to know some of the characteristics of light. These will be taken up one at a time.

## 2.1 Some Characteristics of Light

**a. Light Travels in a Straight Line.** An important characteristic of light is its property of traveling in a straight line. We observe this straight-line travel in the laboratory and on the Earth's surface, and if we assume that light travels in a straight line in free space, beyond the Earth's atmosphere and beyond the solar system, then a star's position in the sky can be determined by noting the direction from which the light comes; that is, by pointing the telescope to the star. Although this may sound very simple, we shall see in later chapters that it is possible for complications to arise even at this point.

**b. Wave Characteristics of Light.** Many experiments lead us to believe that light has wave-like characteristics; in some respects it behaves like a wave caused by a pebble dropped into a smooth pond. A series of these waves on a surface of water is called a wave train (Figure 2-1). Each wave in that train is made up of one crest and one trough. The length of this crest and trough is called the *wavelength*. Waves on the surface of a pond might have a length of 1 centimeter (1 inch equals 2.54 centimeters,

**Chapter Opening Photo**

Water about a telescope steadies the atmosphere permitting better photographs of the sun. (Big Bear Solar Observatory, Hale Observatories, Big Bear City, California.)

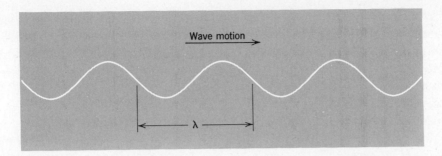

**Figure 2-1**

or 1 centimeter (cm) equals about 0.4 inch). The wavelength of light is much shorter.

Although it is difficult to measure, the wavelengths of light have been measured with great precision. The wavelength of violet light is about $4.0 \times 10^{-5}$ cm* (about $1.6 \times 10^{-5}$ in.); the wavelength of red light is about $7.0 \times 10^{-5}$ cm (about $2.8 \times 10^{-5}$ in.). The wavelengths of all the other colors lie between these two extremes. Since the wavelengths of light are so short a unit of measure, the *angstrom* (abbreviated Å) is generally used. One angstrom equals $10^{-8}$ cm, so the wavelength of violet light is about 4,000 Å; the wavelength of red light is about 7,000 Å.

**c. Polarized Light.** The surface of the water on which a wave train passes moves at right angles to the direction of wave travel—but only up and down, not right and left. The wave travels along the surface; a cork on the surface will bob up and down as the wave passes.

If the medium through which the wave moves (such as water) vibrates in a direction at right angles to the line of wave travel, the wave is called

---

*When dealing with very large and very small numbers, it is much easier to use the exponential system for keeping track of the number of zeros. This system is used throughout this book:

$$1 \times 10^{-10} = 0.000,000,000,1$$
$$4.0 \times 10^{-5} = 0.000,04$$
$$2.1 \times 10^{-2} = 0.021$$
$$10^{-1} = 0.1$$
$$10^{0} = 1.0$$
$$3 \times 10^{1} = 30$$
$$6.1 \times 10^{2} = 610$$
$$10^{3} = 1000$$
$$4.0 \times 10^{6} = 4,000,000$$
$$5.5 \times 10^{9} = 5,500,000,000$$

a *transverse wave*. Light is a transverse wave, but it vibrates in all directions which are perpendicular to the direction of wave travel. A line can be perpendicular to another line, or it can be perpendicular to a plane. A line perpendicular to a plane is perpendicular to all lines in that plane passing through the point of intersection of the plane and line of wave travel (Figure 2-2a).

If the direction of vibration within this plane is restricted, the wave is said to be partially polarized, that is, to vibrate in one direction more than another (Figure 2-2b). A wave on the surface of water is completely polarized, for it vibrates in only one direction (Figure 2-2c). Light emitted from an incandescent lamp is unpolarized, and light from the sun is un-

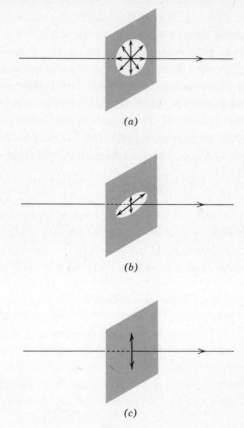

(a)

(b)

(c)

**Figure 2-2**

(a) Unpolarized light; (b) partially polarized light; (c) plane polarized light.

polarized. Light reflecting from a surface is partially polarized; the extent of polarization depends upon the angle of reflection and the material from which the light reflects. Consequently, studies of the extent of polarization of light reflected from the surfaces of the moon and of the planets give us some idea of the nature of their surfaces.

## 2.2 The Velocity of Light

Light travels at a very high velocity; a velocity so high that it becomes of interest to learn how it was first measured. Before it was measured, no one knew whether the velocity of light was finite or infinite. If it were finite, light would require a certain length of time to travel from one point to another. If the velocity were infinite, however, light would need no time at all to travel from one point to another.

The first measurement was made as early as 1675 by an astronomer named Roemer, who at the time was studying the motions of the satellites of Jupiter. Each satellite goes around Jupiter in a set length of time, called a *period*. The period of each satellite had been determined and Roemer was trying to make a more accurate determination. He timed one of the satellites from the moment it emerged from the shadow of Jupiter until it emerged once again, the satellite having made one complete revolution during this time.

Roemer realized that he would be able to determine a period more accurately if he timed many periods and took their average. This involved him in observations over a considerable length of time, in the course of which a curious phenomenon came to his attention. For the first six months the satellite seemed to emerge from Jupiter's shadow somewhat later than he had anticipated. Then for the second six months it emerged a little earlier until it was back on schedule again.

Roemer explained this by a hypothesis that combined the motion of the Earth around the sun with the assumption that light must have a finite velocity (Figure 2-3). When the Earth was in position 1, he obtained a good value for the length of the period. But as the Earth moved on to position 2 and then to position 3 the satellite appeared to emerge a little later because the light that it reflected and by which alone it could be observed had to travel farther each time to reach the Earth as the Earth receded from Jupiter. (Jupiter moves so much more slowly than the Earth that we may consider it stationary for purposes of this explanation.)

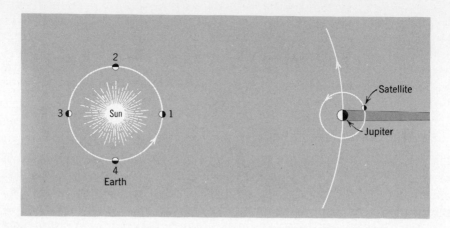

**Figure 2-3**

The relative positions of the Earth, Jupiter, and one of Jupiter's satellites used by Roemer to measure the velocity of light.

Since the satellite appeared to emerge later each time, the length of its period seemed to be increasing.

As the Earth moved from position 3 to position 4 and back to 1 again, the satellite appeared to emerge from the shadow somewhat earlier because the light had a shorter distance to travel each time. Thus during the second six months the period appeared to get shorter until the Earth had returned to the same relative position between the sun and Jupiter, and the period was again what it had been at the beginning of the study.

Although Roemer was able to demonstrate that light has a finite velocity (and at the same time that the Earth revolves about the sun), he had no more than a rough idea of the diameter of the Earth's orbit. Consequently, he could do no more than give an estimation of the speed of light. The speed now accepted is 186,284 miles per second.

From this discussion it is evident that when we observe any celestial object we observe not the object itself as it is at the moment of observation, but as it was at the moment the light that we observe left it. That is, what we see is the object itself at a moment in the past. The interval between that moment and the present is measured by the length of time it has taken the light to travel from the object to the Earth.

## 2.3 Reflection, Refraction, and the Telescope

Light, then, travels at a very high velocity and generally in a straight line. The two most obvious exceptions to the generalization that it travels in a straight line are reflection and refraction. *Reflection* occurs when light is reflected from a mirror or some other object; *refraction*, when light travels from air into some other substance, such as glass or water. This can be seen when a pencil is placed in a glass of water. The pencil appears to bend at the surface of the water. That the pencil itself is not bent can be shown by removing it. It is the light bending as it emerges from the water that creates the illusion that the pencil is bent. Refraction results from a change of velocity when light passes from one medium into another. If its velocity is less in the second medium, the direction of travel is bent toward the perpendicular to the surface. Light travels more rapidly in air than in glass, so it bends toward the perpendicular when passing from air into glass. Conversely it bends away from the perpendicular when it passes from glass into air.

Only because we can change the direction of light's travel are we able to make telescopes. These two ways of changing the direction of travel enable us to build two basically different types of telescopes.

**a. The Refracting Telescope.**   The first telescope was apparently built by a Dutch spectacle-maker, Hans Lippershey. This telescope, employing lenses, was a *refracting telescope*. On October 2, 1608, Lippershey applied to the government officials for a 30-year title of rights and pension on his invention, and the government officials recognized its military value. But news of the invention reached Galileo in Italy in May of 1609, Galileo did not learn any of the details of the instrument, but after reviewing the laws of refraction he was able to construct his first telescope.

The most common refracting telescopes are built according to the simplified drawing in Figure 2-4. The main lens of a telescope is called the *objective lens*. It is the purpose of the objective lens to gather the parallel rays that go to make up a parallel beam of light and to focus them at one point.

The light from a single star (*a* or *b*) enters our figure from the right. Since the star is so far away the light travels in an essentially parallel beam. However, the beam of light from star *a* is not parallel to the beam from star *b*; therefore these two beams will be brought to a focus at two

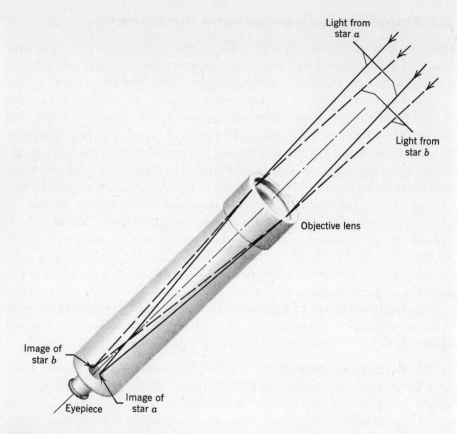

**Figure 2-4**

Schematic diagram of a refracting telescope.

different points. All such points from many stars define what is called the *focal plane*. If the objects observed with the telescope are far away, such as all astronomical objects, then the distance between the focal plane and the lens is called the *focal length* of that lens. If objects closer to the telescope are viewed, their images are located at a greater distance from the lens than the focal length.

It is possible to place a photographic plate at the focal plane and take a picture of a star field or a planet. The telescope is then a very large and specialized camera.

**Figure 2-5**

The 36-in. refracting telescope of the Lick Observatory, Mount Hamilton, California. (Lick Observatory)

If the telescope is pointed toward a planet, the light from any one point on the planet's surface will be traveling in a parallel beam and all the light from this given point will be brought to sharp focus. This is true for every point on the surface of the planet. All these focused points considered together are called an *image* of the planet. The image is real in the sense that a piece of paper could be placed at the focal plane and the image could be seen on the paper. Again, a photographic plate can be placed at the focal plane and a photograph taken of the planet; or an eyepiece, which is essentially a magnifying glass, can be placed just beyond the focal plane and the image, thus magnified, may be seen by the eye.

The larger the objective lens, the more light it can gather and thus the fainter the object that can be seen or photographed. It is because of this that astronomers have built larger and larger telescopes and have seen correspondingly fainter stars. The three biggest refracting telescopes in the world today are the one at Yerkes Observatory in Wisconsin which has an objective that is 40 in. in diameter; the one at Lick Observatory in California with an objective 36 in. in diameter (Figure 2-5); and a 33-in. refractor at the Meudon Observatory in Paris, France.

**b. The Reflecting Telescope.** We have noted that the direction of light can be changed by reflection as well as by refraction. It was Sir Isaac Newton who, in developing the first reflecting telescope, put this principle of optics to use. A schematic drawing of a reflecting telescope is shown in Figure 2-6. Light coming in a parallel beam from a distant star strikes the concave mirror and is converged and focused to a point, *F*. This point is called the *prime focus*. In order to look into an eyepiece placed near the prime focus of this telescope, however, the head of the observer must obstruct the incoming light and thus reduce the amount of light. So Newton placed a small flat mirror in the beam of light converged by the concave mirror and reflected that converging beam to a point, *F'*, called the *Newtonian focus*. The Newtonian focus is outside the telescope tube and therefore readily available to the astronomer. It is true that even this small mirror prevents some light from reaching the objective, but since it is comparatively small little light is lost.

There are many large reflecting telescopes in the world (Figure 2-7). The three largest are so large that instead of a small flat mirror at the prime focus they have a little cage where the observer rides. The two largest are: the 236-in. telescope in the Caucasus Mountains in Russia and the 200-in. of the Hale Observatories on Mount Palomar in California.

c. **Refractors versus Reflectors.** It is evident that many reflecting telescopes are a good deal larger than the biggest refractors. There are good reasons for this. To bring the light to a sharp focus, the lens or mirror must be ground and polished very accurately. Since the lens can be supported only at the circumference, it bends under its own weight as the telescope is placed in different positions. If the diameter of the lens is greater than about 40 in., this bending becomes excessive, the surface of the lens becomes distorted too much, and the image of the star will not be sharply focused. A mirror, on the other hand, can be supported not only on the edge (as the lens) but also on the back, so bending under its own weight can be prevented. Furthermore, since the light must pass *through* the glass of a lens, that glass must be more nearly perfect throughout. No bubbles or uneven streaks can be tolerated. The mirror, on the other hand, has a thin coat of aluminum on the concave surface to give better reflection and less loss of light. Light does not pass through the glass and consequently the glass does not have to be as nearly perfect except over its concave surface. In addition, there are at least two surfaces to be ground and polished on a lens and only one surface on a mirror; and since grinding and polishing is a long tedious job, the mirror is easier to build.

If the light does not even touch the glass of the mirror, why must telescopic mirrors be made of glass? There are several reasons. Since glass is not crystalline but amorphous, and since it can be made homogeneous throughout, its surface can be ground and polished more accurately than that of most other substances. Another factor is that glass, especially Pyrex, expands and contracts less with changes in temperature than most other substances. This is important not only because expansion and contraction change the mirror's *focal length*, the distance between it and its prime focus; but more important, expansion and contraction change the shape of the mirror's surface causing the image to become blurred.

If the shape of the concave surface upon which the aluminum coating is placed is a portion of a sphere, the stellar images will not be sharp points of light. This failure of a spherical mirror is called *spherical aberration*, and is corrected by making the concave surface parabolic in shape. (For the definition of a parabola see p. 163).

d. **Chromatic Aberration.** There is another reason why the mirror is preferred in most telescopes. Unfortunately, refraction of light depends on the color of light so that blue light is brought to a focus which is closer to the lens than the focus for red light. This is seen in Figure 2-8a. If we

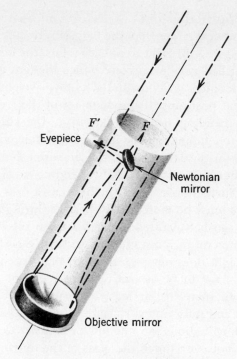

**Figure 2-6**

Schematic diagram of a reflecting telescope.

place an eyepiece at the focus for the blue light, the star will have a blue center with a red ring around it. If we photograph the star and place the film at the blue focus, the picture we take will not be a sharp point but a blurred image. This phenomenon is called *chromatic aberration* and is a great annoyance to astronomers. Fortunately it can be corrected somewhat by placing a second lens directly behind the primary lens. If this second lens is concave (one that diverges a parallel beam of light) and made of a different kind of glass, it will cause the beam of light to converge less rapidly; it causes the blue light in the beam to converge even less rapidly than the red light so that the foci of all colors are brought more closely together (Figure 2-8*b*). Of course an additional lens increases the cost of the telescope.

The correction is never perfect; the refracting telescope always has some chromatic aberration, for only two colors can be brought to the same focus

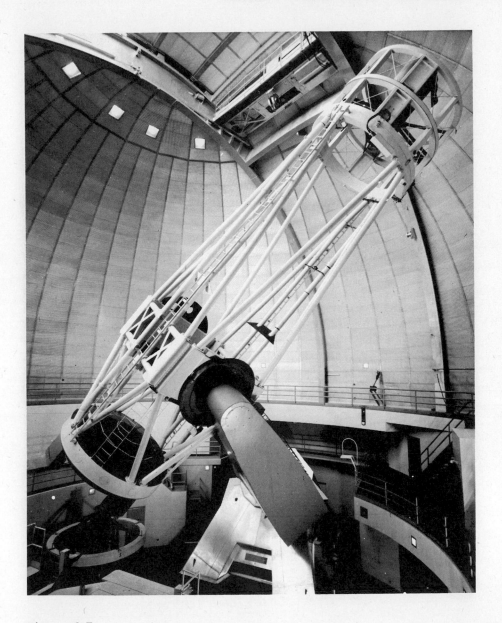

**Figure 2-7**
The 120-in. reflecting telescope at the Lick Observatory, Mount Hamilton.
(Lick Observatory)

by this two-lens combination. Reflection, however, does not depend on the color of the light; consequently, all colors are brought together at one focus.

**e. The Schmidt Telescope.**   Very little astronomical research is done visually, since the eye can be deceived and the memory can be faulty. If we photograph the stars we then have a permanent record that can be studied at any later time. Since astronomers work with very faint light it is desirable to build a telescope that will enable us to photograph very faint objects without too long an exposure. A special type of telescope, called the Schmidt in honor of its inventor, can not only cut down the exposure time but also includes a larger portion of the sky in one photograph. A photograph from a Schmidt telescope may cover a portion of the sky about 7° on each side, whereas a photograph from a reflecting telescope may cover only about 1° on each side.

The Schmidt telescope, invented in 1930, uses a spherical mirror to reflect the light to the prime focus. However, since a spherical mirror causes spherical aberration, Schmidt introduced a *correcting plate* at the upper opening of the telescope tube to correct the aberration.

Each of these major types of telescopes, the refractor, the reflector and the Schmidt, has certain advantages over the others. If the surface detail of the planets is to be studied, a large refractor is best. If a larger portion

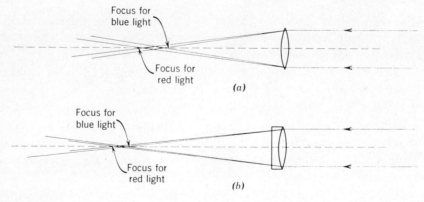

**Figure 2-8**

(*a*) Chromatic aberration results from the fact that a single lens will not focus the light of all colors to the same point. (*b*) With the addition of a second lens chromatic aberration can be partially corrected.

of the sky is to be taken with a relatively short exposure time, a Schmidt telescope is used. If a very faint object is to be examined in detail then a large reflector should be utilized.

To compare the photograph taken by a Schmidt and a large reflector, compare the photographs in Figure 2-9. The top photograph was taken with the 18-in. Schmidt on Mount Palomar and the bottom photograph was taken with the 100-in. reflector on Mount Wilson. These are photographs of the Orion nebula.

The telescope is the astronomer's basic tool and with it he can take pictures of objects in the sky. But the light that arrives from such distant objects can tell us more than just what they look like and where they are located. Again we must refer to Newton's work, which provides the basis for so much astronomy.

## 2.4 The Spectrograph

In discussing refraction we found that the angle through which light is refracted depends on the color of the light. Newton found that by passing light through a prism instead of a lens not only would the direction of the light's travel be changed but it would also be broken up into various colors. With the lens we called this chromatic aberration and considered it a nuisance. But this nuisance is the basis for one of science's very useful tools.

Light from the sun and all celestial objects is not a single color but a mixture. The light from most stars is nearly white. A prism breaks up white light into its component colors, or *spectrum*. A rainbow is a spectrum, the raindrops act as prisms. In studying the spectrum of the sun, another astronomer, Fraunhofer, found that if he put the image of the sun on a narrow slit in front of the prism the spectrum contained a number of dark lines. A narrow slit can be made by placing the cutting edge of two razor blades very close together. This will allow only a very narrow beam of light to pass through (Figure 2-10). The narrow beam will diverge until it passes through a first lens, called the *collimator*. It then becomes a parallel beam of light. The parallel beam is sent through a prism, which disperses the light into its component colors. It then travels through another lens by which it is focused on a photographic plate. Thus the spectrum is photographed. The entire instrument is called a *spectrograph*.

Each dark line that Fraunhofer found is an image of the narrow slit.

**Figure 2-9**

The Orion nebula (*above*) as photographed with the 18-in. Schmidt camera on Mount Palomar. The Orion nebula (*below*) as photographed with the 100-in. telescope on Mount Wilson. (Photographs from the Hale Observatories)

If he had placed the original beam of light on a small circular hole he would have found small dark circular spots on the spectrum rather than dark lines. If the slit is made wider, the lines become wider. Fraunhofer did not know the cause of these dark lines but he did letter them *A, B, C,* etc., according to their position in the spectrum. This lettering is still used. It remained for another man, Kirchhoff, to discover what in the sun caused the dark lines.

## 2.5 Three Kinds of Spectra

**a. Continuous Spectrum.**   Kirchhoff found that a solid object like red-hot iron will give only a *continuous spectrum*, that is, a continuous array of colors from violet through blue, green, yellow, and orange to red. An incandescent gas under pressure or an incandescent liquid will also emit a continuous spectrum.

**b. Bright-Line Spectrum.**   On the other hand, a gas under low pressure, that is, at much less than atmospheric pressure, when caused to emit light (as in a neon sign) will not give a continuous spectrum but a series of bright lines on a dark background. Such a spectrum is called a *bright-line spectrum* and is seen in Figure 2-11*a*. Here again each line is an image of the slit. If a circular hole had been used, a series of differently colored round spots would have been seen. It is easier, however, to work with thin lines.

Each gas has its own distinctive spectrum composed of a certain number

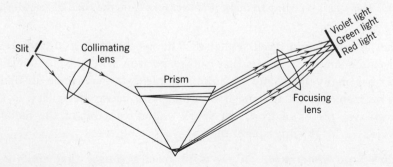

**Figure 2-10**

A schematic diagram of a spectrograph. The spectrograph is attached so that the slit is in the focal plane of the telescope's objective.

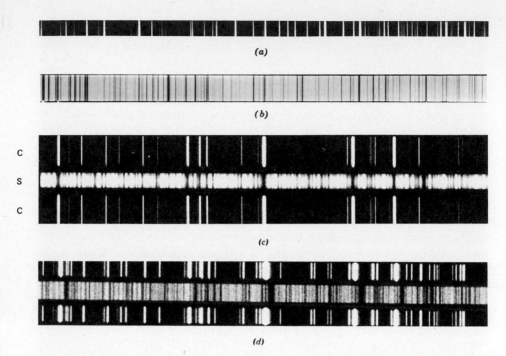

**Figure 2-11**

(a) A bright-line spectrum of iron. (b) A dark-line stellar spectrum. (c) A stellar dark-line spectrum S has a comparison (bright-line) spectrum C above and below it. The iron lines in the stellar spectrum line up with the iron lines in the comparison spectra. (d) The iron lines in this stellar spectrum are each shifted a bit to the right, to the longer wavelength, from the lines in the iron comparison spectra. This star is receding from us. (Photographs from the Hale Observatories)

of lines arranged in a fixed sequence. The spectrum of hydrogen (Figure 3-3) is quite simple, whereas that caused by iron when vaporized (Figure 2-11a) has many lines and is quite complex. It may seem difficult to vaporize iron, but all that need be done is to cause electricity to spark between two pieces of iron; the spark vaporizes a bit of the metal and thus gives a bright-line spectrum.

**c. Dark-Line Spectrum.** If the light from a source that emits a continuous spectrum is allowed to pass through a large container full of a gas that is cooler than the source and under low pressure, the continuous spectrum from the source is crossed by a series of dark lines. These dark lines are in the same position as the bright lines for the same gas when incandescent. This type of spectrum is known as a *dark-line* or *absorption spectrum* (Figure 2-11*b*).

It should be noted that these lines can originate only with a gas under low pressure. Even the smallest dust particles between the observer and the source of a continuous spectrum will not yield absorption lines.

The work of Newton, Fraunhofer, and Kirchhoff enables us to determine what chemical elements are in the atmosphere of a star. An atmosphere acts like a relatively cool gas that absorbs the light from a hotter source which emits a continuous spectrum. Thus the sun's atmosphere caused the dark lines that Fraunhofer first saw.

To determine what chemicals go to make up a star's atmosphere we can take a bright-line source such as a spark between two pieces of iron, and let its light travel through the spectrograph parallel to the star's light. This gives the comparison spectra shown in Figure 2-11*c* and *d*. The star's spectrum (an absorption spectrum) is in the middle while the two comparison spectra (both bright line spectra) are on the top and bottom. We know that the lines in the comparison spectra are caused by iron and we can see that there are lines in the star's spectrum that are located in nearly the same position. This tells us that the star's atmosphere must contain iron.

Since there are more than 100 elements it would not be practicable to form a comparison spectrum for each one of them. There must be an easier means of identifying the dark lines in a stellar spectrum. The wavelengths of the spectral lines of iron can be measured, and by using them as a reference scale we can measure the wavelengths of all the lines in a star's spectrum. The physicists and chemists, too, are interested in the spectra of elements, and they have measured the wavelength of spectral lines for nearly all of them. The astronomer need only take the wavelengths of the star's spectral lines and match them against the known wavelengths of the spectral lines of the elements.

## 2.6 Measuring Wavelengths of Spectral Lines

"Measuring" the wavelength of a stellar spectral line consists in determining its *position* in the spectrum with respect to the positions of the lines in the comparison spectra whose wavelengths are known (see Figure 2-12). In practice, however, this is not quite such an easy matter. There

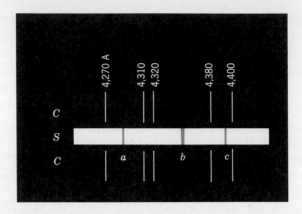

**Figure 2-12**

Estimate the wavelength for the three stellar lines *a*, *b*, and *c*.

are many, many lines in the spectrum of almost every element, and the gases that make up the atmospheres of the stars are not under the same conditions as gases in a laboratory. Consequently, there are always differences that complicate the situation immensely.

There are other difficulties, too. Notice the placement of the stellar iron lines in Figure 2-11*d* with respect to the iron lines in the comparison spectra. The stellar iron lines are all shifted to the right (the red). It might be said that the lines we have attributed to iron are not iron lines but something else very much like iron. But there is no other element that has spectral lines in this same sequence, so there must be some other explanation for the shift.

## 2.7 The Doppler Effect and Radial Velocity

For the explanation of the shifting of spectral lines let us look to sound, which also travels by wave motion. Although the nature of sound waves is quite different from that of light waves the effect we wish to discuss is very similar in both.

The wavelength of light determines its color; the wavelength of sound determines its pitch. A very high sound has a short wavelength and a low-pitched sound has a long wavelength.

Let us consider the sound of the horn of a car as it comes toward us, goes past, and travels away from us. The horn emits a sound made up of a series of wave crests and troughs that results in a characteristic pitch if the car is standing still. Since the velocity of sound can be considered constant it is not affected by the speed of the car. Let us further consider the crests only, one at a time. The horn will emit a certain crest which will travel toward our ear with the velocity of sound. If the horn is approaching us, it will move a little closer to our ear before it emits the next crest. Consequently, the next crest will reach us a little more quickly than if it were emitted when the car is standing still. Not only are these two crests closer together but all the crests emitted will be closer together as long as the car is approaching us. Since these crests are closer together when they reach our ear, the pitch of the horn will be higher. Nothing need be said about how far away the car may be, for we are concerned not with its distance but with its velocity toward us.

When the car is directly opposite it will be neither approaching us nor receding from us, and the crests will be spaced the same as if the car were standing still. The pitch of the sound will be the same as if both the car and the listener were at rest.

As the car recedes from us it will move a little farther away after it emits each crest, and the crests will be spaced a little farther apart, that is, the wavelength will be a little longer. In this case fewer crests will reach our ear each second, and the pitch will become lower.

The change in pitch, called the *Doppler effect*, is often effectively mimicked by the entertainer who assumes the role of a radio sportscaster at an automobile race when he imitates the sound of the racing cars as each passes his broadcasting booth.

The Doppler effect applied to light means that the color—wavelength— of the light will change if there is relative motion between the source and the observer. If the measurements of wavelength of the spectral lines of a stellar spectrum are all a bit too short, we assume that the star and the earth are approaching each other. The entire spectrum is shifted to the short wavelength, which for the visible part of the spectrum is toward the violet.

If the measurements of the wavelength of the stellar spectral lines are all a bit too long, then we declare that that star and the earth are receding from each other. The entire spectrum is shifted toward the longer wavelength, which for light is toward the red.

The amount which the wavelength changes $\Delta\lambda$ ($\Delta$, Greek delta used to designate a change in some quantity; and $\lambda$, Greek lambda, used to represent wavelength) is proportional to the relative velocity between the source (it may be a star or a planet) and the observer. The change in wavelength $\Delta\lambda$ is also proportional to the unaltered wavelength $\lambda_0$. That is, a spectral line with an unshifted wavelength $\lambda_0 = 8,000$ Å will be shifted twice as far as a spectral line with a wavelength of 4,000 Å.

The change in wavelength also depends upon the velocity of light, $c$. These rather lengthy statements can be incorporated into a short simple proportion:

$$\frac{\Delta\lambda}{\lambda_0} = \frac{v}{c}$$

For example, if a spectral line with an unshifted wavelength of 4,860 Å ($4.86 \times 10^3$ Å) is shifted by $+1.5$ Å, we can determine the relative velocity of the source by using this equation

$$v = (c)\frac{\Delta\lambda}{\lambda_0}$$

$$v = (1.86 \times 10^5 \text{ mi/sec})\frac{1.5 \text{ Å}}{4.86 \times 10^3 \text{ Å}}$$

$$v = 57 \text{ mi/sec}$$

where $c = 1.86 \times 10^5$ mi/sec.

With this in mind, we can interpret the spectrum in Figure 2.11$d$. All the iron lines have been shifted to the red, and thus the star must be receding from the Earth.

When the Doppler effect for light was discovered in stellar spectra, it was realized what a powerful tool astronomy had acquired, for by means of the *Doppler shift*, as it is called, an astronomer can determine the velocity with which a star is moving away from or toward us. The velocity can be determined because the faster the star moves the more the spectrum is shifted. The motion of a star either away from or toward the Earth is called the *radial velocity* because it is along a radius (in the line of sight) with the Earth at the center.

The spectrum of a star, then, enables the astronomer to determine not only the chemical composition of the star's atmosphere but also its velocity along the line of sight. Yet a star's spectrum, it must be remembered, is nothing more than starlight after it has been dispersed by a spectrograph.

## 2.8 Photometry

How much light from the star reaches that spectrograph in the first place? Just by looking up at the stars at night we can see that some stars are brighter than others and that all of them are fainter than the sun. If there were not some way of *measuring* the amount of light received, or at least comparing the amount received from one star with that received from another, the astronomer would have little evidence upon which to base his descriptions of the universe.

This question leads us into the field of *photometry*, the measurement of the amount of light. The first method used by astronomers was simply to estimate with the eye, much as we would do if we were to step outside on a clear night. The first astronomer known to have done this was Hipparchus who lived in the second century B.C. He referred to the brightness as *magnitude* and called the brightest stars those of the first magnitude and the faintest those of the sixth magnitude.

As soon as astronomers began to photograph stars, in the second half of the nineteenth century, they found that the brighter stars left a larger area of exposed emulsion than did the fainter ones (Figure 2-13). By arbitrarily establishing certain stars as standards, we can find the brightness, or *apparent magnitude*, of any star by comparing the size and darkness (the astronomer works with negative plates, not with positive prints) of its image on the plate with the size and darkness of the image left by a standard star. This is more accurate than guessing with one's eye, but it involves certain difficulties because the photographic emulsion does not respond in the same manner as the eye to the different colors that make up a star's light.

With the advent of electronics, an instrument called a *photocell* has been developed; the photocell permits precise measurements of the brightness of stars (Figure 2-14). When light strikes a metallic photosensitive surface, electrons are ejected free of that surface; if these electrons are picked up by a wire which is connected first through a meter—called a *galvanometer*—and then to the photosensitive surface, the ejected electrons will flow back to the photosensitive surface through the meter. This flow of electrons constitutes an electric current.

The galvanometer measures the strength of the electric current. The strength of this current is proportional to the number of electrons ejected, which in turn is proportional to the intensity of light. Consequently, the

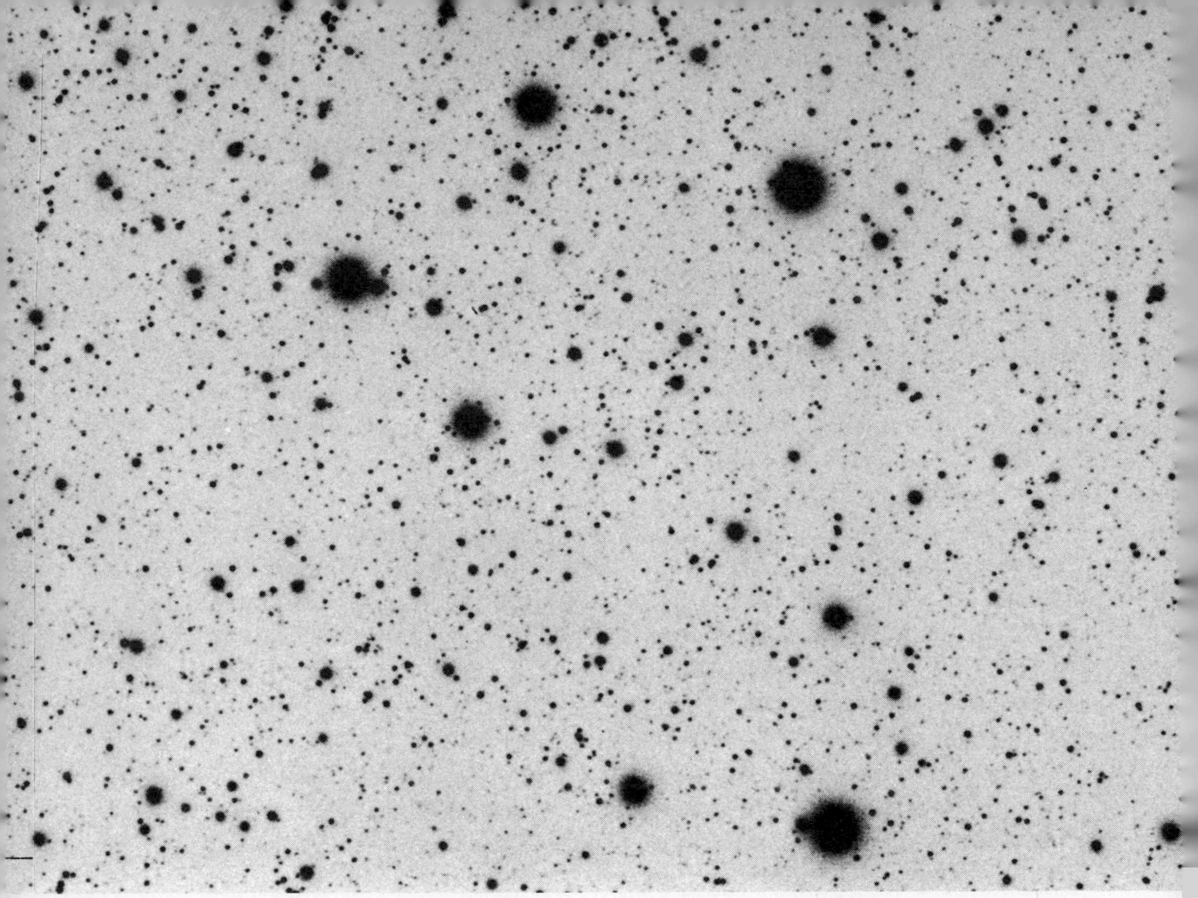

**Figure 2-13**

The brighter stars leave larger and darker images on the photographic plate than the fainter ones. This photograph is a negative, and thus the stellar images appear dark. (Courtesy of Merle Walker, Lick Observatory)

reading of the galvanometer is proportional to the intensity of light.

But even if the photocell is mounted in the focal plane of a large telescope, it receives only a very small amount of light from the very faint stars. Consequently, the electric current is very weak and must be amplified with amplifiers not unlike those used in hi-fi sets. Galvanometers can be made very sensitive to small differences in this amplified current, so very precise measurements of differences in the brightness of stars can be made.

If a photocell is used to study the nature of light, the light appears, startingly enough, to be composed of particles and not waves. These particles are called *photons*. In some experiments light behaves as if it were composed of waves, in other experiments it behaves as if it were composed of particles. Light seems to have a dual nature—a wave-particle nature.

## 2.9 The Electromagnetic Spectrum

The light that we see is actually only a small part of what is called the *electromagnetic spectrum*. The total electromagnetic spectrum includes radio waves, microwaves, infrared, visible light, ultraviolet, X-rays, and gamma rays. The difference between each of these types of radiation lies in their wavelength. The shorter the wavelength, the greater the energy. Figure 2-15 shows the entire electromagnetic spectrum. The wavelengths are expressed in centimeters.

Substances respond differently to each of these radiations. Glass, for example, is transparent to visible light and yet opaque to other radiations such as the ultraviolet. Quartz, on the other hand, is transparent to more of the ultraviolet. Our atmosphere is transparent to visible light and yet is relatively opaque to the far ultraviolet, that is, the region of ultraviolet that is nearest the X-rays. It is also opaque to X-rays and gamma rays when the radiation comes from outside the atmosphere and must pass through the entire atmosphere. All of these radiations will be transmitted over short distances through the atmosphere. There is a region in the radio waves to which our upper atmosphere (the ionosphere) is quite transparent. This region includes the wavelengths that transmit the FM (frequency-

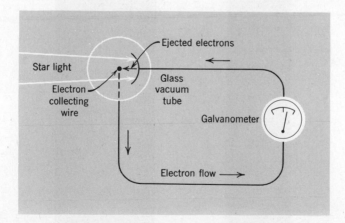

**Figure 2-14**

A schematic drawing of a photoelectric cell. The star's light is focused on the photosensitive surface by means of the telescopic objective, either a lens or mirror.

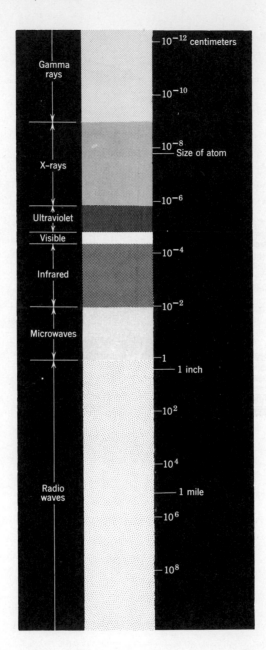

**Figure 2-15**

The entire electromagnetic spectrum.

modulated) and television broadcasts. The ionosphere is opaque to the wavelengths of the standard broadcast band, which it reflects back down to the Earth.

## 2.10 Resolving Power

Galileo recognized that with his telescope he could see things that he could not see with the unaided eye, for example, mountains on the moon. What appears as a single star to the unaided eye, may, in a small telescope (or a pair of binoculars) appear as two stars very close together. In a bigger telescope, these two stars may be recognized as four. This ability to resolve two stars appearing very close together is called *resolving power*. A telescope with high resolving power produces an image in which fine detail can be distinguished, whether that image is of a planet, a nebula, or a star field.

Since distances on the celestial sphere are measured as angles, the separation of objects very close to each other is measured in seconds of arc.* The resolving power of a telescope is defined as the smallest angle separating two stars which can be distinguished as two stars and not merged into one.

Resolving power depends on both the wavelength of radiation used and the diameter of the objective of the telescope. Good resolving power is achieved only when the wavelength is small compared to the diameter of the objective. The resolving power expressed as an angle $\alpha$ (Greek alpha) in seconds of arc is given by the relationship

$$\alpha = 2.1 \times 10^5 \frac{\lambda}{d}$$

where $\lambda$ is the wavelength of the radiation and $d$ the diameter of the telescope. The effective wavelength of visible light is about $5 \times 10^{-5}$ cm, so the resolving power of a 4-in. telescope (4 in. = 10 cm) is

$$\alpha = 2.1 \times 10^5 \frac{5 \times 10^{-5} \text{ cm}}{10 \text{ cm}}$$

$$\alpha = 1.0 \text{ second of arc}$$

*A circle is divided into 360 degrees; the common protractor, a semicircle, has 180 degrees. For small angular measurements, each degree is divided into 60 minutes of arc and each minute of arc is divided into 60 seconds of arc. Consequently there are 3,600 seconds of arc in 1 degree of arc.

The unaided eye has a resolving power of about 60 seconds of arc, or one minute of arc.

## 2.11 The Radio Telescope

In 1931 an employee of the Bell Telephone Laboratories, K. G. Jansky, discovered that celestial objects emit radio waves that can be detected. Not much was done with this startling discovery until the late 1930's when Grote Reber, an electrical engineer from Illinois, built a better instrument to detect radio radiation from the sky. His efforts were stopped by World War II, but out of this war came many new techniques of electronics that prepared the way for the present big radio telescope.

A radio telescope is a reflecting telescope with a reflecting surface not of glass but of either a metallic membrane or a wire mesh. At the prime focus an antenna picks up the radio signals and feeds them into an amplifier, which in turns feeds the signals into a recording device that records the intensity of the radio waves received (Figure 2-16).

A radio telescope with a diameter of 80 ft (2,400 cm) making an observation with radio waves of wavelength 21 cm has a resolving power of

$$\alpha = 2.1 \times 10^5 \frac{21 \text{ cm}}{2.4 \times 10^3 \text{ cm}}$$

$$\alpha = 1{,}800 \text{ seconds of arc}$$

$$\alpha = 30 \text{ minutes of arc}$$

The moon, as seen from the Earth, has an apparent diameter of 30 minutes of arc, therefore, an 80-ft radio telescope observing with 21-cm radio waves could not pinpoint the location of an object on the celestial sphere to an area any smaller than the disk of the moon.

The resolving power of a radio telescope with one reflecting surface is so poor that ways to overcome this disadvantage were investigated soon after the advantages of the radio telescope became evident. For instance, if a source of radio waves lies near the path of the moon, it is possible to obtain a good idea of its position if the moon passes in front of it and eclipses it. Astronomers have very accurate knowledge of the motion of the moon (its location on the celestial sphere at any given time); therefore, the position of an object emitting radio waves can be determined by observing the time the moon first passes in front of the object

**Figure 2-16**

The 300-ft radio telescope at Green Bank, West Virginia. (National Radio Astronomy Observatory)

**Figure 2-17**

The twin radio telescopes used as an interferometer in Owens Valley, California. (California Institute of Technology)

and then, as the moon moves, the time it uncovers the object. However, many objects of interest to the radio astronomer will never be eclipsed by the moon.

The limitations of building a radio telescope with a diameter so large that its resolving power would equal that of an optical telescope are obvious; its diameter would have to be 50 miles or more. But it is possible to achieve good resolution by using only portions of an otherwise mammoth radio telescope. Two relatively small telescopes placed $\frac{1}{4}$ mile apart and electrically connected are as effective in improving the resolving power as one big telescope with a diameter of $\frac{1}{4}$ mile (Figure 2-17). These two smaller telescopes, however, do not have the ability to gather as much energy as the mammoth telescope.

A system of two telescopes placed some distance apart is called an *interferometer*, because in reality it measures the interference effects when the signals from each of the two radio telescopes are brought together. The resolving power of two radio telescopes placed $\frac{1}{4}$ mile apart and observing with 21-cm radio waves is close to 2 minutes of arc.

By special techniques, radio telescopes on different parts of the Earth— such as California and Australia—are being used as interferometers with an effective resolving power of less than 0.01 second of arc.

## 2.12 The Atmosphere as a Filter

The radio telescope is a powerful tool with which to investigate the universe and it is a splendid supplement to the other telescopes used by man, who is still largely Earthbound. As long as we continue to use telescopes at the bottom of the atmosphere, however, we must contend with the absorption properties of that atmosphere. Figure 2-18 shows what portions of the electromagnetic spectrum are transmitted by the atmosphere. It can be seen that there is only a relatively small window in the optical region, but there is a much larger window in the radio region.

The radio astronomer can scan the radio spectrum of performing an operation entirely equivalent to changing a radio from one station to another or a TV set from one channel to another. The optical astronomer scans the visible part of the spectrum by using a prism.

The most important aspect of radio astronomy, however, is that it gives us a form of observation quite independent from optical observations.

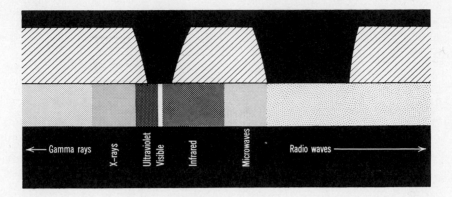

**Figure 2-18**

Regions of the electromagnetic spectrum absorbed by our atmosphere are shown crosshatched. The resulting "windows" enable us to see out.

The stars, so bright to the eye, are (with the exception of the sun) essentially absent from the sky which the radio telescope "sees." Objects which appear bright to the radio telescope are often difficult to see with even the biggest optical telescopes. The radio telescopes have increased our power of observation manyfold and now we have to explain these new observations.

The astronomer would always be at a disadvantage, however, if his observations were made solely from the surface of the Earth. Objects in the universe emit not only radio waves and visible light, but also microwaves, ultraviolet and infrared radiation, X-rays, and gamma rays. Each part of the electromagnetic spectrum contains unique information that the astronomer needs to help answer his many questions. He is aware, of course, that this new information may not only answer old questions but also may ask many new ones.

To make the desired observations, special instruments are being sent above the atmosphere in rockets and space probes. Balloons have taken instruments to heights of 80,000 feet, above 90% of the atmosphere. Instruments carried above the atmosphere have obtained information that would otherwise have been lost through absorption by the atmosphere. The X-ray sky is vastly different from the radio sky or the sky visible to our eyes on a clear night.

## BASIC VOCABULARY FOR SUBSEQUENT READING*

Absorption spectrum .                Focal plane
Ångstrom                             Interferometer
Apparent magnitude                   Objective
Bright-line spectrum                 Prime focus
Chromatic aberration                 Radial velocity
Continuous spectrum                  Resolving power
Doppler effect                       Spectral line
Electromagnetic spectrum             Spectrograph
Focal length                         Wavelength

## QUESTIONS AND PROBLEMS

1. Write the following numbers in the exponential system:
   (a) 93,000,000
   (b) 24,000,000,000,000
   (c) 0.034
   (d) 0.000 000 45

2. Write the following wavelengths in Ångstroms:
   (a) $4.861 \times 10^{-5}$ cm
   (b) $6.0 \times 10^{-3}$ cm
   (c) $3.4 \times 10^{-7}$ cm
   (d) 21 cm

3. If the wavelength of a spectral line is 4,500 Å, what is its frequency (the velocity of light $c = 3 \times 10^{10}$ cm/sec)? (The frequency $f = c/\lambda$.)

4. Determine the Doppler shift for the hydrogen spectral line ($\lambda_0 = 4,860$ Å) for a star whose radial velocity is:
   (a) 10 miles per second
   (b) 100 miles per second
   (c) 1,000 miles per second

5. Determine the radial velocity of an object if the K-line of ionized calcium ($\lambda_0 = 3,930$ Å) is shifted so that it has a measured wavelength of 4,930 Å. Is the object receding from or approaching the Earth?

6. In question 5, the K-line was shifted by 1,000 Å. How far would the 21-cm line be shifted if radio observations were made of the same object?

---

*A list of words will be given at the end of each chapter to indicate to the reader which words are important for further reading. The student should be very familiar with these words.

7. What is the resolving power of:
   (a) the 200-in. telescope at a wavelength of 5,000 Å?
   (b) the 1,000-ft radio telescope in Puerto Rico at a wavelength of 21 cm?

# FOR FURTHER READING

Larmore, L., *Introduction to Photographic Principles*, Dover Publications, New York, 1965.

Mayall, R. N., and M. Mayall, *Skyshooting; Hunting Stars with Your Camera*, The Ronald Press Co., New York, 1949.

Miczaika, G. R., and W. M. Sinton, *Tools of the Astronomer*, Harvard University Press, Cambridge, Mass., 1961.

Page, T., and L. W. Page, ed., *Telescopes*, The Macmillan Co., New York, 1968.

Ruechardt, E., *Light, Visible and Invisible*, University of Michigan Press, Ann Arbor, Mich., 1958.

Steinberg, J. L., and J. Lequeux, *Radio Astronomy*, McGraw-Hill Book Co., New York, 1963.

Struve, O., and V. Zebergs, *Astronomy of the 20th Century*, Crowell Collier and Macmillan, New York, 1962, Chapters I, II and VI.

Texereau, J., *How to Make a Telescope*, Interscience Publishers, New York, 1957.

Woodbury, D. O., *The Glass Giant of Palomar*, Dodd, Mead and Co., New York, 1953.

Blanco, V., "The Inter-American Observatory in Chile," *Sky and Telescope*, p. 72 (Feb. 1968).

Brown, R. H., "The Stellar Interferometer at Narrabri Observatory," *Sky and Telescope*, p. 64 (Aug. 1964).

Kiepenheuer, K. O., "The Domeless Solar Refractor of Capri Observatory," *Sky and Telescope*, p. 256 (May 1966).

Lutsky, V., "The 236-inch Soviet Reflector," *Sky and Telescope*, p. 99 (Feb. 1970).

Small, M., "The New 140-foot Radio Telescope," *Sky and Telescope*, p. 267 (Nov. 1965).

"Giant Mirror Blanks Poured for Chile and Australia," *Sky and Telescope*, p. 140 (Sept. 1969).

"Giant X-Ray Telescope," *Sky and Telescope*, p. 300 (May 1969).

"The Latest Flight of Stratoscope II," *Sky and Telescope*, p. 365 (June 1970).

"Radio Interferometers with Very Long Base Lines," *Sky and Telescope*, p. 143 (Sept. 1967).

"Some Current Programs at Arecibo," *Sky and Telescope*, p. 4 (July 1964); p. 73 (Aug. 1964).

CHAPTER
3
CHAPTER

# THE ATOM

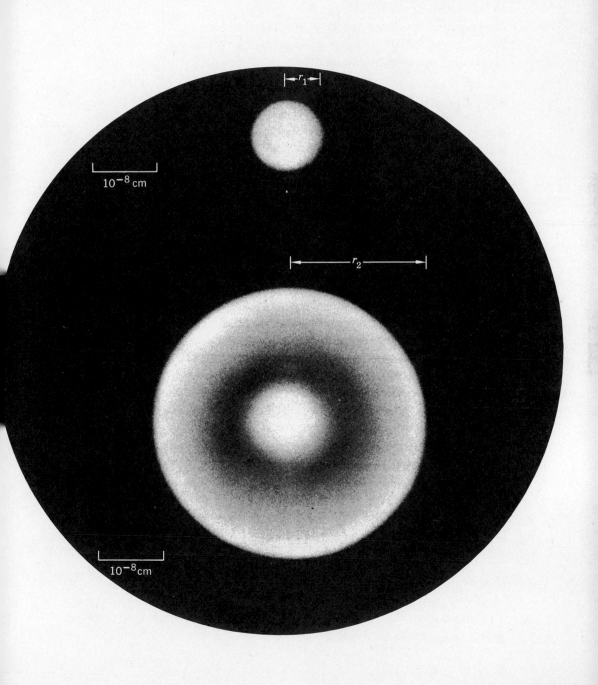

The solar system, the stars, and indeed the entire enormous universe are all composed of atoms. Two extremes confront us: that which is too big to comprehend and that which is too small to see. Yet, to understand the universe, we must understand the atom, for our observations of the universe are the results of actions of atoms.

## 3.1 Atomic Structure

Atoms are composed principally of three basic particles: the *proton*, the *neutron*, and the *electron*. Many other particles are found when we investigate the nucleus of the atom, but we shall limit ourselves to a more simplified description.

A mechanical and oversimplified model of the atom is composed of a *nucleus* about which electrons revolve; at the same time the electrons rotate on their axes. This rotation is called *electron spin*. The number of electrons revolving about a particular nucleus (and thus forming an atom of a particular element) depends on that nucleus and determines how the particular atom will react chemically.

The nucleus is by far the most massive part of the atom and is composed, with one exception, of protons and neutrons. That one exception is the simplest atom of all, hydrogen, which has only a proton as its nucleus. Here, as in any atom, the electron is maintained in its orbit about the nucleus by electrical attraction.

There are two kinds of electrical charges, positive and negative. *Like charges* repel each other, whether they are positive or negative. *Unlike charges* attract each other. The proton has a positive charge and the electron a negative one; these unlike charges attract each other and hold the two particles together to form the hydrogen atom. The hydrogen atom is not only the simplest but also the most common atom in the universe.

The next most prevalent atom in the universe, helium, is only slightly more complex; it is the second atom in a sequence based on the number of protons in the nucleus (the *atomic number*). Its nucleus has two protons rather than one. Consequently the helium atom has two positive charges

### Chapter Opening Photo

An artist's conception of the hydrogen atom in its (*a*) ground state, and (*b*) with the electron in second energy level. (By permission from *Physics*, K. R. Atkins, John Wiley & Sons, Inc., 1965)

and must have two electrons revolving about its nucleus if it is to be electrically neutral.

The mass of the proton is 1,836 times that of the electron, so the mass of an atom (its *atomic mass*) is given as that of its nucleus. The hydrogen atom has a mass of 1. The most common form of helium, however, has an atomic mass of 4 rather than 2. In addition to the two protons there are two other particles in this atomic nucleus, each of which has a mass about equal to that of the proton. These extra particles are called neutrons because they are electrically neutral, that is, they have neither a positive nor a negative charge. The nucleus of the helium atom, then, contains four particles: two protons each of unit positive charge, and two neutrons of zero charge but of about the same mass as the protons. Two electrons revolve about this nucleus to form the complete atom.

There is another variety of helium with an atomic mass of 3. Since it must have two protons (or it would not be helium) it can have but one neutron in its nucleus. These two different types of helium atoms are called *isotopes*, for they have the same (iso) charge on their nucleus and thus fall in the same place (tope) in this sequence, which is based on nuclear charge, even though their masses are different.

The third atom in this sequence is lithium, with three protons in its nucleus and three electrons revolving about it. The most common lithium atom has a mass of 7, which means that there must be four neutrons in its nucleus. There is another isotope of lithium whose mass is 6 and therefore has three neutrons in its nucleus.

The rest of the chemical elements fill out the remainder of the sequence. There are 92 elements found in nature, but by means of particle accelerators (cyclotrons, synchrotrons, etc.) man has created elements that do not occur in nature. Whether they are or were ever formed in nature is another question, but they are not stable and eventually decay radioactively into one of the stable elements. Some natural elements are not stable either, but their rate of decay is so slow that they can still be found on the Earth billions of years after its formation.

This simplified model of the atom is quite easy to imagine; electrons appear as fuzzy bits of negative electricity whirling and spinning about a much more massive nucleus. There are aspects of this oversimplified model, however, which are not supported by observations in physics. Nevertheless, this model is convenient for our purposes and so we will rely on it.

## 3.2 Work and Energy

The idea of the atom is intimately bound to the concept of energy. Consequently, to understand the atom more fully we must first become familiar with the meaning of the term energy. The common usage of the word energy implies the ability to do something. A child has lots of energy; the child is incessantly active. To the scientist, however, a rigorous definition is much more useful. This definition is dependent upon another concept—the concept of work.

The lifting of a weight, the pushing of a car, the accelerating of a rocket, are examples of work—energy is required to do all of these things. Whenever a force is exerted on an object and that object moves as a result of the force, work has been done on that object. We *define work W as the product of the force F exerted and the distance s moved as a result of the force:*

$$W = Fs$$

There is a small catch in this definition, however: both the force and the distance moved must be in the same direction.

Work done on one object increases the energy of that object. The increased energy of that object permits it to do work on a second object. So, if work results in an increase of energy, then energy ought to be able to produce work. In fact, we *define energy as the ability to do work*. Work is action, the act of *changing* the energy of some object; and it is the changes in energy that are significant.

For a 10-lb sack of flour to be lifted from the floor to a greater height, a force must be exerted on it; it then moves the distance to the greater height. Work is done on that sack; its energy is increased. This increase in energy is a consequence of the sack's new position with respect to the Earth; it is energy of position and is called *gravitational potential energy*. The lifting force equals the weight of the sack $w$, and the distance moved is the height $h$ to which the sack was lifted. Since the work done equals the increase in energy, the change in gravitational potential energy $E_p$ is given mathematically as

$$E_p = wh.$$

The gravitational potential energy would increase by 50 ft-lb if that 10-lb sack were lifted to a height of 5 ft.

An electron and a proton have *electrical potential energy* as a result of their mutual attraction. The amount of potential energy depends upon the relative position of the electron and the proton. In order to separate a proton and an electron, work must be done on them; as a consequence of the work, they will have more energy after being separated than before.

Two electrons, however, repel one another, so work must be done on them to force them closer together. The energy of those electrons will increase as the result of being pushed closer together. Once released, they will fly apart faster after having been pushed closer together.

As those electrons fly apart, their electric potential energy decreases. The electrons exert a repulsive force on each other and they move as a result. As the force is exerted they are accelerated. Work is being done on the electrons at the expense of the decrease in potential energy. It would appear, therefore, as if potential energy is being converted into some other form of energy, and since the work done on the electrons increases their velocity, we call this form of energy kinetic energy (Greek, *kinema*, motion, see English *cinema*). *Kinetic energy is defined as energy of motion*. It can be shown that if an object is accelerated from rest, its change in kinetic energy $E_k$ is given by

$$E_k = \tfrac{1}{2}mv^2$$

where $m$ is the mass of the object (see Chapter 5) and $v$ is its velocity.

This example also serves to demonstrate a very fundamental principle in science: the *principle of conservation of energy*. There are many forms of energy and one form may be transformed into another. In all such transformations energy is conserved. The total amount of energy before transformation equals the total amount of energy after transformation.

All forms of energy have the ability to do work. Kinetic energy can do work if a moving object hits something. Chemical energy accelerates an automobile as the gasoline explodes in the cylinder head. Electrical energy permits an electric motor to operate a washing machine. Chemical energy produces heat energy and heat energy boils water which makes the steam engine run.

The concept of energy permeates all aspects of science, whether we consider the solar system, the human body, or the atom. The laws which describe the motions of the planets and stars, Newton's laws of motion and his law of gravity (see Chapter 5), however, do not apply to the atom. For example, meteorites (small chunks of rock revolving about the sun; see Chapter 7) have both potential and kinetic energy with respect to the Earth. Should a meteorite come close to the Earth, it would fall through the Earth's atmosphere to its surface. Its potential and kinetic energy would be converted to heat. Small meteorites do not survive this generation of heat; they melt and burn in the upper atmosphere.

An isolated and moving electron has both potential and kinetic energy with respect to an isloated proton. Should the electron come under the control of the proton the two would attract each other, but they would not fall together as the meteorite which struck the Earth's surface. Instead, the electron falls to within a certain distance of the proton and then "goes into orbit." The proton and its orbiting electron constitute an atom of hydrogen. The potential and kinetic energy lost in the "fall" of the electron is emitted in the form of light.

The size and shape of an electron orbit is predetermined by the nucleus. Each atom has many possible orbits, and each orbit represents a definite amount of energy different from that represented by any other orbit. This means that any electron in an atom will have the amount of energy represented by the particular orbit in which it revolves; that is, it occupies a particular level of energy.

## 3.3 Energy Levels and Atomic Transitions

An *energy level* of an atom can accommodate only one electron at a time. Consequently, if to become electrically neutral an atom requires more than one electron (helium needs two; sodium, eleven) then each electron must occupy a different energy level. But each atom has many more energy levels than it has electrons and certain energy levels are more favored by its electrons than others. The energy levels occupied in an undisturbed, electrically neutral atom are thus set in a pattern predetermined by its nucleus.

In general, the levels of lowest energy in any particular atom are more favored by the electrons. The levels of lowest energy represent those orbits closest to the nucleus, so the electrons tend to hover as close to the nucleus as they can. For example, the hydrogen atom normally has only one electron. If undisturbed, this electron will seek the lowest energy level, also called the *ground state*.

An electron in an atom may, from an outside source, gain energy that will force it up into a higher energy level. This can be demonstrated by introducing hydrogen into a closed tube at low pressure, and passing an electric current through it. Electricity is a flow of electrons, and moving electrons have kinetic energy. If a free electron with enough energy collides with a hydrogen atom, it will give some of its energy to the orbital electron. The bombarding electron will leave the scene of the collision with a reduced velocity because of its loss of energy. The orbital electron will change to a higher energy level as a result of its increase in energy. A change in energy levels by an orbital electron is called a *transition*; in this example it is an upward transition (Figure 3-1). An atom that has had an electron knocked into one of the higher energy levels is said to be *excited*.

Since the hydrogen atom has many energy levels, to which level does the orbital electron go when bombarded by a free electron? This depends on two factors. The first is the amount of energy of the bombarding electron. The more energy it has the higher the orbital electron can go. But when the electron makes a transition it must go from one energy level to another; it cannot stop between them. Since the energy levels are well defined and

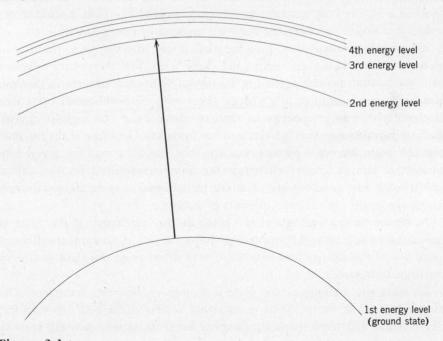

**Figure 3-1**

An electron transition from the ground state to the third energy level of the hydrogen atom.

furthermore, since the electron is permitted to make a transition only from one energy level to another, the orbital electron can absorb only certain discrete amounts of energy from the bombarding electron, namely, the amount necessary to make a single transition between any two energy levels.

The second factor determining which transition an electron will undergo is the probability of making any given transition. Certain transitions are more likely than others, and this helps determine which discrete amount of energy the orbital electron will absorb. It need *not* absorb all the energy made available by the bombarding electron.

The situation is similar to putting a plate on one of a number of shelves in an empty cupboard. The plate must rest on one of the shelves; it cannot be placed between them. To lift the plate from the floor (the ground state) to one of the lower shelves does not require as much energy as to lift it to one of the higher shelves. If the cupboard has five shelves, one of only five discrete amounts of energy is needed to lift the plate to one of the shelves. To move the plate from the floor to any one shelf, or from one shelf to a higher one, amounts to a transition that the plate cannot make unless it is given precisely the required amount of energy.

It is possible, however, to give the plate a toss into the air so that it will go higher than the highest shelf. Since there is no shelf above the cupboard (and we assume no ceiling) it can be tossed with any amount of energy. Similarly, an electron at any energy level may be bombarded by a free electron with enough energy to send it above even the highest energy level. When this happens the electron has been knocked free of the nucleus and the atom becomes *ionized*. For this the electron may be given any amount of energy greater than just the amount required for ionization, and since it has gone beyond discrete transitions, it is no longer limited to the absorption of discrete amounts of energy.

To return to the analogy of the plate in the cupboard: if the plate is permitted to fall, it will drop and perhaps break. In striking the floor it gives off all the energy lent it when it was lifted from the floor to one of the higher shelves.

But here the analogy of the plate and shelves becomes incorrect. The plate will remain on the shelf indefinitely unless disturbed, whereas the electron will fall from the higher energy level of its own accord, usually within $10^{-8}$ second.

**a. Downward Transitions Yield Photons.** When the electron does fall to one of the lower levels it gives off energy. This energy is radiant energy (visible or ultraviolet light, infrared, radio, etc.) and is emitted as a *photon*. A particular photon has an amount of energy that equals the amount of energy given up by the electron making the downward transition. Since an electron in a given atom is permitted only certain transitions in falling, it can give off photons of only certain energies, each representing a particular wavelength of radiant energy. The same transition made downward by electrons in many atoms of the same chemical element gives rise to a bright spectral line.

Many transitions made downward between *different* pairs of energy levels in atoms of the same element produce that element's characteristic bright-line spectrum. We see that in a spectrum each spectral line results from a downward transition between two particular energy levels of an atom.

If the downward transition is short, the amount of energy given off by the electron is small. The photon of radiant energy will travel from the atom with this small amount of energy; but regardless of its energy, it will travel with the velocity of light. The greater the downward transition, the more energy the photon will carry away. Thus the placement of the spectral lines in the spectrum depends on the particular transition made.

The entire electromagnetic spectrum represents a continuous array of wavelengths beginning with the longest, radio waves, and proceeding in order through the microwaves, infrared, visible light, ultraviolet, and X-ray to gamma rays, the shortest. These wavelengths can be translated into energies, and thus the spectrum is a continuous array of radiant energy. The shorter wavelengths have more energy than the longer; blue light, for instance, has more energy than red.

Whenever an electron makes a transition to the ground state of hydrogen, for example, there is given off a photon of sufficient energy to place it in the ultraviolet region of the spectrum. All such transitions from any higher energy level to the ground state of hydrogen constitute a series of spectral lines in the ultraviolet, called the *Lyman series* (Figure 3-2). If a transition is made down to the second level, the atom emits light in the visible region of the spectrum, and all such transitions in hydrogen result in a spectral line sequence called the *Balmer series*. Since hydrogen is the most prevalent atom in the universe and since the Balmer series is in the visible and near-ultraviolet region of the spectrum, this series becomes very important to the astronomer.

If transitions are made from the third to the second energy level of hydrogen, the resulting spectral line is in the red region of the spectrum and has a wavelength of 6,563 Å. Since this is the first line of the Balmer series it is called *Hα* (*H alpha*). If a transition is made from the fourth to the second energy level, more energy will be given off, and this spectral line, called *Hβ* (*H beta*), is in the blue region with a wavelength of 4,861 Å. The third line in this series results from a transition from the fifth energy level to the second, has a wavelength of 4,340 Å, and is called *Hγ* (*H gamma*). There is, theoretically, an infinite number of lines in each of these series; but since each energy level is progressively closer to the preceding one as the outermost level is reached, the lines blend together toward the end of the series (Figure 3-3).

**b. Photons and Upward Transitions.** A bright-line spectrum can therefore be explained by electron transitions from one energy level to any other below it. But how about the dark-line spectrum? It was pointed

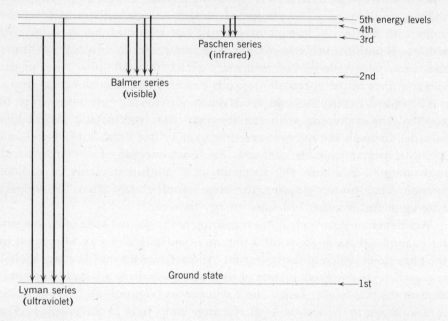

**Figure 3-2**

Electron transitions downward to the ground state emit light that yields the spectral sequence called the Lyman series. All downward transitions to the second energy level give rise to the Balmer series; downward transitions to the third energy level yield the Paschen series. The energy levels have been drawn as straight lines for convenience.

out earlier (see page 32) that a dark-line or absorption spectrum is caused by sending light that consists of all wavelengths (a continuous spectrum) through a low-pressure gas that is cooler than the source of the light. The light in this continuous spectrum consists of photons of all energies within the range of wavelengths emitted. These photons are able to bombard the orbital electrons in the atoms of the gas and cause them to make upward transitions. A given bombarding photon, however, is *completely* absorbed in the process. It disappears as a photon and appears as increased energy in an electron. The electron normally remains in a higher energy level for about a 100-millionth of a second before falling back down to land eventually in the ground state. In making downward transitions, it gives off photons of light which may or may not be of the same wavelength as the energy absorbed in causing the original upward transition. If the electron was originally in the ground state and falls directly back to that state, the photon emitted will be of the same wavelength as that which excited the electron. However, the electron may choose to fall one energy level at a time (*cascading*) and thus the photons emitted will each have less energy (longer wavelength) than the exciting photon. An electron in a higher energy level can be excited to an even higher energy level, but it, too, will eventually wind up in the ground state. If the electron is knocked free of the nucleus by a *photon*, the atom becomes *ionized* and the process is known as *photoionization*.

The absorption spectrum results from this process of absorption and re-emission. But when it happens that the emitted photons are of the same wavelength as the absorbed photons, why should we observe an absorption line?

**Figure 3-3**

Two stellar spectra are contained between the two comparison spectra. The upper stellar spectrum has the lines of the Balmer series in emission, that is, as bright lines. The lower stellar spectrum has the Balmer lines in absorption. (Photographs from the Hale Observatories)

Let us consider only the original light from the continuous source that is heading straight for our spectrograph (Figure 3-4). All the light would reach our spectrograph were it not for the intervening gas. This gas is composed of atoms that absorb light of only certain wavelengths; *which* wavelengths depends on the transitions possible in the atoms. The light that is re-emitted as a result of the electrons' downward transitions is re-emitted in *all* directions. Thus only a small portion of the re-emitted light continues in the original direction toward our spectrograph, and consequently an absorption line is dark. If no energy were to reach the spectrograph, such a line would be completely black. The absorption spectrum, therefore, consists of a continuous background from which selected regions (or lines) have been absorbed by the upward transitions of electrons in the atoms that compose the particular intervening gas.

The transitions for a given atom, whether they are upward (an absorption line) or downward (an emission line), involve the same amount of energy. Thus the two types of lines resulting from transitions between two given energy levels have the same wavelength; for $H\alpha$ of the Balmer series we can have either an emission or absorption line and each has a wavelength of 6,563 Å.

Thus far we have discussed two methods of exciting an orbital electron.

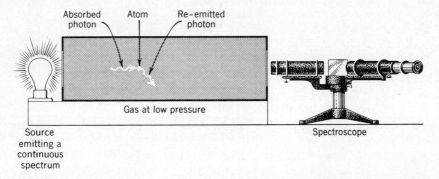

**Figure 3-4**

White light that shines through a gas cooler than the source of light will have certain colors subtracted from it giving rise to a dark-line spectrum. A photon originally heading for the spectrograph is absorbed by an atom in the gas and re-emitted. The re-emitted photon, however, will probably travel in some other direction and thus will not reach the spectrograph. The result is a dark line in an otherwise continuous spectrum.

It may be bombarded by another electron, in which case the *only* visible effect produced is the bright lines emitted when the electron falls back down to a lower energy level. Or the electron may be excited by a photon, in which case the visible effect is an absorption line. This is the only manner in which an absorption line can be produced.

**c. Distorted Energy Levels.**  The electron can be excited in a third manner—by thermal agitation of the particles composing a hot substance. When iron is heated in a forge or crucible the thermal agitation causes the electrons to make upward transitions. The excited electrons fall back, emitting light which is seen as a red glow. Since this is a solid, or perhaps a liquid, why does it give off a continuous spectrum and not the bright-line spectrum of iron?*

To explain this let us imagine some hydrogen inside a closed container that has a piston in one end. When the piston is fully extended the gas is under very low pressure and when excited will emit a bright-line spectrum. If the piston is pushed in, the pressure increases and the energy levels of the hydrogen atoms become distorted by the increased jostling received by the atoms now that they have been forced closer together.

Any electron that makes a transition from a distorted energy level will not give off light of the same wavelength as it would if the energy level were not distorted. Since the energy levels of various atoms are not uniformly distorted, the resulting bright spectral lines become broadened and fuzzy. This is known as *pressure broadening* and provides us with a somewhat simplified explanation of the continuous spectrum. If the pressure of the gas is increased even further the spectral lines eventually become so broadened as to blend into a continuous spectrum. As was noted in Chapter 2, any incandescent solid or liquid emits a continuous spectrum because its atoms (or molecules, for that matter) are close together. Furthermore, each substance emitting a continuous spectrum is opaque.

## BASIC VOCABULARY FOR SUBSEQUENT READING

| | |
|---|---|
| Atomic mass | Balmer series |
| Atomic number | Electron |

*See page 31.

| | |
|---|---|
| Electron spin | Neutron |
| Electron transition | Nucleus |
| Energy level | Photon |
| Ionization | Proton |
| Lyman series | Pressure broadening |

## QUESTIONS AND PROBLEMS

1. A brick is first lifted from the floor to the table top, and then released to fall back to the floor again. Describe its changes in energy.

2. What is the kinetic energy of the Earth if its mass is $6.0 \times 10^{27}$ grams and its velocity about the sun is $3.0 \times 10^6$ cm/second? (The unit of energy is the *erg*.)

3. (a) Explain three ways in which energy can be given an orbital electron.
   (b) What is the natural consequence of the orbital electron's absorbing this energy?
   (c) By what single way can the electron lose this added energy?

4. Explain the process of photoionization.

5. Describe the cause of spectral-line pressure broadening.

6. What are the limitations of electron transitions within a given atom?

7. What are the limitations of energy emitted during the capture of a free electron?

## FOR FURTHER READING

Ruechardt, E., *Light, Visible and Invisible*, University of Michigan Press, Ann Arbor, Mich., 1958.

Semat, H., and H. E. White, *Atomic Age Physics*, Holt, Rinehart and Winston, New York, 1959.

# CHAPTER 4
## CHAPTER CHAPTER

# THE PLANETS

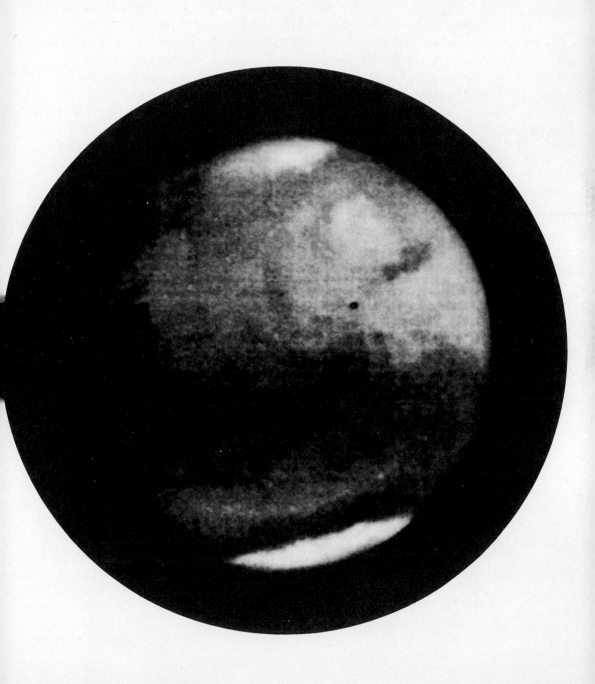

Now that we have some idea of the astronomer's methods, we can better understand the objects of his studies. To the naked eye all the stars (except the sun) and planets appear as points of light. But in the early telescopes those objects closest to the Earth took on a new appearance. The same is true today; planets appear as disks, whereas the stars appear only as points of light even when viewed with the highest magnification available. The early observations led to the first great change in man's concept of the universe—the idea of a *solar system*.

The solar system is composed of the sun, its largest and central member; the nine planets that revolve about the sun; satellites, comets, asteroids, meteors, and something we might call dust.

The sun is really a star and we shall consider it along with the other stars. For the present we are concerned with the planets, which in the solar system are second only to the sun in importance and size.

Within the solar system, the nine planets have at least two different motions. Each of them *revolves* about the sun in a given length of time just as the satellites of Jupiter revolve about Jupiter. The planet Earth, for instance, takes one year to revolve about the sun; in fact, it is because of this revolution that we have the year.

At the same time each planet *rotates* on its own axis like a merry-go-round. The Earth makes one complete rotation on its axis each day. During part of this rotation you, the observer, are facing the sun; during part of it you are facing away from the sun, in the shadow of the Earth.

All planets describe these two basic motions of revolution and rotation, although with differences. But what of their other physical characteristics? Do planets other than the Earth have atmospheres, soil, conditions that would favor life? Conditions on them are largely determined by their distance from the sun because each, like the Earth, receives its heat from the sun. The closer the planet, the more heat it receives.

## 4.1 Mercury

Mercury, less than half as far from the sun as the Earth, is the planet closest to the sun. To put it more briefly, it is 0.4 A.U. from the sun; A.U. is the abbreviation for a measure of length called the *astronomical unit* which is

**Chapter Opening Photo**

Mars as photographed by Mariner 7. (NASA photograph)

used by astronomers because the units of length employed on the Earth are all too short. 1 A.U. is equal to the average radius of the Earth's orbit, about 93,000,000 miles.

**a. Radar and Rotation.**   With a diameter of 3,100 miles and a mass only 0.05 that of the Earth, Mercury is also the smallest planet. It revolves about the sun in 88 days and was long thought to keep the same face toward the sun. However, observations in 1965 made by G. H. Pettengill and R. B. Dyce with the 1,000-ft radio telescope in Puerto Rico by radar (that is, by sending a radio signal to Mercury and observing the echo) indicate that, with respect to the stars, the planet rotates on its axis with a period of 58.6 days. Consequently, during each rotation the entire planet receives sunlight; no face is perpetually dark. But since the direction of rotation is the same as the direction of revolution, the period of rotation with respect to the sun is about 175 days. The sun will shine on one point of the surface for a period of time equal to about 87 $\frac{1}{2}$ days here on Earth.

To measure the period of rotation of a planet by sending a radio signal from the Earth and observing the echo that returns takes a bit of electronic doing. First of all, a very powerful signal must be generated so that it can travel all the way to Mercury and the echo all the way back, and still be detected. A powerful enough signal can, however, be generated for a very brief interval of time.

In order to learn something about the planet from the radar echo, however, we must understand how the planet affects the incident radio waves. Those radio waves that strike the planet near the central part of the disk, region $C$ in Figure 4-1, will reflect back to the Earth. Those that strike the planet near the edge of the disk—the very edge of the disk is called the *limb*—will, for the most part, reflect off into space. But if the surface has mountains and craters then some of the signal will reflect off the sides of the crater walls and mountains, and return directly to the Earth. Since we do indeed receive some (albeit only a small portion) of the signal that strikes the planet near the limb (regions $A$, $B$, $D$, and $E$ in Figure 4-1) we know that Mercury does have mountains.

If the planet rotates, the signal reflecting from part $A$ will have a frequency shift as a result of the Doppler effect. It will reflect from a portion of the planet approaching us, so the returning signal will have a slightly higher frequency. The signal returning from region $B$ will have a slightly lower frequency.

We would not, of course, expect much of a frequency shift, so the signal sent to Mercury must have a single frequency—in reality a very narrow band of frequencies (shown schematically in Figure 4-2a). The echo, greatly reduced in intensity, would have a broader band of frequencies. The signal returning from region A would be shifted to the higher frequency, that from B would be shifted toward the lower frequency (Figure 4-2b). The breadth of the frequency band of the returning signal when compared with the narrow band of frequencies sent enables the radio astronomer to determine the speed of rotation and consequently the period of rotation.

However, the signal returning from region C is so much stronger than that which returns from A, B, D, and E, that it drowns the weaker signal out. Therefore, the returning signal is separated into two parts and only that part returning from regions A, B, D, and E is studied in detail for a Doppler shift.

The signal can be separated because that which reflects from the central part of the disk will reflect first, and thus reach the Earth sooner than the signal reflecting from the crater walls in regions A, B, D, and E. Therefore, if the transmitted signal is a very short time in duration (Figure 4-3a),

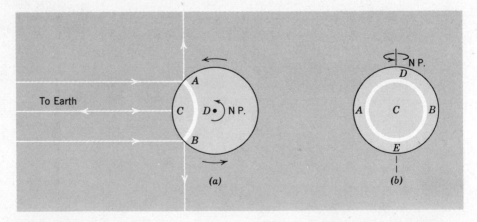

**Figure 4-1**
Radio waves reflecting from A will be shifted to the shorter wavelength (higher frequency), those reflecting from B will be shifted to the longer wavelength (lower frequency). (a) Mercury as seen from above its north pole. (b) Mercury as seen from the Earth.

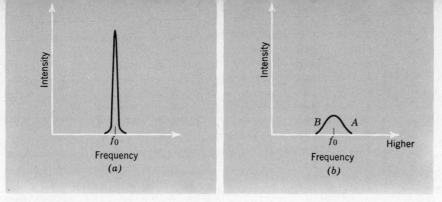

**Figure 4-2**

(*a*) The transmitted radar signal of a very narrow band of frequencies and very high intensity. (*b*) The echo from Mercury with a slightly broader band of frequencies. The intensity of the echo has been exaggerated so that it can be seen on the graph.

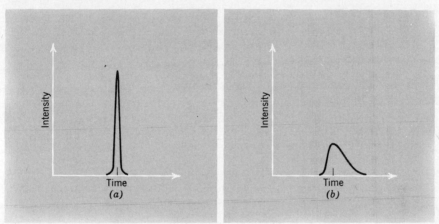

**Figure 4-3**

(*a*) The transmitted radar signal of a very short time interval. (*b*) The echo is spread out over a longer time interval than the transmitted signal.

**Figure 4-4**

(*a*) Only part of the returning signal is studied, the dashed portion is sidetracked. (*b*) The breadth of frequencies in the echo yields the speed of rotation.

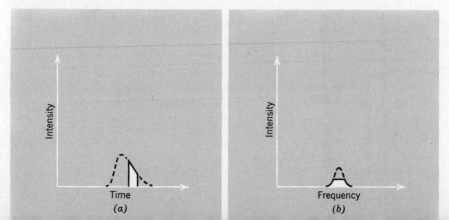

the returning signal will be spread out in time, with the most intense part of the signal returning first (Figure 4-3b).

If the first and last portions of the returning signal are sidetracked (Figure 4-4a), the signal returning from regions A, B, D, and E in Figure 4-1 can be studied in detail. That signal is shown schematically in Figure 4-4b. It is the breadth of this signal that revealed the period of rotation of Mercury.

**b. The Atmosphere.**   If the period of rotation of Mercury seems to be settled, the existence of an atmosphere is not entirely certain. Determining if Mercury has an atmosphere may depend in part upon the way the word "atmosphere" is defined. If the term is applied to a layer of gas about a planet which is not only permanent in nature but which is dense enough to be easily detectable, then the answer is an unqualified no. If, however, the term atmosphere is taken to include very rarified gases which may or may not be permanently associated with the planet, then Mercury may have an atmosphere.

Mercury, like every other planet, shines by reflected sunlight. Therefore, its spectrum must be the same as the sun's spectrum—unless it has an atmosphere of its own. (Surface materials do absorb some light but not as spectral lines.) If Mercury has an atmosphere, the sunlight would have to pass through it to reflect from the surface, and then pass back through the atmosphere again before reaching our telescopes. Since this atmosphere would act as a cool gas at low pressure, it would add absorption lines to the solar spectrum. The wavelengths of these lines would indicate the chemical nature of the atmosphere.

There is slight evidence that the spectrum of Mercury does contain spectral lines other than those of the sun. N. A. Kosyrev of the Pulkovo Observatory in Russia reports that he has observed spectroscopic evidence that Mercury does contain a very slight atmosphere of hydrogen gas. But this atmosphere would not remain were it not replenished by protons and electrons (the two constituent parts of the hydrogen atom) from the sun.

Whether a planet can retain an atmosphere or not depends upon the mass and radius of the planet, its temperature, and the kind of gas. The mass and radius of the planet determine the escape velocity at the surface. The *escape velocity* is the velocity an object must achieve in order to escape from that planet. The escape velocity at the surface of Mercury is 2.6 miles per second. By comparison the escape velocity at the surface of the Earth is 7.0 miles per second.

The velocity with which the atoms of gas travel depends upon the temperature of the gas and upon the kind of gas. Even if an atmosphere is composed of only one kind of gas, its atoms, because of the atomic collisions, will be traveling at different velocities. As a result of a collision, one atom may acquire a greater velocity, another a lower velocity. The number of atoms at each particular velocity depends upon the temperature of the gas. As a result, the fastest atoms will escape from the planet if they acquire the escape velocity. If the atmosphere contains several kinds of gases, the atoms of the lightest gases will have the greatest velocity. The atoms of those gases with the greatest mass will have the least velocity. Consequently, a planet is more apt to retain gases with more massive atoms; for example, the heaviest gases of all which might comprise a rarified atmosphere are the inert gases krypton and xenon. A planet is more apt to retain oxygen, nitrogen, and carbon dioxide than it is hydrogen and helium, for the latter two gases are the lightest gases of all. It is possible, therefore, that Mercury might have a slight, but a very rarefied, atmosphere composed of the heaviest inert gases.

However, it is well known that the sun expels protons and electrons which when combined, form atoms of hydrogen. Consequently, it is possible that Mercury retains a slight atmosphere of hydrogen which is always changing; gases leave because the atoms acquire the escape velocity, but these gases are replenished by those expelled by the sun and captured by Mercury. It is also possible that some gases are still erupting from volcanic-like activity on the surface of Mercury.

c. The Surface.   We are led to believe that the surface of Mercury is similar to the surface of the moon (see Chapter 6). Mercury reflects 6% of the light incident on it. This can be determined because we know how much light the sun emits; we know how far Mercury is from the sun and thus how much of the sun's light it intercepts. We also know, at the moment of observation, how far we are from Mercury. With the added knowledge that the intensity of light radiating from the sun and reflecting from Mercury decreases as the square of the distance, we can calculate how much light we on Earth would receive from Mercury if it were a perfect reflector. We receive about 6% of that quantity. Astronomers use the term *albedo* to describe the reflectivity of a surface; the albedo is the fractional part of the incident light which reflects from a surface. The albedo of Mercury is 0.06; the albedo of the moon is 0.07.

The fact that the albedoes of these two bodies are nearly the same and that other properties of the reflected light are also similar (for example, the color and the extent of polarization), leads us to conclude that the surface of Mercury is similar to that of the moon.

The temperature of the surface of Mercury has been measured by a number of methods. Measurements of the temperature of the sunlit side using the infrared region of the electromagnetic spectrum indicate the sunlit face, at its hottest, may reach a temperature of 650°F. More recently, observations of Mercury with radio telescopes using radio waves with a length of about 3.5 cm indicate that the *average* temperature of the sunlit side is not far from 220°F. As expected, the temperature of the surface must drop in those regions where the sunlight strikes the surface more obliquely. Measurements of the nighttime side using radio waves with a length of about 3.4 millimeters (mm) indicates a temperature of about − 100°F.

## 4.2 Venus

The second planet from the sun (0.7 A.U.) is Venus. With a diameter of 7,522 miles and a mass 0.89 times that of the Earth, the planet is only a little smaller than the Earth. The optical telescope reveals only a layer of clouds which are opaque to visible light. Occasionally these clouds show markings, although these markings are not permanent (Figure 4-5*a*).

**a. Radar and Rotation.** Since the markings are only temporary, astronomers cannot determine the period of rotation by watching the planet rotate. The period of rotation has, however, been determined by radar to be about 243 days, but in the direction opposite to the direction of revolution. The planet rotates in *retrograde motion*, that is, in the reverse direction. The period of rotation with respect to the sun must, therefore, be about 122 days. Consequently, the sun shines on one location for about 61 days.

Retrograde rotation is a bit disconcerting, for it is not the easiest situation to explain. Present theories of the origin and evolution of the solar system (see Chapter 8) attempt to explain only those planets which rotate in the same direction in which they revolve about the sun. At this stage, we simply do not know enough of the past history of the solar system to

| Aug. 13 | Aug. 14 | Aug. 15 |

### Figure 4-5 (a)

Three photographs of Venus taken on consecutive days to show the changes in its cloud cover. (Lick Observatory)

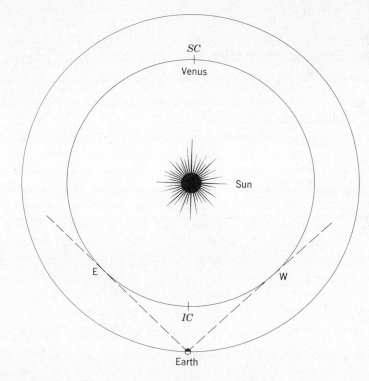

### Figure 4-5 (b)

From the Earth we cannot observe either Mercury or Venus in full phase, for at full phase the planets are at *superior conjunction (SC)* behind the sun. We can only see a thin crescent at most when either planet is at *inferior conjunction (IC)*. As viewed from the Earth the planets are farthest from the sun at *eastern elongation (E)* or *western elongation (W)*, and are most easily observed at these times. Radar studies can be made near inferior conjunction, however.

explain retrograde rotation. We must wait until we have gathered much more information, particularly about the very early history of our solar system. Space exploration appears to be the only method by which we will be able to gather such information.

**b. Other Techniques of Observation.**   Among the most exciting observations made of Venus have been the space probes. Especially valuable were those made by Mariner 5, sent by the United States, and the Russian's Venus 4. Mariner 5 passed behind Venus only 34 hours after Venus 4 made a landing on the surface.

Measurements were made by Venus 4 as its egg-shaped capsule parachuted through the atmosphere. It radioed its findings during this passage, and it was assumed that its radio would stop sending when it struck the surface. Observations from Mariner 5, however, indicate that Venus 4 stopped sending signals at a height of about 15 miles above the surface. Apparently the high temperature of the atmosphere caused an electronic failure in the equipment.

Mariner 5, which passed within 2,500 miles of the surface of the planet on October 19, 1967, gave us information on the atmosphere by sending radio signals through it. Before the space craft went behind the planet, its radio signal was uninterrupted; but as it went behind the planet, its radio signals passed first through the upper atmosphere, then through the full thickness of the atmosphere before it was finally blocked by the rocky surface of the planet itself (Figure 4-6). As the space craft emerged from behind the planet, 21 minutes later, its signal again passed successively through all levels of the atmosphere. That radio signal was altered by Venus' atmosphere, and the manner in which it was altered depended upon the characteristics of the atmosphere. Consequently, a study of that radio signal revealed useful information about the atmosphere of Venus.

**c. The Surface.**   Radar observations, similar to those made of Mercury, indicate that Venus' surface is as rough as surfaces of Mercury and the moon, and probably made of similar material. But this finding is not really a surprise.

**d. The Atmosphere and the Greenhouse Effect.**   The space probes, Venus 4 and Mariner 5, did turn up some surprises with the atmosphere, however. Information radioed back from both space craft can only be interpreted to mean that the atmosphere of Venus is between 80% and 90% carbon dioxide. This is a striking difference from the atmosphere

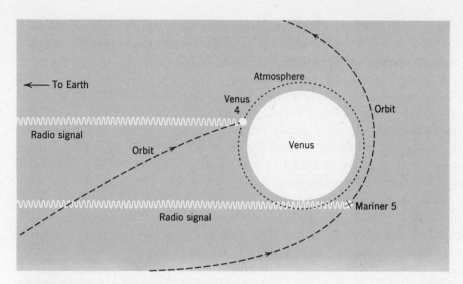

**Figure 4-6**

The signal sent to Earth from Mariner 5 passed through Venus' atmosphere only 34 hours after Venus 4 had crashed onto the planet's surface.

of the Earth, which is 78% nitrogen, 21% oxygen, and only 0.03% carbon dioxide. Information returned from Venus 4 indicates that the atmospheric pressure on Venus is about 20 times that on the surface of the Earth. However, since its radio apparently stopped sending signals at a height of 15 miles, the atmospheric pressure at the surface of Venus must be nearly 100 times that of the Earth.

It is reasonable to ask why there is such a difference between the atmospheres of Venus and the Earth, since both planets are about the same size and reasonably close together in the solar system. But the answer to this question should wait until we have considered the atmosphere of Mars, for its composition should help us draw more general conclusions.

Observations made by radio waves of 2–3 cm in length indicate that Venus radiates energy as if its surface temperature were near 800°F. Observations from Mariner 2 (sent in 1962) and Venus 4 confirm that the surface is indeed hot, perhaps 700°F. A more precise temperature will certainly be determined with future observations.

That the surface of Venus is hot enough to melt not only mercury but also such metals as lead, tin, and zinc is at first puzzling. Observations indicate that the surface of Mercury is 650°F. Why should the surface of

Venus, nearly twice as far from the sun as Mercury, be hotter? Mercury has no atmosphere and Venus does, so perhaps Venus' atmosphere is the cause of the high temperature.

The atmosphere apparently acts as a heat trap. The term describing this action is the *greenhouse effect*: it is recognized that radiation of one wavelength—the short wavelengths of visible and infrared light—can travel through the glass of a greenhouse and warm up the interior, but that since the radiations emitted by the interior are of longer wavelength, they are blocked by the glass. An automobile parked in the summer sun with all the windows rolled up demonstrates this principle only too well.

From studies here on Earth it is known that carbon dioxide is a particularly effective producer of the greenhouse effect. So although the cloudy atmosphere of Venus does prevent some of the sun's radiation from reaching the surface, apparently an ample supply of radiation does reach the surface to warm it up. If heat is then trapped by the atmosphere, the temperature will rise until the planet is giving off as much to outer space as it receives from the sun. At that point the surface temperature becomes stable. Planets radiate heat out into space just as the sun does, by radiant energy. Since the planets are a good deal cooler than the sun, however, they radiate less energy, and that energy has a much longer wavelength. A soldering iron and a clothes iron, even though they do not glow, radiate energy in the infrared region of the spectrum.

Venus' atmosphere is also known to contain at least some water vapor. It is difficult to determine how much, because the light that leaves Venus must pass through our own atmosphere, which certainly contains water vapor. It is not easy to differentiate spectral lines caused by the water vapor in our atmosphere from those originating in the atmosphere of Venus.

In 1964, a spectrometer designed and operated by astronomers at Johns Hopkins University was carried aloft by an unmanned balloon to a height of 87,500 ft—above all but 0.03% of the water vapor in the Earth's atmosphere. A photoelectric cell scanned the spectrum of Venus formed by the spectrometer in the infrared region, for it is in this region where spectral lines of water vapor appear. These observations indicate that the atmosphere of Venus does indeed contain some water, but that much of it may be frozen into ice crystals. The temperature at the top of the layer clouds, as determined by studies in the infrared, is $-38°F$, so it would not be surprising to find the water there frozen into ice.

Spectroscopic studies to detect molecular oxygen in the atmosphere of Venus have yielded negative results. There appears to be very little oxygen, at least on top of the cloud layer. It may well be that most of the oxygen is tied up in the carbon dioxide molecules.

Observations at radio wavelengths in the order of 2–5 mm generally yield a temperature lower than 600°F. It may be that radiations of this wavelength are emitted by a layer of gas in the atmosphere which is colder than the surface. Such layers exist in the Earth's atmosphere.

One other interesting finding of Mariner 5 is that Venus has only a very weak magnetic field, no stronger than 0.03% the strength of the Earth's magnetic field. The big question, however, is not why doesn't Venus have a magnetic field, but why does the Earth? We will consider the question again after we have discussed more of the planets and can draw more general conclusions.

## 4.3 Earth

The third planet out from the sun is the Earth (Star Plate 1). It is special to us because it is our means of life, and it is to our present knowledge unique because it does have life. As our investigations of the universe become more precise and more penetrating, however, it is logical to expect that we will find many, many stars with planets about them and many of these may well have life on them. We have advanced to the stage of sophistication where we do not believe that the universe was made for the sole benefit of mankind here on Earth.

We know more about the Earth's surface and its interior than we do about any other planet; consequently, in our study of the solar system we can take advantage of our knowledge of a particular planet. This knowledge will add greatly to our understanding of all the planets and perhaps help us understand how they were all formed.

**a. The Interior.** Most of the ideas about the Earth's interior are based on information gained from the study of earthquake waves. As these waves travel through the Earth, their speed and direction are changed according to the type of wave and the material through which they are traveling.

These studies show that it is convenient to consider the Earth as four concentric spheres: the inner core, the outer core, the mantle, and the crust (Figure 4-7). The core is composed principally of iron and nickel; the inner core is believed to be solid, the outer core molten. Above the core is the mantle, composed chiefly of the denser rockforming substances. Atop the mantle is the crust, the rocky surface of the Earth. It will be seen in Chapter 7 that knowledge of the Earth's iron–nickel core has influenced our thinking about the formation of the solar system.

The separation of chemicals, as seen in the Earth, can take place in a solid astronomical body only if the interior of that body becomes heated to temperatures in the order of 1,000°F. Only then can the heavier chemicals slowly sink to the center. The most plausible explanation for such high temperatures is that radioactive substances trapped inside act like a furnace, converting nuclear energy into thermal energy.

**b. The Atmosphere.**   The Earth's atmosphere has influenced the surface of the Earth a great deal. The atmosphere carries the water vapor which forms the rains, and the rains have been the chief agent in erosion and transportation of material from one location on the crust to another. The

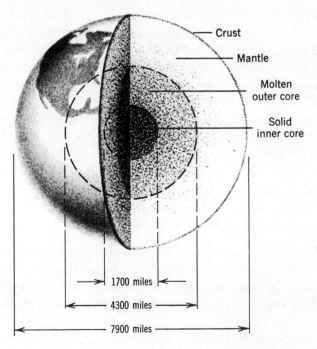

**Figure 4-7**

A cross section of the Earth.

crust has responded to this relocation of material by sinking where the material was deposited and rising in that location from which the material was taken, changing the patterns of erosion.

The atmosphere actually contains very little water vapor. Nitrogen is the dominant gas in the atmosphere (78%), oxygen is next (21%), with only 0.03% carbon dioxide. The rest of the gases comprise the very small percentage that remains.

As valuable as the atmosphere is, the astronomer is anxious to transport telescopes above and beyond it. Earthbound telescopes permit the astronomer to examine the light and radio waves of celestial objects only after they have passed through the atmosphere and its disturbing influence. Severe restrictions are placed upon the terrestrial astronomer because the atmosphere acts like a filter, and permits only certain regions of the electromagnetic spectrum to pass through, namely the visible region and large regions in the radio wavelengths. It blocks out all the other radiations such as X-rays, and much of the ultraviolet and infrared, which contain a great deal of information about the universe. Consequently space explorations have made major contributions by observing the sun and the celestial sphere from above the Earth's atmosphere in X-ray and ultraviolet radiation. These observations (like most completely new observations) tell us not only what we have been missing but also that we have much yet to learn.

The light that does reach our terrestrial telescopes has some alterations imposed upon it by the atmosphere. The atmosphere acts somewhat like a lens, for by refraction it changes the direction of light passing through it (unless that light is traveling vertically). Thus the position of each object in the sky is shifted slightly toward the *zenith* (the point directly overhead) from its true position. The shift is zero for stars in the zenith and increases for stars as they approach the horizon. At the horizon the shift amounts to about 30 minutes of arc, or nearly the angular diameter of the sun and the moon. Therefore when you see the sun (and the moon) sitting on the horizon it has actually already set and is just below the horizon.

Not only does refraction cause a star's position to change in this systematic manner, but also the differential refraction caused by convection currents in the atmosphere (similar to those visible above a hot road) causes a star's image to move about slightly and erratically. To the eye this is twinkling, which may be poetic and romantic. The astronomer, however, calls it "bad seeing," for the unpredictable movement of a star's image can only result in a blurred photograph, or a decrease in the amount of light passing through the slit of a spectrograph.

It is the atmosphere that gives rise to the blue sky which outshines the stars in the daytime. Sunlight that enters the atmosphere is intercepted not only by the gases that give rise to absorption lines but also by tiny dust particles. These dust particles, as well as the gas molecules, act somewhat like reflectors in that they can change the direction of light's travel. The process, which is different from reflection, is called *scattering* and is selective in that molecules and particles scatter the blue light (the shorter wavelengths) more effectively than they scatter red light. Since the blue is scattered more thoroughly, our sky is blue for it consists of this scattered light; the setting sun is red and the harvest moon orange because we see them after their light has had much of the blue taken out. When the sun and the moon are on the horizon their light travels through a longer path in the atmosphere, with the result that more of the blue is scattered out of the direct beam of light.

c. **The Earth's Magnetic Field.** From above the Earth's atmosphere, rockets and space probes have made observations and measurements of features not dreamt of before. Two of the first discoveries of the space age are the Van Allen radiation belts and the extended magnetic field of the Earth. Radiation counters aboard satellites and space probes sent to both Venus and Mars have revealed that the Earth is surrounded by a large magnetic field, a portion of which traps high-energy electrons (Figure 4-8). These electrons travel in corkscrew paths from the region of one pole to the other, along the magnetic lines of force. In the polar regions, these electrons dip into the upper atmosphere, causing it to glow in much the same manner that a gaseous tube glows. This glow is observed as the *aurora*, both the aurora borealis (northern lights) and the aurora australis (southern lights) (Figure 4-9).

The magnetic field of the Earth, however, is not free from external disturbing influences. Space probes have detected the existence of what is called the *solar wind*, a stream of charged particles, mostly electrons and protons, emitted by the sun and flowing past the Earth. Since these particles are charged, they carry a magnetic field with them. This magnetic field reacts with the magnetic field of the Earth in a way which appears to "blow" the magnetic field away from the Earth.

On the sunward side, the Earth's magnetic field extends to a distance of about 10 Earth's radii. Just beyond 10 Earth's radii, a region of magnetic turbulence has been observed, with an outer boundary consisting of a shock wave (Figure 4-10). The shock wave is the encounter between the particles of the solar wind and the Earth's magnetic field. It resembles a shock wave set up by a rock projecting above the surface of a river, or the wave set up by the bow of a ship as it cuts through the water. Outside this shock wave, the magnetic field is that of the solar system.

On the "leeward" (nighttime) side of the Earth the magnetic field forms a "magnetic tail." The Earth's magnetic field has been detected in this direction as far away as 25 Earth's radii, and it appears that the tail is about 40 Earth's radii in diameter. The magnetic lines of force in the northern part of the magnetic tail point toward the Earth, that is, the north end of a compass needle if placed there would point to the Earth. The magnetic lines of force in the southern part of the tail point away from the Earth. The boundary between these two regions of oppositely directed magnetic fields within the tail shows very little—if any—magnetic field and is called the *neutral surface*. The similarity between the shape of the Earth's magnetic tail, with its encircling region of magnetic turbulence, and the shape of a comet's tail is striking.

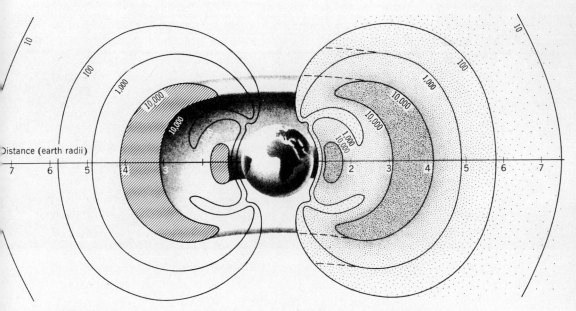

**Figure 4-8**

Van Allen radiation belts. (Reprinted by permission from *Scientific American*)

**Figure 4-9**
Aurora Borealis in Alaska. (Photograph by V. P. Hessler)

## 4.4 Mars

Mars, the "red planet of mystery," is the fourth planet we investigate as as we go farther away from the sun. We are able to study Mars better than we can any of the other planets because of its proximity to us, and because when it is closest to the Earth we see its fully lighted face.

Mars, 1.5 A.U. from the sun, rotates on its axis in 24 hours, 37 minutes, and revolves about the sun once every 687 days. Its diameter, 4,200 miles, is about $\frac{1}{2}$ that of the Earth; its mass is 0.11 times that of the Earth. It has an atmosphere, but one which is not so thick that we cannot see through it. It has permanent surface markings and polar caps. The markings change color during the Martian year and we can see cloud formations that move across the surface of the planet. Most of the early telescopic observations encouraged ideas of life on Mars.

**a. Techniques of Observation.** Mars has been studied extensively with telescopes for more than 100 years. The first reliable maps of its dark markings were made in the middle of the 19th century. That those markings are basically the same today indicates that they are permanent (Figure 4-11).

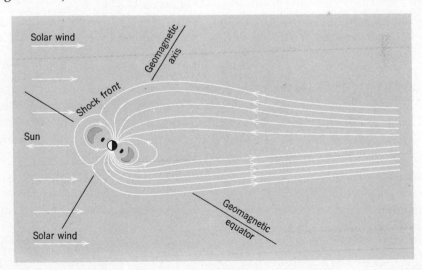

**Figure 4-10**
The Earth's magnetic tail caused by the solar wind.

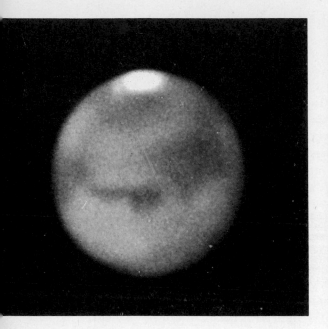 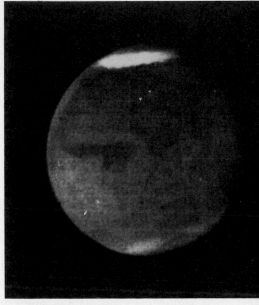

**Figure 4-11 (a)**

Mars photographed by the 24-in. refractor at Flagstaff, Arizona, in 1924. (Lowell Observatory)

**Figure 4-11 (b)**

Mars as photographed by Mariner 6 when within only 460,000 miles of that planet in 1969. (NASA photograph)

Telescopic visual observations of Mars are exciting to make, but they are subjective and rely on the observer's ability to jot down or sketch what he sees. Photographic observations permit a record that can be studied later and by other people. However, the disturbing effect of the Earth's atmosphere, that is, *bad seeing*, causes the image of Mars to move about on the photographic plate and to blur out any distinctness of the features. Although considerable reliance must be placed on good visual observations, thousands upon thousands of good photographs have been made of this well-marked planet.

Photoelectric cells have been attached to telescopes to ascertain the brightness of Mars, and to study any changes in brightness as that planet rotates. The proper use of filters with the photocells has yielded reliable measurements of colors.

Measurements in the infrared and radio region of the spectrum have permitted good estimates of the temperature of the Martian surface to be made.

In 1965, however, a new and rewarding approach to the study of Mars was introduced when Mariner 4 passed behind the planet and not only took photographs of the surface, but also made observations of temperature and sent radio waves back to the Earth through the Martian atmosphere, as Mariner 5 did two years later of Venus. Mariners 6 and 7 flew behind Mars on July 31 and August 5, 1969. These coordinated observations produced excellent photographs, reliable measurements of the temperature, and observations of the atmospheric content and pressure.

**b. The Surface.** The permanent dark markings on the surface of Mars have commanded the strongest interest and aroused the most controversy. Many reliable observers reported dark lines on Mars that came to be regarded as canals, some thought canals made by an intelligent life. The three Mariner flybys, however, ended that long standing controversy. The Martian surface, like the moon, is pockmarked with craters (compare Figure 4-12a with 6-6).

As with the moon, it is strongly suspected that these craters are, for the most part, a result of large chunks of rock falling from outer space onto the surface. Before impact, each of these rocks had presumably been revolving about the sun in an orbit. The Earth, too, has some impact craters (see Figure 7-9), and there must have been many more that have been erased by persistent erosion and catastrophic upheavals of the Earth's crust.

Although only a very small portion of Mars' surface has yet been photographed in detail, it does appear that the craters on Mars are, in general, more eroded than those on the moon. But then Mars has an atmosphere; the winds do blow and the dust forms huge clouds. All this motion of atmosphere and dust must cause erosion on Mars. The moon has no atmosphere.

From a study of the photographs transmitted back from Mariners 6 and 7, it has been concluded that there are three distinct types of topography on Mars. The first is the *cratered* topography; the lunar-type topography seen in Figure 4-12a. The second is called the *chaotic* topography and is shown in Figure 4-12b. The region at the top appears to have resulted from a cave-in or slumping of a vast area; at its widest the slump area is about 60 miles wide. Neither the Earth nor the moon have any known

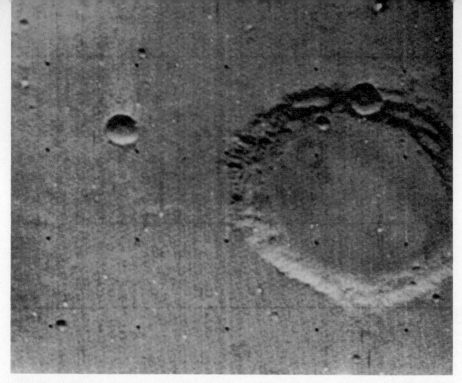

**Figure 4-12 (a)**

Craters on the surface of Mars were photographed in 1969 by Mariner 6.

**Figure 4-12 (b)**

Chaotic terrain photographed on Mars by Mariner 6. (NASA photographs)

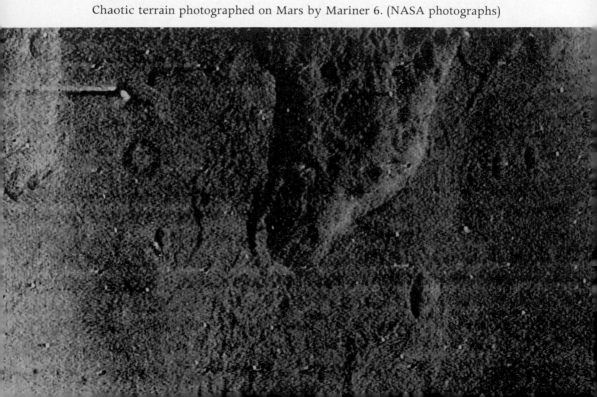

topographic features that resemble this. The third type of Martian topography is the *featureless plain,* a flat area without mountains, ridges, or craters (Figure 4-13). It seems like a vast desert of sand or dust. If that is what it really is, then why should only certain regions be so thoroughly dust covered and not others?

1. *Surface Soil.* The basic red color of Mars is caused by the surface material. Studies of both the color and the polarization of the sun's light reflected from Mars have led astronomers to suspect that the surface soil is composed largely of limonite, an oxide of iron. But whatever the material is, it must exist to a large extent in the form of fine powder, for at times dust storms can cover an entire hemisphere of the planet. Sometimes these dust storms are local, progressing at a rate of up to 30 miles per hour or more, and lasting a day or two. When they are so large as to cover a good part of the planet, as the dust storm in 1956 did, it may last for a month. The question remains, however, where does the dust go when it settles

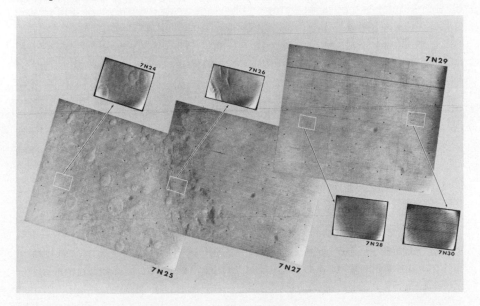

**Figure 4-13**

Photographs of Mars taken by Mariner 7. Photographs 7N25, 27, and 29 were taken with a wide angle camera and show portions of the featureless plane. Photographs 7N24, 26, 28, and 30 were taken with a telescopic lens to reveal detail, if there is any to see. (NASA photographs)

out and falls on the surface? Why aren't the dark areas completely covered with dust? When the dust settles not even the polar caps are covered. Maybe they aren't really dust storms.

2. *The White Polar Caps.* The polar caps of Mars change with the seasons, and are every bit as intriguing as the craters and dark markings. Mars has seasons as does the Earth, except they are nearly twice as long. During the winter in one hemisphere, clouds form over that pole and appear to rise above the limb of the planet (Figure 4-14). Once established, the clouds remain throughout the winter, but dissipate in the spring to reveal the whiter polar cap with its sharper boundary (Figure 4-15).

As the polar cap begins to shrink in the spring, a faint band of darkening develops immediately about its circumference. With the advance of the season, the polar cap shrinks still more and the band of darkening travels toward the equator as a "wave of darkening." The wave of darkening is not a sharply outlined belt or band, it is a slight increase in the darkness of the permanent dark markings. The fact that the wave of darkening travels at the fairly steady rate of about 20 miles per day indicates that it might be caused by winds or some change in the atmosphere which in turn influences the dark markings. The rate of travel is the same season after season.

If the polar caps are assumed to be snow or frost, then it can be presumed that the wave of darkening is water vapor in the atmosphere that diffuses away from the melting snows of the polar cap. Such a theory would have to presume that the water would progress from pole to pole season after season, becoming trapped at the pole which has winter, then moving to the other pole for its winter. It is difficult to imagine a mechanism that would cause such a seasonal migration of water. Yet spectroscopic observations have been made at both the Lick Observatory in California and the MacDonald Observatory in Texas to indicate that the amount of water vapor in the Martian atmosphere is greater in the hemisphere which is in winter and spring, than the amount in the hemisphere that is in summer or autumn.

c. **The Temperature.** The temperature of the surface has been determined by both ground-based telescopes, radio telescopes, and from observations from Mariners 6 and 7. The maximum temperature recorded at noon on the equator is close to 82°F; the coldest, recorded at the south polar cap by Mariner 7, is −190°F. The usual daytime temperatures at the

**Figure 4-14**

Three photographs of Mars taken on three different dates in 1939—July 23, July 25, and July 30—to show the formation, disappearance, and reformation of clouds over the north polar region in just a few days. (Lowell Observatory)

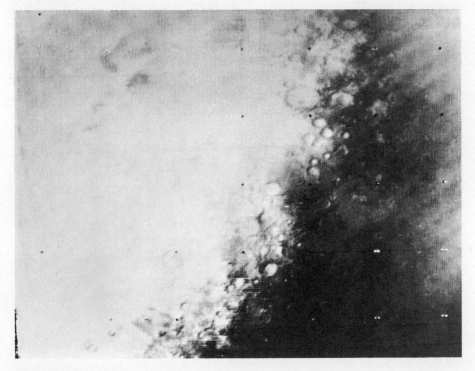

**Figure 4-15**

The edge of Mars' south polar cap photographed by Mariner 7. (NASA photograph)

equator, however, vary from a high of about 70°F at noon to a low of perhaps −100°F just before sunrise. This results in a daily change in temperature of about 170°F. It has been observed that the dark markings are warmer than the rest of the planet.

**d. The Atmosphere.** One of the reasons for the extreme daily changes in temperature is the rarified atmosphere. Observations made with Mariners 6 and 7 indicate an atmospheric pressure at the surface of Mars of only 6 millibars; the atmospheric pressure here on the surface of the Earth is about 1,000 millibars. A pressure of 6 millibars is reached in the Earth's atmosphere only at a height of about 20 miles.

The main constituent in the Martian atmosphere is apparently carbon dioxide. Mariner 6 detected some hydrogen and oxygen in the upper atmosphere. Water vapor has definitely been detected, but in small amounts. There is, surprisingly enough, no evidence of nitrogen. It would appear, therefore, that the atmospheres of Venus and Mars are similar in their chemical makeup; it is the Earth's atmosphere that is unique with its high percentage of nitrogen and oxygen. It seems likely that the composition of the Earth's atmosphere is a result of life. It may be that life on Earth has altered the original atmosphere here—a sort of natural pollution. Detailed studies of other planetary atmospheres, therefore, will certainly help us understand our own atmosphere better. It may well be that such studies could help avoid a catastrophe with manmade pollutants.

**e. The Clouds.** Clouds also appear in the atmosphere of Mars. The dust storms are the best explanation of the yellow clouds, even if we are not yet able to answer all of the questions about the storms. It has been observed that the area covered by a dust storm will be a lighter color for 2 or 3 weeks before returning to its normal color.

White clouds also appear from time to time. They are often seen along the surface at sunrise, and they dissipate as the morning progresses (Figure 4-16). They also form in the late afternoon, so they may persist throughout the night. From 140 nights of observation of Mars in 1964–1965, when the Earth and Mars were on the same side of the sun and consequently close together, the morning clouds were observed on two-thirds of the Martian days, the evening clouds on about one-half.

Whether the clouds are water vapor or not is not yet known. Spectro-scopic studies indicate that if all of the water vapor in the Martian atmo-sphere were to precipitate out onto the surface, the water level would be only 0.01 mm (or 0.004 in.) thick. If all the water in the Earth's atmosphere were to precipitate out (and not sink into the soil) it would cover a layer of water about 1 mm thick on the deserts and about 10 mm or 1 cm thick in regions of high humidity. This means that the desert atmosphere of the Earth has nearly 100 times more water than the atmosphere of Mars.

Because of the dryness of the Martian atmosphere and because of the very high percentage of carbon dioxide, it has been proposed that both the polar caps and the clouds are crystals of carbon dioxide, more commonly called dry ice. Carbon dioxide gas should crystallize out in the atmosphere of Mars at about $-190°F$, the same temperature that Mariner 7 recorded of the south polar cap.

Although many questions have been answered by the extended observations of the Mariner flybys, many new exciting questions are being asked. Progress in research leads to further progress; discoveries lead to still newer discoveries.

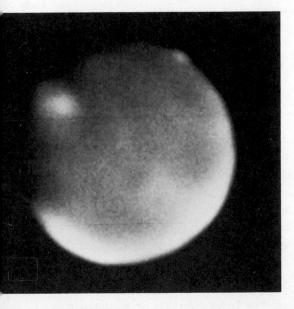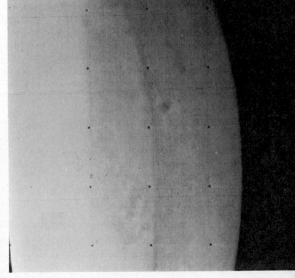

**Figure 4-16**
(*a*) White clouds seen along the terminator of Mars (*left*). (Lick Observatory photograph. (*b*) A haze or dust in the atmosphere of Mars was photographed by Mariner 7 (*right*). This haze is not the same as the white clouds in (*a*). (NASA photograph)

## 4.5 Jupiter

The fifth and largest planet is Jupiter. Its mass is 318 times that of the Earth and its diameter is 86,700 miles, over 10 times the Earth's diameter. It is about 5.2 A.U. from the sun. Because of its enormous distance from the source of heat its temperature is quite low, approximately $-200°F$. There are markings which enable us to determine its period of rotation near the equator at 9 hours, 50 minutes (Figure 4-17). The period of rotation increases at latitudes nearer the poles, to about 9 hours and 56 minutes. From this observation we conclude that the surface visible to us is not solid and that solid material does not occur until a considerable depth under the visible cloud layer.

Jupiter is not only the largest planet but also rotates on its axis in a shorter period than any other planet. The material at the equator whirls about the axis at about 30,000 miles per hour. (The Earth's equator whirls at about 1,000 miles per hour.) As a result, Jupiter is flattened at the poles to the extent of presenting a slightly elliptical rather than a circular disk.

**a. Techniques of Observation.** Jupiter has been investigated much as the inner planets, except that, as yet, no space probe has been sent for a closer look. Nevertheless, the telescope, both optical and radio, has compiled a large store of information from which we can form ideas about the characteristics of Jupiter.

For example, W. A. Baum and C. D. Code of the Hale Observatories observed a star that was in the path of Jupiter's motion. As the star was *occulted* by Jupiter (an occultation is the term applied when a planet or the moon passes in front of a star even if the term eclipse would•apply), its light passed through the upper atmosphere of Jupiter. This observation is similar to the Mariner space probes sending radio signals through the atmospheres of Venus and Mars. The star's light was altered by Jupiter's upper atmosphere, and the way it was altered yields information on the characteristics of that atmosphere.

**b. The Atmosphere.** The atmosphere of Jupiter is completely different from that of any of the inner planets. The main constituent is hydrogen, making up perhaps three-fourths of the atmosphere. Of the remaining one-fourth, perhaps half is helium gas, with the remaining one-eighth being composed of other gases such as ammonia ($NH_3$), water ($H_2O$), and methane ($CH_4$). It may be presumed that since there is so much hydrogen on Jupiter it has combined with all the carbon, nitrogen, and oxygen available.

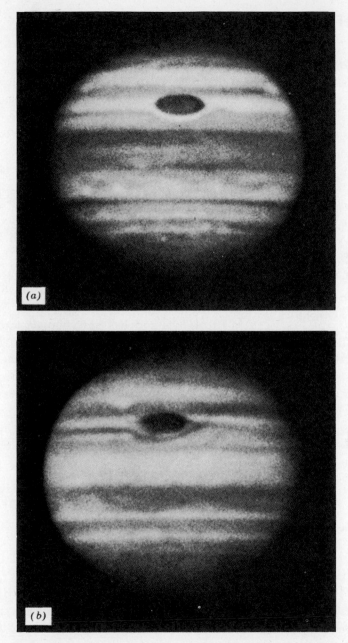

## Figure 4-17

(a) Photograph of Jupiter taken on October 23, 1964 shows the Equatorial Belt dark. The Great Red Spot is obvious. (b) This photograph was taken on February 21, 1966 and the Equatorial Belt is bright. Even the belt containing the spot has changed in the intervening months between these two photographs. (New Mexico State University Observatory)

The temperature at the top of the cloud layer is about −227°F, so cold that most of the ammonia and water are frozen into crystals, but the methane is gaseous. Some of the ammonia exists in the gaseous state however, since spectral lines absorbed by that molecule have been detected. Absorption lines of methane and hydrogen have also been detected.

The dark and light bands that distinguish Jupiter from the other planets are formed, in part at least, by the ammonia and methane crystals. Yet colors such as yellow and shades of red appear from time to time. These may be caused by molecules that arise from beneath the top of the cloud layer, circulate for a time, and then sink again. As a brief study of Figure 4-17 makes evident, the colors of the clouds change, and, in fact, can change drastically in just over one year.

There is one notable feature in these clouds that does remain fairly constant. It is a large spot, often a dull brick red in color, which is 7,000 miles wide and 30,000 miles long. It is shown in Figure 4-17. The Great Red Spot was first seen in the 1660's and has not changed appreciably since, although its brightness does vary as well as its color. It has been observed to move through the surrounding clouds up to 8,000 miles in one (terrestial) day, which indicates that it probably is not attached to any solid surface below the clouds. The true nature of the Great Red Spot is a complete mystery.

White spots, too, appear from time to time; they also move in relation to the clouds but are rather short-lived. They generally smear out into a belt or band.

**c. The Interior.**   The motions of the clouds indicate that they are not bound by any underlying solid surface. This gives us a clue in our study of Jupiter's interior. Although Jupiter has a mass 318 times the mass of the Earth, its volume is 1,312 times that of the Earth. Consequently the material of the Earth, volume for volume, must have a mass that is about 4 times as great as that of Jupiter. This can be expressed in terms of *density*.

The *density* of a substance is its mass per unit volume. The density of water may be taken as 1 and all other materials can then be compared to water. The average density of the Earth is 5.5 and the density of Jupiter is only 1.3. Consequently, volume for volume, the Earth has a mass that is 5.5 times as great as water, whereas Jupiter is only 1.3 times as massive. This low density indicates that Jupiter is about three-fourths hydrogen.

According to one theory, the atmosphere of Jupiter may be less than 500 miles thick, and may rest upon and gradually merge into a sea of liquid hydrogen and helium. With increased depth, the corresponding increase in pressure must compress the hydrogen into a solid. In the very center may be a rocky core not much larger than the Earth.

This concept of Jupiter's structure is supported by the oblateness of its surface. The diameter from pole to pole is 83,200 miles, and the equatorial diameter is 88,700 miles. This is about 5,500 miles out of about 85,000, considerably more than the Earth's oblateness of 27 miles out of about 8,000. Jupiter's greater oblateness can be accounted for not only by its more rapid rotation but also by the theory mentioned above that Jupiter must be much denser at the center than near the surface.

**d. Radio Signals.** Besides reflecting the sun's light, Jupiter emits its own radiation in the form of radio waves. This discovery, made in 1955 by K. L. Franklin and B. Burke, then with the Carnegie Institute, opened up a whole new field of planetary radio astronomy, and gave us a new tool with which to probe the secrets of our solar system. Jupiter has been studied intensively since 1955, not only with radio telescopes, but with increased optical interest. The attempt is to correlate as many observations as possible.

There are two basically different radio emissions from Jupiter: the *microwaves*, with a wavelength between 3 and 68 cm, and the *decameter waves* with wavelengths between 10 and 50 meters. The decameter range is more often referred to by its frequency between 5 and 25 megacycles per second. These will be considered one at a time.

The microwaves are difficult to observe and there is some question as to their nature. For example, by using the radio energies in this range to compute the temperature of Jupiter, discordant calculations result. At a wavelength of 3 cm, the surface temperature is calculated to be −210°F, in close agreement with optical studies. But as longer and longer wavelengths are used, the temperatures become more appropriate for a star than for a planet. Therefore it is necessary to assume that this radio energy is emitted by nonthermal sources, perhaps by electrical discharges, or more regular electron motion. Through the use of radio interferometers to more nearly pinpoint the source of the microwaves, it has become clear that they originate not on the surface but in radiation belts similar to the Van Allen radiation belts around the Earth.

The radiation belts about Jupiter are evidence that this planet, too, has a magnetic field, and one much stronger than the Earth's. The radiation belts about Jupiter extend out to a distance of more than 130,000 miles, far enough to include the innermost satellite, Number V. (It was the fifth satellite to be discovered and happens to be closer to the planet than the four which Galileo discovered.)

The decameter waves which arrive are much more intense than the microwaves and consequently are more easily studied. These waves may appear in bursts lasting only a few seconds, or in "noise storms" lasting many hours. One source of difficulty in studying the decameter waves is that the Earth's atmosphere makes them scintillate or twinkle just as stars twinkle at night. Observations made by two radio telescopes placed 40 miles apart were quite different. Each radio telescope may pick up "bursty" type signals, but there is no correlation between the bursts received from each of the telescopes. Again the need to place telescopes above the Earth's atmosphere becomes apparent.

It has become clear, however, that the decameter waves are associated more closely with the surface than the microwaves. In fact, the decameter waves exhibit a periodicity close to the period of rotation of the planet itself. They would appear to originate from a definite region on the planet.

Since the period of rotation of Jupiter depends upon latitude, two periods of rotation were given in the past. The first is called system I and relates to the equatorial rotation. Its period is given as 9 hours, 50 minutes, 30.003 seconds. The second is system II and is associated with the rest of the planet; its period is 9 hours, 55 minutes, 40.632 seconds. The period of rotation observed by decameter radio waves is 9 hours, 55 minutes, 29.37 seconds and is called system III.

A mysterious correlation between the radio bursts in the 10-meter range and the motion of the second satellite, Io, was detected in 1964. It appears that radio bursts in this wavelength range are much more apt to occur when Io is on one side of the planet as seen from the Earth, and not when it is on the other side. This correlation is not understood at all. More observations and theoretical study are needed.

## 4.6 Saturn

Saturn, 9.5 A.U. from the sun, is perhaps the most spectacular astronomical sight in a telescope, because of the fantastic rings which encircle it, and which have long been used by cartoon artists to set the scene out in space. Saturn revolves about the sun in $29\frac{1}{2}$ years. Its oblateness is more extreme than that of Jupiter; its polar diameter is 67,900 miles and its equatorial diameter is 75,100 miles. The period of rotation at the equator of Saturn is 10 hours, 14 minutes, but with increasing latitude its period of rotation increases even more rapidly than with Jupiter. The density of Saturn is more extreme than that of Jupiter; at 0.68 it is even less than that of water— even less than that of butter! Its mass, however, is 95 times that of the Earth.

Saturn's atmosphere is similar to that of Jupiter except that since it is colder on Saturn ($-290°F$) some of the ammonia has crystallized out of the gaseous state. Therefore Saturn's atmosphere has in the gaseous state relatively more methane and less ammonia. (The solid particles, of course, do not add any dark lines to its spectrum.)

Saturn also has belts or bands somewhat like Jupiter's, though they are much less pronounced. These also change from time to time. Occasionally, white spots appear which merge into a white belt and then disperse into the atmosphere.

The multiple rings are what make Saturn unique in our solar system. They have a maximum diameter of 170,000 miles, are concentric in a single plane, cannot be much thicker than 10 miles and may be thinner. Looking through a large telescope one can see gaps between them. The

**Figure 4-18**

A photograph of Saturn with its ring system showing Cassini's division. (Lick Observatory)

outer gap, called Cassini's division, is seen in Figure 4-18. There is also an inner division.

The composition of the rings was a major puzzle for astronomers until the advent of the spectrograph. Around the turn of the century an astronomer named Keeler placed the image of the planet formed by his telescope across the slit of a spectrograph in order to find the rate of rotation of the rings. If the rings are solid then the outer part should rotate faster since it has farther to go in one rotation. If they are not solid, however, then the inner portion should have a higher velocity than the outer. This is derived from the same principle (discussed in Chapter 5) that explains why Mercury revolves about the sun faster than the other planets, and why Pluto, the planet most distant from the sun, revolves the slowest in its orbit.*

Keeler determined the velocities of approach and recession of the limbs of both Saturn and its rings by the Doppler principle. This critical test revealed that the inner portion of the rings rotates more rapidly than the outer portions and therefore the rings are not solid.

But if the rings are not solid, what are they? So far as we know they yield no spectral lines of their own and therefore cannot be gaseous. Since they do reflect the sun's light, however, the spectrum of the light reflected can be examined. As two paints can be matched by comparing their reflection spectra, so the material on the surface of particles composing Saturn's rings can be identified by comparing the reflection spectrum of Saturn's rings with materials in the terrestrial laboratory. Studies of this nature reveal that the reflection spectrum of the rings closely matches the reflection spectrum of ice at $-190°F$. The conclusion has been drawn, therefore, that the particles are either small bits of ice, or ice-coated particles.

As viewed from the Earth, the rings present different aspects because of the orientation of the three planes: the plane of the Earth's orbit, the plane of Saturn's orbit, and the plane of the rings. At most times, these planes are such that the rings of Saturn appear tilted as seen from the Earth. But at times the rings appear edge on, and since they are so thin, they are then scarcely visible.

*The orbital velocity of Mercury is 29.7 miles per second whereas Pluto's orbital velocity is only 3.0 miles per second.

## 4.7 Uranus

The planets discussed thus far are bright enough to be seen with the naked eye and consequently have been known since prehistoric times. Uranus, the first planet discovered in recorded history, is the next to be seen as we go out from the sun. One of the greatest astronomical observers of all times, William Herschel, discovered it in 1781. The story of the discovery indicates the value of better instruments and experience in observing.

Herschel built his own telescopes as an amateur and devoted his life to observing. Those were days when but few of the spectacles of the sky were known, and it was Herschel who discovered many of the objects that today are being investigated so that we may better understand how the universe is put together and why it stays together—if it is staying together.

Herschel spent much of his time scanning the skies looking for objects that were not stellar in appearance, objects such as the large nebulae we discuss in a later chapter. One of these objects presented a disk-like structure which made it appear different not only from the stars but from the nebulae as well. On magnifying it further Herschel thought it was a comet, but later it was recognized as the first discovered planet. Moreover, it had already been seen by one other astronomer who at that very time was preparing a star map of the sky and had in fact recorded the planet's position twelve times. But he had thought of it only as a star, partly because his telescope was not as large as Herschel's and partly because it was stars with which he was concerned.

Uranus is about 19.2 A.U. from the sun, with a mass 14 times as large as that of the Earth. It revolves about the sun in 84 years and rotates on its axis in 10 hours, 49 minutes. Its diameter is about 30,000 miles. Because of its great distance from the sun its surface temperature, at $-300°F$, is even less than that of Saturn. Its atmosphere is similar to that of Jupiter and Saturn, except that even more of the ammonia has crystallized, leaving an atmosphere with an even higher percentage of methane gas. Molecular hydrogen has been discovered on Uranus as well as on Jupiter.

Uranus is so far from the Earth that any surface markings it may have are very difficult to see, so its period of rotation had to be determined by the spectrograph.

The most unusual thing about Uranus is the inclination of its axis of rotation. The Earth's axis of rotation is not perpendicular to the plane of its orbit but is tipped $23\frac{1}{2}°$ from this perpendicular. The axis of rotation of Uranus, however, is tipped 98° from the perpendicular to its orbit. Or, to put it another way, its axis is tipped by only 82°, if we consider it as rotating in retrograde motion (Figure 4-19). Just how Uranus got this way is a puzzle which must be answered by any satisfactory theory of the origin of the solar system.

After Uranus was discovered the astronomers observed its position carefully over a period of years in order to determine its orbit, that is, where it goes and how fast. Newton's laws of motion and his law of gravity (see Chapter 5) describe the motion of the planets about the sun very well and can be used to predict the position of a planet or comet once its orbit has been determined, but Uranus did not move as expected. Two astronomers, Adams and Leverrier, noted this discrepancy and independently both concluded that Uranus was not following its prescribed path because an assumed planet, still more distant from the sun, was disturbing its motion. By Newton's law of gravity they both calculated where such a planet should be.

Leverrier made contact with an observatory that had been preparing a star map. By comparing the position of the predicted planet on the star map with the actual sky, the new planet was seen within a matter of minutes. This was a tremendous triumph for Newton's law of gravity; not only could it account for the motions of the planets but it enabled the

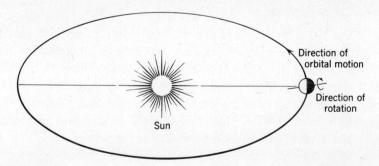

**Figure 4-19**

The axis of rotation of Uranus is tipped by 98° from the perpendicular to the plane of its orbit. To compare with the Earth see Figure 5-5.

astronomer to detect heretofore unknown planets by their gravitational effect on already known planets, *and* to determine their positions!

## 4.8 Neptune

The new planet was named Neptune. It is about 30 A.U. from the sun, has an equatorial diameter of 28,000 miles (it is not as oblate as Jupiter), and revolves about the sun in 165 years. The mass of Neptune is 17 times that of the Earth. It has a period of rotation a little longer than 15 hours and a surface temperature of about − 330°F. Aside from knowing that Neptune has stronger bands of methane in its spectrum than any other planet, astronomers know little about the planet.

Uranus was discovered in 1781 and Neptune was discovered in 1846. These planets travel so slowly (their average orbital velocities are respectively 4.2 and 3.4 miles per second), that it takes a long time to determine an accurate orbit for them. By the early 1900's Lowell, the founder of the Lowell Observatory in Flagstaff, Arizona, had reobserved Uranus and concluded that the discovery of Neptune could not account for all the discrepancies in the orbit of Uranus. He therefore predicted that there must be another planet even farther away than Neptune. The new planet was discovered by Clyde Tombaugh in 1930 at the observatory in Flagstaff and named Pluto.

## 4.9 Pluto

Pluto travels around the sun in 248 years and with an average velocity of 3.0 miles per second. It is about 40 A.U. from the sun and its surface temperature seems to be about − 348°F. Because it is so very far away, it is extremely difficult to observe. Nevertheless, its period of rotation has been determined by photoelectric measurements by Robert Hardie of the Dominion Observatory, Ottawa. Apparently its surface is not uniform, for the light it reflects varies with a consistent period of 6 days, 9 hours, 17 minutes. This is assumed to be its period of rotation.

The diameter of Pluto remains a bit of a mystery, although its maximum size has been estimated to be 3,600 miles. This estimation is based on the knowledge of which stars Pluto's tiny apparent disk does *not* occult (pass in front of) as it moves across the sky.

Since Pluto's spectrum does not contain any molecular absorption

Moon

Mercury · Venus · Earth · Mars

Jupiter · Saturn · Uranus · Neptune · Pluto

Portion of the sun

## Figure 4-20

The largest members of the solar system drawn to scale. The sun's diameter is nearly 10 times that of Jupiter, which in turn is nearly 10 times that of Earth.

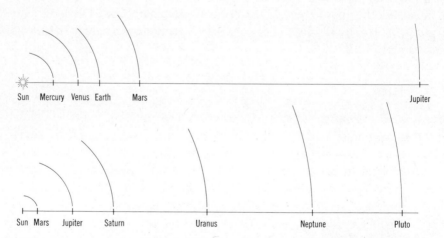

Sun   Mercury  Venus  Earth      Mars                                    Jupiter

Sun  Mars   Jupiter  Saturn        Uranus              Neptune          Pluto

## Figure 4-21

The Jovian planets are so much farther from the sun than the terrestrial planets that two scale drawings are needed to show both. The planets themselves would be microscopic on either scale.

bands, we assume that it is quite different from the giant planets between Mars and Pluto. In fact, it has been speculated that Pluto was once a satellite of Neptune that escaped and began circling the sun in an orbit of its own. (Since Pluto now revolves about the sun and not about Neptune, it must be classified as a planet and *not* as a satellite.) The main justification for this hypothesis is Pluto's eccentric orbit. The planet actually travels inside of Neptune's orbit for a portion of its journey around the sun.

## 4.10 Sizes and Distances

We have spoken of the sizes and distances of the planets, but it is difficult to grasp the dimensions involved because they are so large. Figure 4-20 shows the planets with respect to their sizes. The sun is added in the figure to show how much larger it is than any of the planets (the sun's diameter is nearly 10 times that of Jupiter). This is fairly easy to imagine, but conceiving of their distances is more difficult. Mercury is 36 million miles from the sun and Pluto is about 100 times as far away or 3,662 million miles. This can be drawn to scale, but to fit such a model of the solar system on a page the size of this book it has been necessary to break the solar system into two parts and to draw each to a different scale. This is shown in Figure 4-21. On both scales each planet is too small to be seen except by

**Table 4-1 ● Data on the Planets**

| | Mean distance from sun | | | | | | Den-sity in | Escape veloc-ity, |
| | Astro-nomical units | Million miles | Period of revolution | Period of rotation | Diameter in miles | Mass in terms of Earth's mass | terms of water | miles per second |
|---|---|---|---|---|---|---|---|---|
| Mercury | 0.39 | 36 | 87.96 d | 58.6 d | 3,100 | 0.054 | 5.3 | 2.6 |
| Venus | 0.72 | 67 | 224.69 | − 243 d | 7,600 | 0.82 | 5.0 | 6.5 |
| Earth | 1.00 | 93 | 365.24 | 23 hr 56 min | 7,917 | 1.00 | 5.5 | 7.0 |
| Mars | 1.52 | 142 | 686.95 | 24 hr 37 min | 4,200 | 0.11 | 4.0 | 3.2 |
| Jupiter | 5.20 | 484 | 11.86 yr | 9 hr 50 min | 88,700 | 317.8 | 1.3 | 37.1 |
| Saturn | 9.56 | 889 | 29.56 | 10 hr 14 min | 75,100 | 94.2 | 0.7 | 22.3 |
| Uranus | 19.2 | 1,782 | 83.95 | 10 hr 49 min | 29,000 | 14.5 | 1.6 | 13.9 |
| Neptune | 30.1 | 2,784 | 163.9 | 15 hr 40 min | 28,000 | 17.2 | 2.3 | 15.4 |
| Pluto | 39.4 | 3,662 | 247.3 | 6 d 9 hr 17 min | < 3,600 | 0.2? | ? | ? |

a good microscope. The differences in the sizes and distances from the sun of the terrestrial planets (the rocky Earth-like planets) and the Jovian planets (the Jupiter-like planets) is clear.

The Earth seems large to us; and so, upon reflection, do the other planets. All are complex in their structure and fascinating in their details. But what we know of them, even of those we have studied the most, is little compared to the complete picture we would like to have. It is space and the distance between the planets that keeps us from this knowledge— distances so vast that even Jupiter, 87,000 miles across, is microscopic in comparison to the 480 million miles that separate it from the sun, or the 3,700 million miles that separate the sun from Pluto.

## BASIC VOCABULARY FOR SUBSEQUENT READING

| | |
|---|---|
| Albedo | Revolution |
| Astronomical unit | Rotation |
| Density | Scattering of light |
| Escape velocity | Solar system |
| Limb | Terminator |
| Mass | Zenith |

## QUESTIONS AND PROBLEMS

1. Draw to scale the sun and its nine planets. (One possible scale is to let 8,000 miles be represented by 0.5 centimeter.)
2. Explain how the period of rotation of a planet can be determined by the use of radar.
3. Compare the surface of Mars with that of the Earth and that of the moon.
4. Describe the radio signals received from Jupiter.
5. What are the significant differences between the terrestrial planets and the Jovian planets?
6. Discuss the possibility of life on each of the other planets of our solar system.
7. Explain how the period of rotation of Pluto was determined.

# FOR FURTHER READING

Glasstone, S., *Sourcebook on the Space Sciences*, D. Van Nostrand Co., Princeton, N.J., 1965, Chapters 10 and 11.

Glasstone, S., *The Book of Mars*, National Aeronautics and Space Administration, Washington, D.C., 1968.

Jackson, J. H., *Pictorial Guide to the Planets*, Thomas Y. Crowell Co., New York, 1965.

Page, T., and L. W. Page, ed., *Neighbors of the Earth*, The Macmillan Co., New York, 1965.

Petrie, W., *Keoeeit—The Story of the Aurora Borealis*, Pergamon Press, New York, 1963.

Slipher, E. C., *A Photographic Study of the Brighter Planets*, Lowell Observatory, Flagstaff, Arizona, and The National Geographical Society, Washington, D.C., 1964.

Smith, A. G., and T. D. Carr, *Radio Exploration of the Planetary System*, D. Van Nostrand Momentum Book, Princeton, N.J., 1964.

Struve, O., and V. Zebergs, *Astronomy of the 20th Century*, Crowell, Collier and Macmillan, New York, 1962, Chapter VIII.

Whipple, F. L., *Earth, Moon, and Planets*, 3rd ed., Harvard University Press, Cambridge, Mass., 1968.

Akasofu, S.-I., "The Aurora," *Scientific American*, p. 55 (Dec. 1965).

Cahill, L. J., "The Magnetosphere," *Scientific American*, p. 58 (March 1965).

Chapman, C. R., "The Discovery of Jupiter's Red Spot," *Sky and Telescope*, p. 276 (May 1968).

Cross, C. A., "A Mariners' 1969 closeup Map of Mars," *Sky and Telescope*, p. 16 (July 1971).

Eshleman, V. R., "The Atmospheres of Mars and Venus," *Scientific American*, p. 78 (March 1969).

Franklin, K. L., "Radio Waves from Jupiter," *Scientific American*, p. 35 (July 1964).

Goodman, J. W., "The Edgewise Presentation of Saturn's Rings," *Sky and Telescope*, p. 128 (Sept. 1965).

Leighton, R. B., "The Surface of Mars," *Scientific American*, p. 26 (May 1970).

O'Brian, B. J., "Radiation Belts," *Scientific American*, p. 84 (May 1963).

Parker, E. N., "The Solar Wind," *Scientific American*, p. 66 (April 1964).

Shapiro, I. I., "Radar Observations of the Planets," *Scientific American*, p. 28 (July 1968).

Strong, J., "Infrared Astronomy by Balloon," *Scientific American*, p. 28 (Jan. 1965).

"Bright Flares on Mars," *Sky and Telescope*, p. 83 (Feb. 1970).

"The Composition of Saturn's Rings," *Sky and Telescope*, p. 14 (Jan. 1970); p. 80 (Feb. 1970).

"First Findings of the Mariner Flybys," *Sky and Telescope*, p. 232 (Oct. 1969).

"A Jovian Atmospheric Feature of Special Interest," *Sky and Telescope*, p. 118 (Feb. 1965).

"Mars Pictures from Mariners 6 and 7," *Sky and Telescope*, p. 212 (Oct. 1969).

"Pluto's Diameter," *Sky and Telescope*, p. 213 (Oct. 1965).

"Some Highlights of the Current Apparition of Mars," *Sky and Telescope*, p. 72 (Aug. 1969).

# PLANETS
# IN MOTION

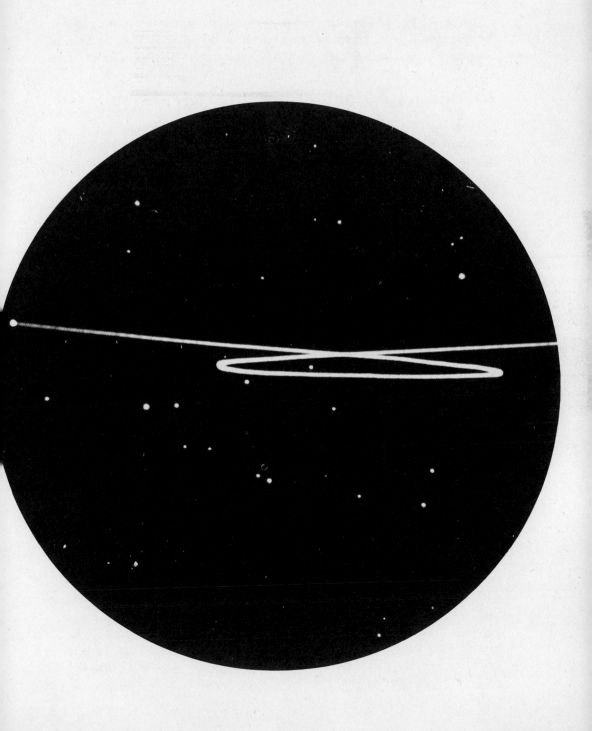

Chapter 4 gave us some idea of the physical characteristics of the planets. But this is not a complete description of the solar system, or even of the planets. To complete their story it is necessary to discuss their motions.

## 5.1 Observations and a Frame of Reference

"Planet," derived from a Greek word meaning "wandering," is an appropriate name for these objects in the sky that appear to be continually moving stars. Their motions were observed many thousands of years ago and can be observed today by looking at the sky with some care.

Few people, however, pay much attention to the sky. If they did they would realize that the planets move against the background of stars over the weeks and months. They would also be able to see that the moon moves against the same backdrop in just an hour's time, and that even the sun moves differently in the sky than do the distant stars.

It should be noticed that each of these motions has been related to the background of stars, for these apparent motions become more obvious when compared to the "fixed" system of stars. But that to which we must ultimately refer all these motions is the Earth itself, for the sun, the moon, and the planets would appear to move as they do if the stars were not there. In other words, the Earth (the local horizon) is the *frame of reference* against which we measure all these apparent motions.

Every motion must be referred to some frame of reference. The motion of a car is tacitly referred to the Earth when we say that it is traveling 60 miles per hour, for it is traveling much faster than this when referred to celestial frames of reference.

The apparent *diurnal* (daily) motion of the stars is from east to west. The moon moves differently from the stars in that it moves westward more slowly than they do. Consequently it appears to move eastward relative to the stars (that is, it lags behind in the "diurnal race") approximately one of its diameters ($\frac{1}{2}°$) every hour. Thus the moon sets on the average 49 minutes ($1° = 4$ minutes of time) later each day.

### Chapter Opening Photo

The retrograde motion of Mars during August 1971 as photographed with a time exposure in the Medford (Oregon) Schools Planetarium. Since the speed of the planet changes, its brightness in a time exposure also changes. The path is brighter where the planet moves more slowly. (Courtesy Jack Fink, Director, Medford Schools Planetarium)

The sun, too, moves eastward relative to the stars, but this motion is only about 1° each day. To put it another way (since we reckon civil time by the sun), the stars move west relative to the sun about 1° each day. Thus each star sets about 4 minutes earlier on succeeding nights.

The planets, generally speaking, also move eastward through the field of stars, but there are times when each planet moves westward. Since this motion is contrary to the usual motion it is called *retrograde motion*. The photograph that opens this chapter and Figure 5-1 show schematically how Mars may move during retrograde motion when it is brighter than at any other time.

## 5.2 Kepler's Laws

Since the time of Plato, in ancient Greece, there have been a number of theories to explain the motion of the planets. Plato, in effect, challenged the mathematicians to explain the apparently erratic motion of the planets, and he went so far as to specify that the motion must be circular, at a

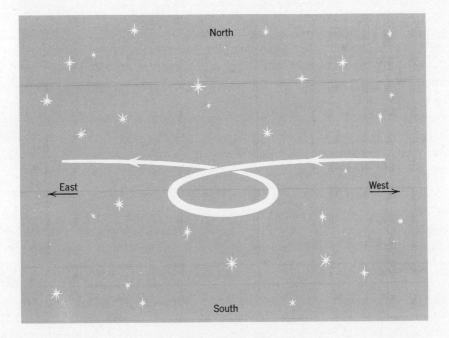

**Figure 5-1**

The westward motion of Mars (or any planet) is called retrograde motion.

steady speed, and *geocentric* (from the Greek, *ge*, Earth). The circular orbits and the steady speed were selected essentially because of their basic symmetry and "perfection." The geocentrism, however, was undoubtedly favored because it is more comforting to believe that the Earth is *not* moving, and more pleasing to man's ego to think that the Earth is at the very center of the universe. However, in the seventeenth century, the *heliocentric* (from the Greek, *helio*, sun) theory became established once and for all, but not without a struggle and not without men saying, "Prove it to me!" But theory must precede the proof.

In the first two decades of the seventeenth century, Johannes Kepler published what have become known as Kepler's three laws of planetary motion. The first two laws describe the shape of a planetary orbit and the motion of the planet in that orbit. The third law compares the motion of any two of the planets revolving about the sun. Although the laws were derived specifically for the planets revolving about the sun, they apply to satellites revolving about the Earth and planets revolving about any star.

**a. First Law.** The shape of each planetary orbit is described by the first law as an *ellipse*. An ellipse is a figure that can be drawn by taking a piece of string, 8 in. long, for example, tacking the ends down by tacks that are placed 6 in. apart, and running a pencil inside the string, always keeping the string taut. The position that each tack occupies is called a *focus*. *The sun occupies one of the foci of each planetary orbit*.

Since the sun is at one of the foci, the distance of each planet from the sun varies (Figure 5-2). When the planet is in position *A*, *aphelion*, it is at its farthest from the sun. When it is in position *P*, *perihelion*, the planet makes its closest approach to the sun. The corresponding terms in the orbit of a satellite of the Earth are *apogee* and *perigee*.

**b. Second Law.** The second law is a little more difficult, but it is important to understand the motions of objects about a central body when under gravitational attraction. The line joining the planet with the sun is called the *radius vector* and, according to Kepler's second law, *the radius vector sweeps out equal areas in equal lengths of time*. This means that while the planet goes from point *a* to point *b* (both near perihelion) in a certain length of time, say one month, its radius vector will sweep out the area that is shaded (Figure 5-2). Starting from point *c*, its radius vector will sweep out the same area in the same length of time, but the planet will move more slowly since the radius vector is longer near aphelion. Thus when a planet is near perihelion it travels faster than when it is near aphelion. The average velocity of the Earth in its orbit is 18.6 miles per second; at perihelion it travels 18.9 miles per second and at aphelion it travels 18.3 miles per second.

c. **Third Law.** Kepler's third law describes how the periods of the planets depend on the average radii of their orbits. It can be most easily expressed in the form of a proportion:

$$\frac{R^3}{r^3} = \frac{P^2}{p^2}$$

where $R$ is the average orbital radius of one planet, $r$ the average orbital radius of the other, and $P$ and $p$ their respective periods of revolution about the sun. In effect, Kepler's third law tells us that the closer a planet is to the sun the faster it travels. Mercury's average orbital velocity is 30 miles per second, that of Pluto is 3 miles per second.

Kepler knew the periods of all the planets visible to the naked eye and

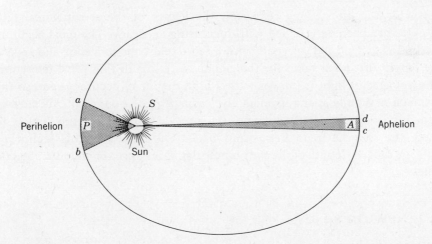

**Figure 5-2**

According to Kepler's second law, the line joining the sun with any planet sweeps out equal areas in equal intervals of time. If the planet takes the same length of time to move from $a$ to $b$ as it does from $c$ to $d$ then the areas $Sab$ and $Scd$ must be equal.

he had only to find the length of the orbital radius of one planet to calculate the orbital radius of each planet and thus determine the size of the solar system.

The average distance of the Earth from the sun is 1 A.U., so Kepler's third law permits us to find the distance of any planet in astronomical units by comparing that planet with the Earth. Jupiter's period of revolution is nearly 12 years, the Earth's is 1 year; the distance of the Earth from the sun is 1 A.U. We can now find $R$, the average distance of Jupiter from the sun:

$$\frac{R^3}{1} = \frac{12^2}{1}$$

$$R^3 = 144$$

$$R = 5.2 \text{ A.U.}$$

Although Kepler never learned the size of the solar system in comparison to a standard distance on the surface of the Earth (such as a mile), he did know that Jupiter was about 5.2 times as far from the sun as the Earth is.

During the intervening centuries, the length of the astronomical unit in miles has been determined with increasing accuracy. The most accurate method is to send radar signals with a radio telescope to Venus and record the length of time it takes for their echo to return to that same telescope. The velocity of light is known with a high degree of accuracy, and so the distance of Venus at the moment the radar signals bounce off its surface depends upon our ability to time the interval it takes for the signals to make the round trip to Venus and back. The results of this technique indicate that the length of the astronomical unit is approximately 92,956,000 miles.

## 5.3 Newton's Laws

Newton sought the answer to the questions "Why does the moon travel about the Earth in an elliptical orbit? Why doesn't it fly off into space and travel in a straight line at a constant speed?"

**a. First and Second Law of Motion.** Newton recognized that any object which has no unbalanced force acting on it will either remain at rest or, if it is in motion, will continue its motion in a straight line at a constant speed. This is Newton's *first law of motion*. We sometimes call it the law of inertia, for inertia is that property of matter which resists change from either its position of rest or its motion in a straight line at a constant speed.

Any change in motion we call an *acceleration* whether it is an increase or decrease in speed, or a change in the direction of travel. Newton realized that for an object to accelerate, an unbalanced force has to act on that object. It is Newton's *second law of motion* which explains that the cause of acceleration is an unbalanced force. The larger the unbalanced force acting on an object, the greater the acceleration. But not all objects have the same inertia, that is, the same resistance to acceleration. The amount of inertia that an object has is called its *mass*, and the greater the mass (for the same force) the less the acceleration. This can be expressed mathematically as

$$\text{Force} = \text{mass} \times \text{acceleration}$$

$$F = ma$$

Therefore, since the moon (or an artificial satellite or a planet) is continually accelerating by continually changing its direction of travel, there must be an unbalanced force acting on it (Figure 5-3). Since the moon travels in nearly a circular orbit this force must act so as to pull the moon toward the center of that orbit.

**b. Law of Gravitation.**   But the Earth is at the center of the moon's orbit, and this fact led Newton to the conclusion that it is the Earth which exerts the required force on the moon. Likewise, he felt, the moon must exert a force on the Earth. But if the moon and the Earth exert an attractive force on each other, then so must the sun and the Earth, the sun and Saturn, the Earth and Saturn, etc. Newton then derived the expression relating the force of gravitational attraction between two bodies to the mass of each body and to the distances separating them. This relationship is *Newton's law of gravity: every object in the universe attracts every other object in the universe with a force that is proportional to the product of the masses of the two objects, $m_1$ and $m_2$, and inversely proportional to the square of the distance d between them.*

$$F = G \frac{m_1 m_2}{d^2}$$

where $G$ is the universal gravitational constant which changes the proportionality expressed above into an equality. In effect, $G$ tells us something about the universe—about the strength of the gravitational forces when compared with other forces such as electrical forces.

But how does this gravitational force operate? Newton was not able to say, but the fact that it does operate as described by his law of gravity and that it alone accounted for the motion of the planets (until the introduction of the principle of general relativity some 300 years after Newton) makes it a bold and decisive step in man's description of the universe.

But even if Newton's laws of motion and his law of gravity were generally accepted within a few decades after he published them in 1687, men still looked for observational *proof* that the Earth rotates on its axis and revolves about the sun.

## 5.4 A Proof and Consequence

A formal proof of the heliocentric theory was long in coming, for the observations necessary are not easy; they require either refined instruments or an advanced analysis. As a result, the geocentric theory, out of

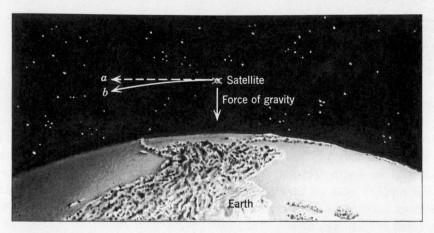

**Figure 5-3**

Any object, such as a satellite, if not acted on by a force, would continue traveling in a straight line. The Earth's force of gravity, however, causes a satellite to follow an elliptical path. The satellite is always falling but is moving ahead with such a high velocity that it falls into an orbit which misses the Earth.

sheer egotism, nourished man's beliefs of his superiority. The geocentric system placed mankind in the auspicious position of being in command of the Earth which was at the very center of the universe. This made man the most important thing in the universe; the universe was made for man. It was not so much a geocentric system as it was an *anthropocentric* (man-centered) system.

The problem revolves about the questions: "What constitutes *a proof* of the Earth's motion?" and "What constitutes *a consequence* of that motion?"

**a. A Proof of the Earth's Motion.** It was realized rather early that if the Earth does revolve around the sun and if the stars are at different distances, then as a consequence of changes in the Earth's position there should be a shifting back and forth of the nearest stars with respect to those more distant. This apparent annual shift of position for the nearest stars is called *heliocentric parallax* (Figure 5-4).

When the Earth is in the position shown in *a* the nearer star appears in the direction indicated, and a photograph of the region might appear something like star plate 1. In *b*, 6 months later, the Earth has changed its position and the nearer star is seen from a different direction. The corresponding photograph might appear as star plate 2.

The result is that the nearest stars will appear to oscillate back and forth against the background stars in a period that is equal to the period of the Earth's revolution. This fact, since it cannot be explained independently of the Earth's revolving, constitutes in itself sufficient proof that the Earth does revolve.

Since the stars are so far away in comparison with the radius of the Earth's orbit, the resulting change in their positions is extremely small and thus was not detected until the nineteenth century. Other observations before the nineteenth century, however, helped convince many of the philosophers and scientists that the Earth does revolve about the sun. One such observation was the measurement of the velocity of light by Roemer, which was done in 1675 (see page 19).

A second proof of the Earth's revolution can readily be made from the study of spectroscopy. The Earth has an average velocity in its orbit of 18.6 miles per second, which is enough to cause a measurable Doppler shift in the spectrum of a star that is in or nearly in the plane of the Earth's orbit. This has not only been observed but it becomes a nuisance when

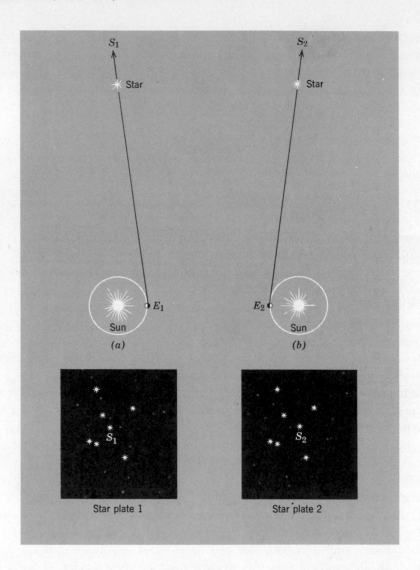

**Figure 5-4**

At one time of the year a nearby star (*a*) will be seen in one direction, $S_1$; 6 months later (*b*) the same star will be seen in a slightly different direction, $S_2$. The difference in direction is highly exaggerated, for in reality it is always less than 1 second of arc. Star plate 1 shows the position $S_1$ against the more distant stars. Star plate 2 shows the new position of the nearby star.

an astronomer wants to measure the radial velocity of a star. To obtain consistent radial velocities for any star it is convenient to use the sun as a frame of reference rather than the revolving Earth, so the Earth's motion must be subtracted out of each measurement of radial velocity.

**b. A Consequence of the Earth's Motion.** It is common knowledge today that the Earth revolves about the sun and as a *consequence* we have our seasons. It was pointed out that the axis of rotation of the Earth is not perpendicular to the plane of the Earth's orbit (see page 100). For purposes of this discussion only, the axis of rotation may be said to maintain the same direction with respect to the stars (Figure 5-5). When the Earth is in position 1 the north pole is tipped toward the sun, and the south pole is tipped away from the sun. The darkened portion of the Earth is in the shadow and therefore in nighttime. It can be seen that not only is the north pole receiving sunlight, but, as the Earth rotates on its axis, the sun will not set. This is called the midnight sun. At the same time, the south pole is in continual night.

The sun's rays are shining more nearly perpendicularly on the northern hemisphere and so their heating effect is stronger. At the same time they are shining more obliquely on the southern hemisphere and their heating effect is accordingly diminished. Thus when the Earth is in position 1 the northern hemisphere is beginning summer while the southern hemisphere is beginning winter.

Three months later the Earth has moved over to position 2 while its axis has remained parallel to the direction in which it pointed in position 1. At this time both hemispheres are equally illuminated by the sun. The sun is on the horizon of both poles, that is, it is setting on the north pole and rising on the south pole. This is the beginning of autumn for the northern hemisphere and spring for the southern hemisphere. Similarly, the seasons follow in sequence as the Earth moves on to positions 3 and 4.

Referring to position 1 we can see that the sun is north of the Earth's equator, and in position 2 it is in the plane of the Earth's equator. In position 3 the sun appears to be south of the equator, and in position 4 it is again directly overhead at noon for those on the equator.

As viewed from the Earth, therefore, the sun appears to move north and south in the sky as well as eastward. The celestial sphere is bisected by the *celestial equator* just as the Earth is bisected by the terrestrial equator; in fact, the projection of the Earth's equator onto the celestial sphere defines the celestial equator. The sun appears to travel throughout the year from the northern celestial hemisphere to the southern celestial hemisphere and back again, crossing the celestial equator twice (Figure 5-6). When the sun crosses the celestial equator on its apparent journey north, about March 21, the northern hemisphere begins spring and the southern hemisphere begins autumn. This point of intersection of the ecliptic with the celestial equator is called the *vernal equinox*. The other intersection of the ecliptic with the celestial equator, that is, where the sun crosses the celestial equator going from the northern to the southern hemisphere, is called the *autumnal equinox*. The crossing occurs about September 21.

The point in the sky where the sun reaches its most northern position, about June 21, is called the *summer solstice*. Its most southerly point, the *winter solstice*, is reached about December 21.

It should be pointed out that since these points were named by inhabitants of the northern hemisphere, they naturally refer to the seasons in that hemisphere. The seasons are exactly opposite for people in the southern hemisphere. When the sun is in the vernal equinox, autumn begins for those south of the equator, and so forth.

The seasons are a consequence of the Earth's motion and not a proof, for they can be—and were—explained by the geocentric theory formulated by the ancient Greeks. Since they can be explained by either theory, they

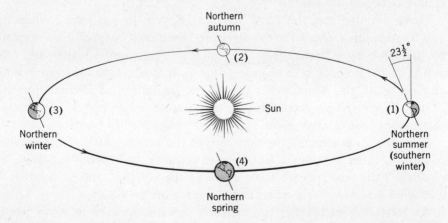

**Figure 5-5**

The seasons result from the fact that the Earth's axis of rotation is not perpendicular to the plane of the Earth's orbit, and because this axis of rotation points to essentially the same point in the sky during the year.

cannot act as a proof of one. We therefore classify the seasons as a consequence of the Earth's motion.

For the astronomer, and even the casual observer of the skies, two other consequences of the Earth's motion are important.

## 5.5 The Sun's Apparent Motion

As the Earth revolves about the sun we see the sun appear to revolve about us. This is the apparent motion of the sun eastward against the background of stars referred to in the first part of this chapter. The sun's apparent path through the field of stars is called the *ecliptic*, which is the intersection of the plane of the Earth's orbit with the celestial sphere. The sun, like all celestial objects, appears projected on this sphere.

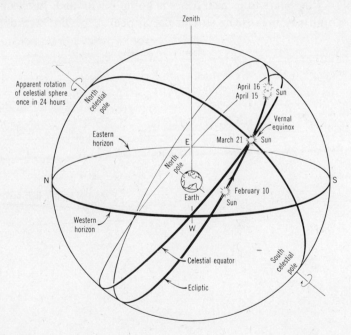

**Figure 5-6**

The Earth is drawn at the center of a greatly decreased celestial sphere. The Earth's axis of rotation intersects the celestial sphere at the north and south celestial poles. A projection of the plane of the Earth's equator on the celestial sphere is the celestial equator. The apparent path of the sun against the background stars is called the ecliptic. The sun moves eastward along the ecliptic about 1° each day.

Since the Earth moves in an elliptical orbit it must rotate more than 360° with respect to the stars in order to make one complete 24-hour period with respect to the sun (Figure 5-7). The Earth in position *a* must rotate more than 360° with respect to the stars in order for the same region of the Earth to return to midnight in position *b*. Since the Earth revolves about 1° each day, it must rotate 361° with respect to the stars in a 24-hour period. The rotation with respect to the stars yields the *sidereal day*; the rotation with respect to the sun gives us the *solar day*. The length of the sidereal day is about 23 hours, 56 minutes solar time.

Another consequence of revolution is that puzzling retrograde motion of the planets. Since Mercury travels faster than Venus, when they are on the same side of the sun Mercury will overtake Venus. Similarly Venus will overtake the Earth and the Earth will overtake Mars, etc. When we overtake Mars it appears to us as if Mars is going backward, just as passing a car on the highway makes the slower car appear to go backward.

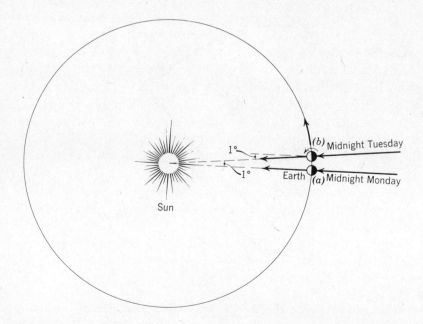

**Figure 5-7**

During any 24-hour period the Earth moves along its elliptical orbit so that in order for midnight to return to the same locality the Earth must rotate about 1° more than one complete rotation, or about 361°.

## 5.6 Great Circles in the Sky

It becomes essential for astronomers to have a coordinate system for defining the position of each star in the sky, just as the position of a given locality on the Earth is defined by its latitude and longitude.

Defining a position on the Earth requires two dimensions: one north or south of some reference circle on the Earth, and one east or west of some other reference circle. The reference circle for the first is quite obvious, for the Earth is divided into two equal hemispheres by the equator, which therefore has become a prime reference circle.

Any circle on the Earth parallel to the equator is called a *parallel of latitude* and defines a given locality as being so many degrees north or south of the equator. The latitude of San Francisco, for example, is about +38° (the plus sign referring to northerly latitudes). The latitude of Sydney, Australia, is about −34° (the minus sign signifying a southerly latitude) and the latitude of the north pole is +90°.

The reference circle for the east and west measurement is not so obvious. Any circle on the Earth that passes through both poles is called a *meridian*. It had to be agreed that one of the meridians should be chosen as the reference and called the *prime meridian*. The one selected by international agreement goes through the Greenwich Observatory in England. Any locality on the Earth can, therefore, be referred to by its longitude as being so many degrees east or west of the prime meridian. San Francisco has a longitude of 122° west; the longitude of Sydney is 151° east. East and west longitudes meet on the 180th meridian which goes north and south through the Pacific Ocean just west of Midway Island and east of New Zealand.*

In the sky, positions of the stars also require two dimensions and are most conveniently stated in terms similar to latitude and longitude. As has been already mentioned, the projection of the plane of the Earth's equator intersects the celestial sphere in a circle called the celestial equator. A star's position north or south of the celestial equator is given as so many degrees of *declination*, this being analogous to latitude. A star with a declination of +38° will pass directly overhead for the people in San Francisco. A star with a declination of −34° will pass through the zenith of Sydney.

---

*With the advent of manmade satellites, it has become convenient to give the longitudes in only one direction from the prime meridian, namely east. The longitude of Sydney would remain the same on this system, but that of San Francisco would become 238°.

The choice of the celestial equator as a reference circle is as obvious as that of the terrestrial equator, but, again, we are faced with the choice of a reference circle for the east–west direction in the sky. Circles on the celestial sphere analogous to meridians on the Earth pass through both celestial poles and are called *hour circles*. Since hour circles are to stars in the sky what meridians are to cities on the Earth, the hour circles must rise and set with the stars. The hour circle chosen as the reference is the one that passes through the vernal equinox. A star's position with reference to this prime hour circle is given as *right ascension* and is measured only to the east in hours instead of in degrees (24 hours equals 360°).

For example, the bright star named Vega has a position of right ascension = $18\frac{1}{2}$ hours, declination = $+39°$. It is therefore located 39° north of the celestial equator and will cross the meridian of San Francisco 1° north of its zenith. It is also located $18\frac{1}{2}$ hours east of the vernal equinox, that is, it will cross the meridian of a given locality $18\frac{1}{2}$ hours after the vernal equinox crosses the same meridian. Antares has a right ascension of $16\frac{1}{2}$ hours and a declination of $-26°$. It will cross the meridian of Sydney $16\frac{1}{2}$ hours after the vernal equinox and be 8° north of the zenith of Sydney.

Since the astronomer is concerned with the stars and since the stars have an apparent rotation about the Earth different from that of the sun, he uses clocks whose rate is different from those of the clocks used for ordinary timekeeping.

It has been mentioned that the stars set 4 minutes earlier each day according to solar time (see page 111), and therefore the clocks set to *sidereal time* (star time) must run 4 minutes faster each day. The solar day begins at midnight when a point in the sky exactly opposite the sun crosses the meridian of a given locality—or in actual practice crosses the midpoint of a time zone. The sidereal "day" begins when the vernal equinox crosses the meridian of a particular locality (time zones not being used or being of any use in this case).

All this is further complicated by a third motion of the Earth that prevents the vernal equinox from being fixed in the sky.

## 5.7 Precession

The vernal equinox, being one of the intersections of the celestial equator with the ecliptic, depends on the location of both of these circles with respect to the stars. If we use the ecliptic and the stars as a frame of reference, we find that the two equinoctial points move against the backdrop of stars.

The celestial equator is the circle that is everywhere 90° from the *celestial poles* and thus depends on the location of these poles which are the intersections of the axis of rotation of the Earth with the celestial sphere. It is the motion of the Earth's axis itself (not of the Earth about its axis) about a line perpendicular to the plane of the Earth's orbit that causes the celestial equator to move in the sky. This motion, called *precession* (Figure 5-8), is caused by the gravitational pull of the sun and the moon on the equatorial bulge of the Earth. This pull tries to make the plane of the equator more nearly coincident with the ecliptic, but the Earth's rotation will not permit this. As a consequence, the direction in which the axis points continually changes. The axis cuts out a cone in space as it

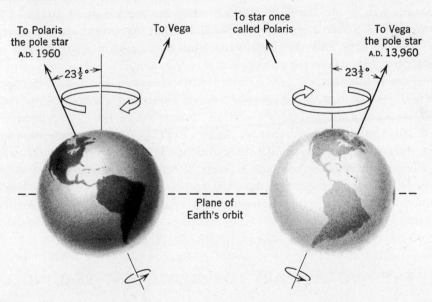

**Figure 5-8**

The Earth's axis of rotation points nearly to the star we now call Polaris, because of its unique position in the sky. As the Earth's axis of rotation precesses (about a line perpendicular to the plane of the Earth's orbit) the celestial poles trace out circles on the celestial sphere. In about 12,000 years the north celestial pole will be about 5° from the star Vega.

makes one complete rotation in its period of 26,000 years. Since precession would not exist if the Earth did not rotate, the fact that it does precess constitutes proof of the Earth's rotation.

The star we call Polaris is now the north star because the Earth's axis nearly lines up with it, but because of precession it will not always be the north star (see page 7). In 12,000 years the Earth's axis will point to another star, the star Vega referred to above (right ascension = $18\frac{1}{2}$ hours and declination = $+39°$). It will then be the north star; its declination will be about $+85°$. Polaris, however, will again be the north star 26,000 years from now. The result of precession, then, is that the positions (right ascension and declination) of all the stars are continually changing. Hence when a position is given the date or *epoch* must also be specified.

These three motions of the Earth are rather difficult to conceive, and may be more easily pictured by likening the Earth to a spinning top placed on the edge of a merry-go-round. The top's spinning represents the rotation of the Earth on its axis, and the top's motion around the center of the merry-go-round represents the revolution of the Earth around the sun. This top, however, is not spinning upright; its axis of rotation is tipped from the perpendicular to the floor of the merry-go-round, with the result that the axis of rotation itself rotates about that perpendicular (as all tops do, especially as their spinning slows down). This motion is precession.

It must be pointed out that all planets experience these three motions to a greater or lesser degree. All three of them may be conveniently referred to the supposedly fixed stars as a frame of reference. Moreover, we have assumed so far that the sun does not travel through space among these "fixed" stars, just as a merry-go-round remains on the same piece of ground.

## BASIC VOCABULARY FOR SUBSEQUENT READING

| | |
|---|---|
| Aphelion | Declination |
| Autumnal equinox | Diurnal |
| Celestial equator | Ecliptic |
| Celestial pole | Ellipse |
| Celestial sphere | Frame of reference |

| | |
|---|---|
| Heliocentric parallax | Retrograde motion |
| Latitude | Right ascension |
| Longitude | Summer solstice |
| Meridian | Vernal equinox |
| Perihelion | Winter solstice |
| Precession | |

## QUESTIONS AND PROBLEMS

1. Explain why Mercury and Venus are always seen relatively close to the sun.

2. If the sun moves east relative to the stars, why does it set in the west?

3. Draw the planetary distances from the sun on a scale such that the radius of Mercury's orbit is 0.5 cm. Can you draw the planets to the same scale?

4. Describe one proof of rotation and one of revolution of the Earth.

5. Explain why the Earth has seasons.

6. If a planet revolved about the sun in an orbit with a radius vector of 2.8 A.U. what would be its period of revolution?

7. Describe the appearance of the sky both day and night through each of the four seasons as seen by (a) explorers on the North Pole; (b) people on the equator.

8. Calculate the gravitational force of attraction (in dynes, 1 dyne $= 2.2 \times 10^{-6}$ pound) between the Earth and the moon from the following information:

   mass of the Earth $= 6.0 \times 10^{27}$ grams
   mass of the moon $= 7.3 \times 10^{25}$ grams
   radius of moon's orbit $= 3.8 \times 10^{10}$ cm
   $G$, gravitational constant $= 6.7 \times 10^{-8}$ dyne-cm$^2$/gram$^2$

9. Trace the ecliptic in the sky by following the moon and as many planets as you can see.

10. Plot the location (right ascension and declination) of Mars throughout the year from its positions as listed in the current *American Ephemeris and Nautical Almanac*. Its motion can readily be seen by plotting positions for the 10th, 20th, and the last day of every month.

11. An artificial satellite was launched with a period of 100 minutes. By using the moon as a comparison find the radius of the artificial satellite's orbit. (Radius of moon's orbit is $2.4 \times 10^5$ miles; its period is $2.4 \times 10^6$ seconds.) Now find the velocity of this artificial satellite in its orbit.

12. Explain why there is both a sidereal and a solar day. Which is longer and why?

## FOR FURTHER READING

Cohen, I. B., *The Birth of a New Physics*, Doubleday Anchor Books, New York, 1960.

Geymonat, L., *Galileo Galilei*, McGraw-Hill Book Co., New York, 1965.

Glasstone, S., *Sourcebook on the Space Sciences*, Van Nostrand Co., Princeton, N.J., 1965, Chapter 2.

Hawkins, G. S., *Stonehenge Decoded*, Doubleday and Co., New York, 1965.

Keston, H., *Copernicus and His World*, Roy Publishers, New York, 1945.

Koestler, A., *The Watershed*, Doubleday Anchor Books, New York, 1960.

Kuhn, T. S., *The Copernican Revolution*, Modern Library Paperback, New York, 1959.

Munitz, M. K., *Theories of the Universe*, The Free Press, New York, 1957.

Page, T., and L. W. Page, ed., *Wanderers of the Sky*, The Macmillan Co., New York, 1965.

Santillana, G. de, *The Crime of Galileo*, The University of Chicago Phoenix Books, Chicago, Ill., 1959.

Christianson, J., "The Celestial Palace of Tycho Brahe," *Scientific American*, p. 118 (Feb. 1961).

Franklin, K. L., "The Astronomer's Odd Figure 8," *Natural History*, p. 8 (Oct. 1962).

Franklin, K. L., "The Gravitational Forces and Effects," *Natural History*, part I, p. 12 (Oct. 1963); part II, p. 44 (Nov. 1963).

Muul, Illar, "Day Length and Food Caches," *Natural History*, p. 22 (March 1965).

Rawlins, D., "The Mysterious Case of the Planet Pluto," *Sky and Telescope*, p. 160 (March 1968).

Ronan, C. A., "Phoenix of Astronomers," *Natural History*, p. 52 (Jan. 1965).

Rothrock, G. A., "Steps to New Astronomy," *Natural History*, p. 64 (May 1965).

"Galileo Galilei," series of articles, *Sky and Telescope* (Feb. 1964).

CHAPTER CHAPTER

6

# SATELLITES

## THE MOON

## OTHER SATELLITES

Our discussion of the planets has covered one phase of the description of the solar system. But of the nine planets, six are known to have satellites circling them as each pursues its complicated motions through space.

Every satellite, since it revolves about a planet, in turn has one more motion than those characteristic of the planets. For us here on the Earth the most obvious satellite in the universe is our own natural one, the moon, which has an average distance from us of 239,000 miles, and which we see going through its phases during the month.

## THE MOON

It is not completely correct to say that the moon revolves about the Earth, for in reality they both revolve about a common point called their *center of mass*. The Earth is about 81 times as massive as the moon, and the center of mass is therefore about 3,000 miles ($\frac{1}{81}$ of 239,000 miles) from the center of the Earth on a line joining that center with the center of the moon. Since the center of mass is inside the Earth, the Earth's motion about this center is barely perceptible. The corresponding motion of the other planets is even less noticeable, for all other satellites are much smaller when compared with their mother planets.

## 6.1 Phases of the Moon

Since we see the moon only by reflected sunlight, it is understandable that its phases result from its revolution about the Earth (Figure 6-1). Starting with the new moon and following it through the lunar month, the phases progress in order: new moon, first quarter, full moon, and third quarter (Figure 6-2). Starting with each new month, the length of time within the lunar month is reckoned by the "age" of the moon. That is, the full moon is said to be 14 days old, etc. (Figure 6-3).

Just before and after the new moon, the sunlit side of the moon visible from the Earth assumes a crescent shape; at first quarter it assumes the half-moon shape. Just before and after full moon, the lighted portion assumes a gibbous shape.

### Chapter Opening Photo

The crater Goclenius with a diameter of 45 miles and a graben running through the wall and central mountain peak. Photographed by the Apollo 8 crew. (NASA)

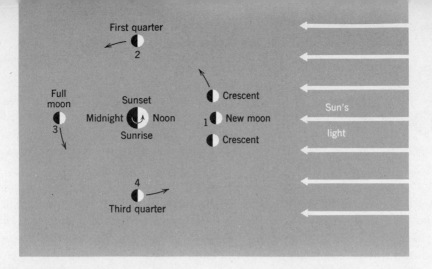

**Figure 6-1**

The phases of the moon.

**Figure 6-2**

(*a*) 26 days old, (*b*) 23 days old, (*c*) 4 days old, (*d*) 3 days old. (Photographs from the Hale Observatories)

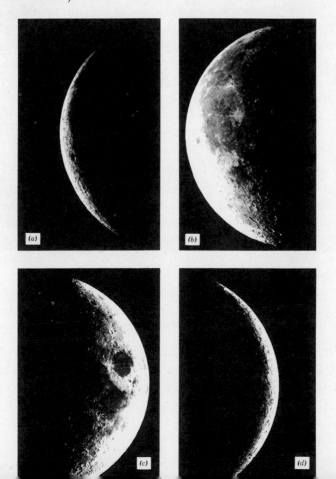

**Figure 6-3**
The 14-day old moon showing the impact craters Tycho, Copernicus, Kepler, and others. (Lick Observatory photograph)

The first-quarter moon differs from the third-quarter in that the western portion of the near side of the moon (as viewed from the Earth) is visible during first quarter; the eastern portion is visible in the third-quarter moon. The first-quarter moon rises about 6 hours after sunrise, so it is high in the sky at sunset. The third-quarter moon rises 6 hours after sunset, so it is high in the sky at sunrise.

The crescent moon has one unique feature. Most of the side that faces the Earth is dark. But the side of the Earth that faces the crescent moon is fully lighted; consequently the moon is faintly illuminated by the Earth's reflected light, or *earthlight*. It is by this reflected light that astronomers have been able to estimate the albedo of the Earth at about 0.4. The fact that this is less than the albedo of Venus (0.76) is in line with what we might expect, since the atmosphere of the Earth has fewer clouds than that of Venus; on the other hand it is more cloud-laden than the atmosphere of Mars, with its albedo of 0.15.

## 6.2 The Moon's Atmosphere

What about the moon with its albedo of 0.07, which has already been mentioned in our discussion of Mercury (see page 72)? The spectroscopic test, when applied to the moon, reveals no more atmosphere than there is on Mercury. But the moon's closeness to the Earth enables us to employ a still more conclusive test.

As will be seen, the moon moves eastward through the field of background stars one of its diameters every hour. In the course of its motion it passes in front of stars, blocking them from our view; this is called an *occultation*. An occultation of a star by Jupiter, it will be recalled, permitted us to learn something of that planet's atmosphere. If the moon has any atmosphere it would absorb and scatter some of a star's light. Thus just before a star becomes occulted it would appear slightly fainter than it does when not disturbed by the moon. But with the aid of a sensitive photocell, it has been demonstrated that the star's light does not vary until the solid edge of the moon's limb cuts it from our view. This very sensitive test has been made with radio telescopes as well. The conclusion is that the moon has no permanent atmosphere. There may be small pockets of gas that form from time to time; but these, like those that might form on Mercury, would be dissipated into space, for the moon's escape velocity is only 1.5 miles per second (even less than that of Mercury). The only molecules that could possibly form an atmosphere at all would be those of some of the heavier inert gases such as xenon and krypton, but these elements are very rare.

It is the moon's lack of an atmosphere that enables us to distinguish its features so clearly. Even when seen by the naked eye, it does not appear to have a uniform surface. Its darker portions not only provide children with an image of the "man in the moon," but they also indicate differences in light-reflecting properties of the surface. When the full moon is viewed or photographed through a telescope these differences may be clearly seen. When the first- or third-quarter moon is viewed through a telescope irregularities not only in surface materials but also in depth are visible along the *terminator*, the boundary between day and night.

## 6.3  Techniques of Observation

As Galileo's telescope opened up a new era for astronomical investigation, so have electronics, computers, and rocket propulsion opened up a new and exciting era. The successful space voyages that have put both men and equipment on the moon have—as did Galileo's telescope—answered old questions but asked new ones as well. The identification and study of lunar features are now following lines vastly different from the studies made as recently as the middle of the 20th century.

A series of Ranger space craft made hard (destructive) landings on the moon and took photographs as they descended (see Figure 6-8). Surveyor space craft made soft (nondestructive) landings, took close-up photographs of the lunar surface, and dug into the soil a bit (Figure 6-12). Lunar orbiters carried cameras into orbits around the moon, taking wide-angle photographs (Figure 6-5) and narrow-angle photographs with a telescopic lens to record detail (Figure 6-10). Finally the Apollo rockets have sent men to the moon to photograph surface features, to set up equipment for special studies, and to bring surface material back to the Earth for detailed studies in our laboratories.

From Earth-based telescopes, studies are still being made in the infrared, microwave, and radio regions of the spectrum. Radar observations are also being made. From all of these studies a great deal of information is being compiled which will increase our understanding of the moon, its origin, and history. These studies will, we hope, add to our understanding of the solar system and even of the universe.

## 6.4  The Lunar Surface

**a. Mare.**  Each of the large dark areas of the moon is called a *mare* (plural *maria*). These have been studied intensively. Observations of the color of the maria have been made by photographing them with film sensitive to the ultraviolet and again with film sensitive to the infrared. From the differences in color it can be concluded that the maria are lava flows. In fact, some of the maria are composed of several lava flows.

To substantiate the lava nature of the maria, some of the material brought back from the moon by the Apollo astronauts has been dated by using radioactive techniques (see Chapter 8) as $3.5 \times 10^9$ years old. This is the length of time it has been since those rocks solidified. Other rocks seem to be older. At least one appears to be $4.0 \times 10^9$ years old and others are $4.6 \times 10^9$ years. The moon, itself, is $4.6 \times 10^9$ years old (see Chapter 8), so some of the rocks brought back by the Apollo crews were molten and on the moon's surface one billion years after the moon was formed. During the first one-quarter of its lifetime, the moon was not a "dead" body.

**b. Masscons.** Further evidence of past changes on the moon in the region of the maria has been obtained from an unsuspected source: the variations in the orbits of the Lunar Orbiter and Apollo space craft. These orbits depend on the mass of the moon as well as the height of the space craft. Very small changes in the Doppler shift of the radio signal sent back to Earth indicate clearly that the space craft occasionally made slight and unexpected dips in their orbits. Each dip was about 3 to 4 feet and each occurred over a mare. We are left to conclude that the gravitational field above at least some of the maria is greater than the gravitational field above the rest of the moon.

The best way to account for this increased gravitational field is to assume that there is material under the maria whose density is greater than the material just under the surface of the rest of the moon. This greater density means that there is more mass per unit volume and consequently a greater gravitational pull on the space craft. These concentrations of mass have, quaintly enough, become known as *masscons*. We conclude that they are large high-density rocks that collided with the moon during the first billion or so years of the moon's existence. The maria are certainly circular, and, in fact, the Apennines (Figure 6-4) form part of a circular wall about Mare Imbrium.

The Orientale Basin (Figure 6-5) seems to be the youngest of the large circular maria on the moon. The outer circular ring of mountains is some 600 miles in diameter, with peaks rising to more than 20,000 feet above the surrounding plains. There seems little doubt but that this is an impact crater, and it lends support to the idea that all the maria are impact craters with their rims greatly eroded.

**Figure 6-4 (a)**

A mosaic of the first and third quarter moon (Lick Observatory photograph)

**Figure 6-4 (b)**

A map of the moon. 1. Clavius; 2. Tycho; 3. Arzachel; 4. Alphonsus; 5. Ptolemaeus; 6. Albategnius; 7. Abulfeda; 8. Copernicus; 9. Eratosthenes; 10. Archimedes; 11. Autolycus; 12. Aristillus; 13. Cassini; 14. Plato; 15. Julius Caesar; 16. Flamsteed; 17. Kepler.

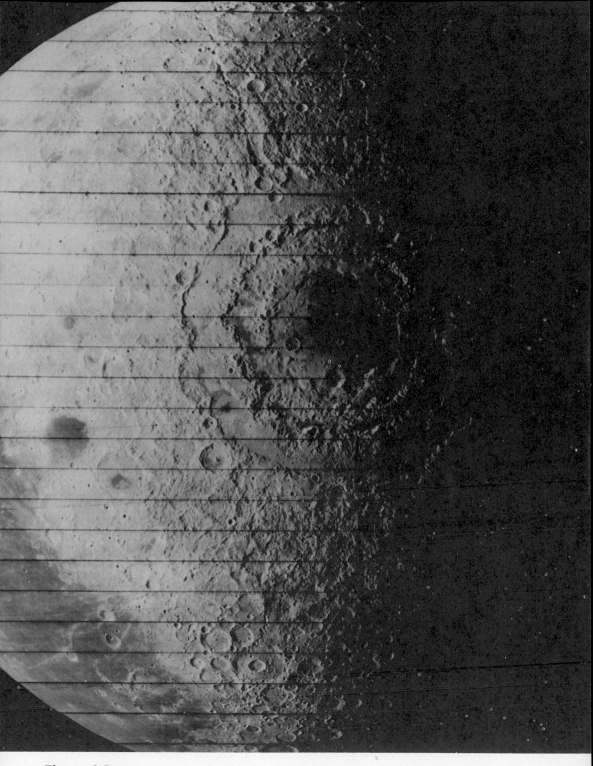

**Figure 6-5**
The Orientale Basin, photographed by Lunar Orbiter 4. (NASA photograph)

c. **Craters.**   The craters that make up so much of the topography of the moon are probably both from interplanetary rocks colliding with the moon's surface and from volcanic activity. However it is suspected that volcanic activity has produced only some of the smaller craters; most of the craters are impact craters.

The crater Tycho (see Figures 6-4b, 6-3, and 6-6) has extensive *rays* emanating from it, as do Copernicus, Kepler, and many smaller ones. These rays are felt to be debris that was ejected from the explosion that resulted when the kinetic energy of the rock colliding with the moon was converted into thermal energy. The pattern of secondary craters along some rays has been photographed by some of the Ranger space craft. That pattern supports the idea that the rays are indeed ejected material from the impact explosion.

Since not every crater suspected of being an impact crater has rays about, it must be assumed that the rays fade and disappear with time. Ranger photographs have been made of two intersecting ray systems, and the ray system which overlies the other is the brighter of the two.

But then so do craters fade and change with time. An examination of the Crater Tycho (Figure 6-6) reveals that it is clearly much younger than the craters that surround it. Not only do its rays lie on top of other craters, but its features are sharp. The features of some of the other craters have been rounded and smoothed considerably (Figure 6-7). The agents of erosion on the moon, however, are vastly different from the agents of erosion on the Earth.

d. **Erosion.**   The erosion of the surface features on the moon is caused by a number of different factors. It does suffer a large change in temperature between its day and night. The highest temperature measurement is +215°F, the lowest temperature is probably about −240°F. This large change in temperature causes the surface of rocks to expand and contract more than the interior, which causes a cracking and flaking off of material.

The electrons and protons that make up the solar wind must cause changes on the surface of the rocks on the moon. These changes can only result in a weakening of the crystalline structure of the rocks. Ultraviolet radiation, X-rays, and gamma rays from the sun must also cause a breakdown of the crystalline structure of the surface of the rocks.

Another agent of erosion is cosmic rays. *Cosmic rays* are the nuclei of atoms that are accelerated somewhere in the universe to velocities very close to that of light. Protons, the nuclei of hydrogen atoms, make up most

**Figure 6-6**

The Crater Tycho in the lunar highlands. (Lick Observatory photograph)

of the cosmic rays, but nuclei of other atoms, such as helium and iron, are also found in cosmic rays. When these very high energy particles strike the surface of rocks on the moon, they leave little streaks, slightly altering the structure of the rock. These cosmic ray tracks have been found in rocks brought back by the Apollo astronauts. It may well be that the record of cosmic ray tracks in rocks on the moon will tell us a great deal about the history of the sun (it ejects some low energy cosmic rays) and of the universe.

Perhaps the biggest cause of erosion on the moon, however, is the continual bombardment of tiny bits of material that permeate the inter-planetary spaces. As large chunks of rock colliding with the moon have

**Figure 6-7**
This photograph taken from Lunar Orbiter 5 gives direct evidence of lunar erosion; two rocks have rolled down a crater wall. The larger rock is about 75 feet in diameter. (NASA photograph)

left large craters, so small particles colliding with the moon leave small impact marks. These small impact marks amount to a wearing away, a slow chipping.

The small chunks of rock and grains of sand that circle the sun are called *meteorites*. There are vastly more small meteorites than large ones. When a small one strikes the Earth it is altered by the atmosphere into the common streak of light called a shooting star or, preferably, *meteor*. Its speed is so great that when it is slowed down by the atmosphere it is heated to incandescence; it burns up in the atmosphere. Its small size is recognized by the length of time it takes it to burn up—only a fraction of a second. Larger ones make it through the atmosphere and strike the Earth's surface.

Meteorites strike the surface of the moon since it has no atmosphere to shield it. Each time a particle strikes the moon it chips a little bit off a lunar rock. And bit by bit, over the millions and billions of years, these small particles do an effective job of sandblasting the surface of the moon.

**e. Rills.**   The *rills* are another common feature of the moon. They are frequently found on the floor of craters, such as those in Alphonsus (Figure 6-8). These were photographed by Ranger 9 as it plunged to the surface. It can be seen from this figure that the rills are actually a series of craters resulting from a collapse of the surface material. These *collapse craters*, as they are called, could have resulted from lava flowing out from under the surface reducing the support of the surface. This lack of support permitted the surface to collapse. Similar features are found in lava formations on the Earth: when not collapsed they are called lava caves.

**f. Graben.**   The *graben* are features of the moon which result from a sinking in of a portion of the surface (Figure 6-9 and the photograph that opens this chapter). These features have been named after similar features on the Earth which have been shown to be the result of earthquakes. If the crust of the Earth cracks along two parallel seams, and if the portion in the middle sinks, a graben is formed. At least one graben on the moon's surface passes right through the rim and central peak of a crater.

**g. The Alpine Valley.**   A striking feature of the moon's surface has been named the *Alpine Valley* and has long puzzled astronomers (Figure 6-10). Lunar Orbiter 5 photographed this feature with a telescopic lens and revealed that the long straight valley which cuts through the lunar Alps just south and east of the crater Plato has what appears to be a

**Figure 6-8**

(*a*) The crater Alphonsus and the edge of Mare Nubium to the left. Ranger 9 struck at the head of a rill marked by the circle. Note the "dark-haloed" craters about the periphery of the floor of Alphonsus. Taken 265 miles above the surface. (*b*) The white circle again marks impact point of Ranger 9 and the rill is now seen as a string of collapse craters. Taken 12.2 miles above the surface, 8.09 seconds before impact. (NASA photograph, Jet Propulsion Laboratory)

**Figure 6-9**

A photograph, taken with the 120-in. telescope at Lick Observatory, of a graben extending from the "highlands" into Mare Tranquilitatis. (Lick Observatory photograph)

meandering stream running down the middle. The cause of this valley with its crooked rill running down the middle is the subject of speculation right now, but we do not expect to find stream-worn pebbles along its banks.

## 6.5 Recent Lunar Activity

It has been suspected from Earthbound observations that there has been recent activity of some sort or another around the crater Alphonsus.

**Figure 6-10**
The Alpine Valley photographed by Lunar Orbiter 5. (NASA photograph)

ograms taken by the Russian astronomer Kosyrev indicate the
nal presence of luminous gases, for *bright* spectral lines appeared
mposed on the solar spectrum reflected from the lunar surface.
pposed that the gas is occasionally emitted from the surface and
. to emit its own light by solar radiation. Ranger 9 photographs
that gases are probably not, however, emitted by the central peak
ing 3,500 feet above the surrounding floor, for this peak has no
through which gases might escape. The peak appears to be very
ı, with little or no detail—at least no gross detail. It is possible, of
, that the surface has many small holes through which gases could

material has been ejected from the surface in the not-too-distant
made evident by a half-dozen or so craters all on the periphery of
or of Alphonsus. These unique craters are all surrounded by dark
(Figure 6-8a). Fortunately, Ranger 9 obtained a good photograph
e of these dark-haloed craters. Figure 6-11 is a photograph of the
edge of the floor of Alphonsus on the side opposite Mare Nubium.
tion of the photograph makes it clear that the dark area immediately
nding the crater at the bottom of the photograph is smoother than
st of the surface which is fairly broken up with small craters. It
obvious that material, perhaps volcanic ash, ejected from this crater
ed in the older craters and even part of the rill which this crater
les.

## ∟unar Explorers

ost exciting phase of lunar study has been the achievement of man-
exploration. Man, by his very natural sense of curiosity and ability
on, no longer lives in caves and hunts for food with a club. Whether
lay are better off may remain a philosophical or ethical question,
e still retain that natural sense of curiosity; man will walk the moon,
erhaps other astronomical bodies, in search of something new.
and a manned space craft, such as the Apollo 12, within a few
ed feet of a previously launched space probe, Surveyor 3 (Figure
is an exciting technical achievement. But to bring back selected
from the surface to be studied in terrestrial laboratories in many
ies throughout the world is more rewarding for scientific knowledge
ternational cooperation.

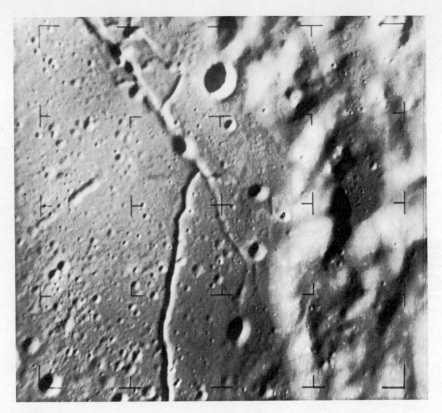

**Figure 6-11**
A closer view of one of the "dark-haloed" craters taken 107 miles above the surface by Ranger 9. Note the "lava lake" in the smooth walls of the crater Alphonsus. (NASA photograph, Jet Propulsion Laboratory)

The rocks that have been brought back have consisted of a variety of specimens, among them are: breccias, basalts, glasses, and others (Figure 6-13). Breccias are a composite rock with angular pieces bound by a matrix material. One type of basalt rock is the basic material of the Earth's crust. The granite continents on the Earth, being less dense, actually float on the basaltic crust.

Some bits of the glass brought back from the moon appear as tiny sphericules, others appear jagged. It would seem that the tiny sphericules of glass may have been formed by the impact of a meteorite on the moon causing a crater. If some material was melted during the resulting ex-

**Figure 6-12**

One of the Apollo 12 astronauts took this photograph of their lunar module (on the left) close to the Surveyor 3 craft (on the right). (NASA photograph)

plosion, the tiny sphericules could be bits of the ejected material that solidified before striking the surface of the moon. Some of the rocks found on Mare Tranquilitatis are suspected to have been thrown there by the impact that caused the crater Tycho.

The granular nature of the material on the surface of the moon is shown in Figure 6-14a. How that material packs together (at least in the Oceanus Procellarum not far from the crater Copernicus) is shown in Figure 6-14b. The Apollo 12 astronauts found a greater amount of dust on the Oceanus Procellarum than was found by the Apollo 11 crew on Mare Tranquilitatis.

It is significant that the average density of the rocks from the moon is about 3 times the density of water. The overall density of the moon is only 3.35 times the density of water, so apparently the moon does not have a dense iron core as does the Earth. It would seem that the inside of the moon never became hot enough to permit the heavier chemicals to settle to the center.

With the advent of actual travel to the surface of the moon, interest in our nearby neighbor has increased. New studies have been initiated at a number of observatories. Amateurs have been asked to help by observing the moon in the hopes of recording local color changes which might indicate volcanic activity of some kind. It is hoped that studies of the lunar surface will help answer questions not only about its past, but also about the origin and evolution of the entire solar system. Because of the lack of atmosphere and water, the moon should abound with records of the past.

## 6.7 The Moon's Motion

We have mentioned that the moon revolves about the Earth, but its motion needs to be described in greater detail. Although the moon's diurnal motion is westward, it actually revolves in its orbit from west to east (see page 110). Its orbit is elliptical, as are the orbits of all celestial bodies. The point of its closest approach to the Earth is called *perigee*, and the point farthest from the Earth is called *apogee*.

Our evidence for the moon's eastward motion lies in the fact that it sets on the average 49 minutes later every time it sinks below the western horizon. If the stars are used as reference, the moon has a period of $27\frac{1}{3}$ days, during which time it completes a circuit of $360°$ in the sky. But at the same time the Earth has continued its motion about the sun; thus the moon needs more than 2 additional days, $29\frac{1}{2}$ days in all, to complete its period of phases from new moon through full and back to new moon again. The difference is seen in Figure 6-15. One complete revolution of the moon with respect to the stars (the $27\frac{1}{3}$ days) is called a *sidereal*

**Figure 6-13**

Tiny lunar rock specimens brought back by the Apollo 11 crew. (Courtesy of J. A. Wood, Smithsonian Astrophysical Observatory)

**Figure 6-14**

(*a*) Close-up photograph of the lunar soil taken by the Apollo 12 crew. (NASA photograph). (*b*) Close-up photograph of the imprint of a boot showing how the lunar soil compacts. (NASA photograph)

(a)

(b)

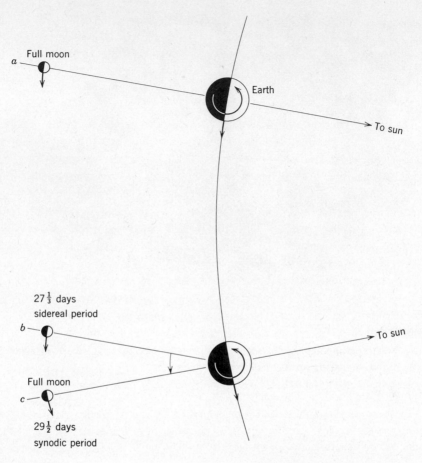

**Figure 6-15**

The sidereal period of the moon, $27\frac{1}{3}$ days, is the time required for the moon to make one complete revolution of 360° with respect to the stars. The synodic period, $29\frac{1}{2}$ days, is the time required for the moon to revolve from one phase to that same phase again, such as from full moon to full moon.

*period,* while one complete revolution with respect to the line joining the sun and the Earth (the $29\frac{1}{2}$ days) is called a *synodic period.* This latter is the basis for the month.

Since the moon moves 360° around the Earth in about 30 days, it must move 12° each day, or about $\frac{1}{2}$° each hour. Its apparent diameter, however, is about $\frac{1}{2}$°, so the moon must move eastward relative to the stars one of its diameters each hour.

The plane of the moon's orbit is coincident neither with the plane of the Earth's orbit (the ecliptic) nor with the plane of the Earth's equator. It is closer to the ecliptic, however, from which it is tipped by some 5°.

## 6.8 Eclipses

It is because the moon's orbit is not coincident with the ecliptic that the moon does not go into the Earth's show every full moon to give us a *lunar eclipse*; nor does the new moon pass in front of the sun every month to give us a *solar eclipse*. Eclipses can occur only during two seasons of the year.

In order for the moon to pass in front of the sun they both must be in the same part of the sky. This means that the moon must be on or very near the ecliptic because, by definition, the sun is always on the ecliptic. The two points in the sky where the moon's path crosses the ecliptic are called the *lunar nodes*: at the ascending node the moon crosses the ecliptic going north, and at the descending node the moon crosses it going south. The moon passes through each node once a month, but the sun passes through each of the nodes only once a year. Thus, a *solar eclipse* can occur only during those two seasons of the year when the sun is at or near one of the nodes during the new moon.

In order for the Earth to cast its shadow on the moon, a *lunar eclipse*, the moon must be at one of its nodes and the sun at the other. Again this can happen only during two seasons of the year. The nodes, however, are not fixed in the sky, for the moon's orbit is precessing much as is the Earth's axis.

The number of eclipses (lunar and solar) each year varies. It cannot be less than two nor more than seven, although the usual number is four. Solar eclipses are more common than lunar eclipses, but since a lunar eclipse can be seen from nearly half of the Earth's surface, more people have seen a lunar eclipse than have seen a solar eclipse.

A total solar eclipse is a beautiful sight. It occurs when the disk of the moon completely covers the disk of the sun (Figure 9-16). But a total solar eclipse is seen only over a small area of the Earth; over a larger area it appears as a partial eclipse. The reason for this is apparent from Figure 6-16*a*. (It has been necessary to draw the sizes and distances of the sun, moon, and the Earth out of proportion to each other.)

In the black region, the sun is completely hidden from view by the moon. This part of the shadow is called the *umbra*, meaning shade or cover

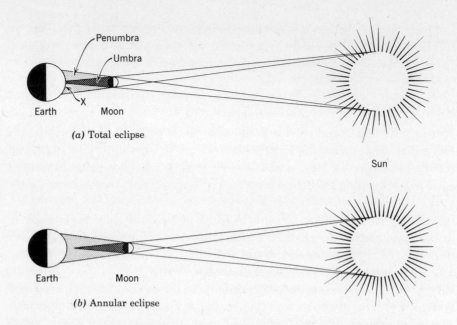

(a) Total eclipse

Sun

(b) Annular eclipse

**Figure 6-16**

(a) The umbra of the moon just reaches the Earth when the moon is near perigee.
(b) When the moon is near apogee the umbra does not reach the Earth, and so a total eclipse cannot occur, only an annular eclipse.

(see our word "umbrella"). In the part of the shadow that is gray the sun is not completely hidden. For example, a person standing on the Earth at X would be able to see part of the sun. Where this portion of the shadow, called the *penumbra*, strikes the Earth, a partial eclipse may be seen. If an observer is near the umbra, a large portion of the sun's disk is covered by the moon, but if he is near the outer edge of the penumbra, only a small section of the sun will be obscured from view. This small section is a silhouette of part of the moon.

The umbra is actually a cone, and it is sheer coincidence that the length of this cone is very nearly the same as the distance from the Earth to the moon. When the moon is near perigee and the Earth near aphelion, the cone of the umbra projects a dark circular shadow about 167 miles in diameter on the face of the Earth. When the moon is near apogee, however, the tip of the cone does not even reach the Earth (Figure 6-16b). The resulting visible effect makes the disk of the moon appear slightly smaller than the disk of the sun. Consequently the moon does not cover the sun completely. The ring of the sun that overlaps the moon's disk gives rise to the term *annular eclipse*.

Since the umbra follows the moon in its motion and is projected on a nearly spherical rotating Earth, it traces a rather complicated path, called the *path of totality*, on the Earth's surface. Anyone located in this path may see a total eclipse that can at best last 7 minutes, 40 seconds. The penumbra may stretch for a distance of several thousand miles on either side of the path of totality. The path of totality for the eclipse of October 12, 1958, is seen in Figure 6-17.

At the time of totality the outer regions of the sun become visible. (Because they are fainter than the blue sky they cannot normally be seen.) The pearly white area that extends farthest from the sun is called the *corona*. Very close to the surface of the sun is a region called the *chromosphere* because it is reddish. Prominences that appear to be gases erupting from the sun may be seen standing out from its surface. The nature of these is discussed in Chapter 9.

A total lunar eclipse, although not as spectacular as a total solar eclipse, is a sight long remembered. Even though the moon's disk may be entirely

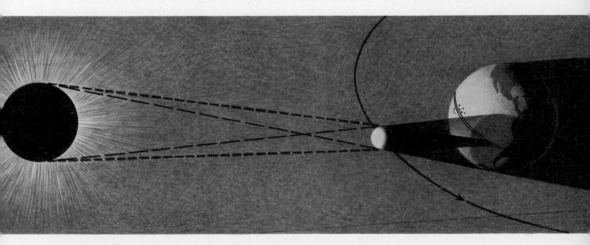

**Figure 6-17**

As the moon revolves in its orbit and the Earth rotates on its axis, the umbra traces the path of totality across the surface of the Earth. (Courtesy of *Sky and Telescope*)

covered by the umbra of the Earth's shadow it is never completely black. The Earth's atmosphere acts like a lens and focuses some of the sunlight on to the moon. Since the red light can travel through a long path in the atmosphere better than the blue, the moon takes on a deep reddish or copper color. The blue light has been scattered out into the surrounding atmosphere.

Eclipses, both lunar and solar, have been beneficial to the historian and archeologist as well as to the astronomer. In the records of ancient peoples there are many references that not only enable the historian to date an event, but also give the astronomer observations extending over many centuries of time. With the help of historic eclipses he has been able to describe the motion of the moon with greater accuracy.

## 6.9 Tides

Not only through eclipses does the moon affect us here on the Earth; the moon is also the main cause of the ocean tides. These tides result from the fact that the gravitational pull of any body, including the moon, decreases with increasing distance from that body. If there were no moon (or sun) the water layer over the Earth would be uniform as shown in Figure 6-18a. When the moon is put back in place (shown by the arrow in Figure 6-18b) the Earth responds to the additional gravitational force. Those portions of the Earth nearest the moon are subjected to a stronger

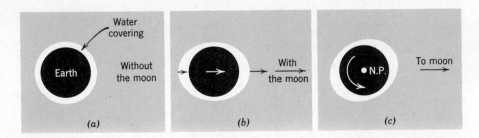

**Figure 6-18**
(a) The layer of water about the Earth in the absence of the moon (and the sun) would be uniform about the equator. (b) The moon is there, however, and pulls the Earth toward it. Those parts of Earth closer to the moon are pulled with a stronger force. (c) Because the Earth rotates faster than the moon revolves about the Earth, the continents of the Earth drag the tides with them.

gravitational pull, and thus the top of the ocean near the moon is pulled closer to the moon than is the bottom of the ocean. On the far side of the Earth, the bottom of the ocean, being nearer the moon than the top, is pulled closer to the moon. Thus the ocean bulges on two sides: the side facing the moon and the side directly opposite the moon. It is in these regions that we experience high tides. Halfway between the high tides are the low tides, the regions which supply the water for the high tides.

The net effect of this, and all tidal forces, is that the moon tries to pull the Earth apart but fortunately never succeeds. The Earth tries the same with the moon, and were the moon closer the Earth might succeed.

The Earth rotates underneath the tidal bulges to give us two high and two low tides every day. During one rotation of the Earth, however, the moon has moved eastward in its orbit; consequently the Earth must rotate farther so that the cycle of tides may be completed. Thus there are two high and two low tides every 24 hours, 49 minutes.

The variation in the level of the water differs with locality. Tides of 3 to 6 ft are fairly common on the shores of the oceans, but in some places, because of funneling, the water level may differ by 50 ft between high and low tide. In such places (for example, the Bay of Fundy, which is the gulf between Nova Scotia and New Brunswick) special harbors must be built.

It is bays such as these and shallow waters such as the Bering Strait that cause friction between the tides and the Earth. As a consequence the tidal bulge is dragged by the Earth, which rotates faster than the moon revolves. Hence the tidal bulge does not line up with the moon, but precedes it (Figure 6-18c).

The sun, too, influences our tides, but since it is so much farther away than the moon its effect is not as great. However, at new or full moon the gravitational pulls of the sun and the moon complement each other to give us the *spring tides* (Figure 6-19a). During the spring tides the high tides are higher and the low tides are lower than usual. The other extreme, called the *neap tides*, occurs when the moon is at first- or third-quarter (Figure 6-19b). During the neap tides the lines joining the sun and the moon with the Earth form a right angle. The moon attracts a high tide from the side of the Earth directly beneath and the opposite side, whereas the attraction of the sun operates to leave only a low tide in these same places. The result is that the high tides are not as high, or the low tides as low as the average.

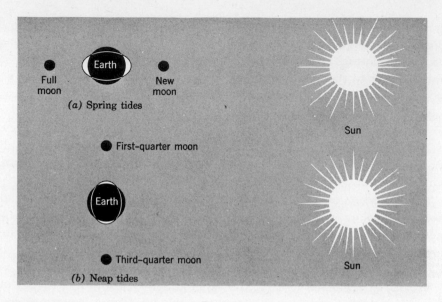

**Figure 6-19**

(a) Spring tides occur when the tidal forces of the moon and the sun work together to produce the high and low tides. (b) Neap tides occur when the high tides caused by the sun are superimposed on the low tides caused by the moon.

Since the Earth is not a rigid body it, too, gives slightly with the tides. Although small, this movement has been measured—evidence that having your feet on "solid ground" means something quite different, for that ground is moving up and down every 12 hours, 25 minutes.

## OTHER SATELLITES

### 6.10 Satellites of Other Planets

Many of the characteristics of our natural satellite can be generalized to apply to satellites of the other planets. Of the nine planets, only Mercury, Venus, and Pluto do not, to our knowledge, have satellites. Those planets that do will be considered next, with emphasis on the unusual satellites.

After the Earth the first planet from the sun to have satellites is Mars with two, Phobos (the inner one) and Deimos. They are small, and their dimness indicates that they are no more than 10 miles in diameter; Deimos is perhaps only 5 miles in diameter.

Phobos is only 5,820 miles from Mars, so close that its period of revolution is only 7 hours, 39 minutes. Since the period of rotation of Mars is 24 hours, 36 minutes Phobos makes a little more than two complete circuits around the planet while the latter is making one complete rotation. This means that Phobos, whose synodic period is 11 hours, 7 minutes, not only rises in the west and sets in the east but may rise twice in one night.

Deimos has a sidereal period of 30 hours, 18 minutes, only a little longer than Mars' period of rotation. Consequently it rises in the east and progresses very slowly toward the western horizon, where it sets about $2\frac{2}{3}$ days later. During this long stay in the sky above any one location it will have gone through its phases twice. But it is doubtful that its phases can be seen. The only indication would be a change in brightness from new to full and back to new phase again.

Jupiter, because of its gigantic size, has more satellites than the sun has planets. The first four to be discovered were discovered by Galileo in 1609. The fifth satellite was discovered in 1892 and has a smaller orbit than the innermost Galilean satellite. The twelfth satellite was not discovered until 1951 by the late Seth B. Nicholson at the Mount Wilson Observatory.

The four satellites discovered by Galileo are called the Galilean satellites and are visible with binoculars. By watching them from night to night an interested observer can follow the motions of these largest Jovian satellites. Two of them are larger than our moon and one, with a diameter of 3,080 miles, is only slightly smaller than the planet Mercury (3,100 miles in diameter).

Since the four Galilean satellites revolve in planes that are very nearly parallel to the plane of Jupiter's equator, they not only go behind Jupiter during part of their cycle but also pass in front of it. During a transit of one of these satellites its shadow is visible on the surface of Jupiter (Figure 6-20).

Of the ten satellites of Saturn (we do not count the tiny particles that compose its rings), Titan is the most interesting because of all the satellites in the solar system, it alone has a spectrum with absorption lines indicating an atmosphere. Its atmosphere is composed of at least methane, though there may be other gases present but simply not yet detected.

Uranus, whose axis of rotation is nearly parallel to the plane of its orbit, has five satellites revolving about it. Their planes of revolution are nearly parallel to the plane of Uranus' equator. Since these planes are more or less fixed in space, we on the Earth can at certain times (as in 1944) see

**Figure 6-20**

One of the Galilean satellites casting its shadow on the cloud covering of Jupiter.
(Photograph from the Hale Observatories)

the satellites moving in circles about the planet; that is, at certain times our line of sight is nearly perpendicular to the planes of their orbits. By the time Uranus moved in its orbit one-quarter the way around the sun (by 1967), the satellite orbits were seen edge on and the satellites appeared to oscillate in straight lines.

Neptune has two satellites, one of which when compared to our moon illustrates the dependency of the period of revolution on the mass of the central body. Neptune is 14.5 times as massive as the Earth. One of its satellites, Triton, which has an orbital radius of 220,000 miles (nearly the same as our moon's orbital radius of 239,000 miles), has a period of revolution of only 6 days as compared with our moon's $27\frac{1}{3}$ days.

The sun rotates, the planets all revolve and rotate, and most of the satellites revolve and rotate all in the same direction. Triton, however, like the four outermost satellites of Jupiter and the outermost satellite of Saturn, moves in retrograde motion, that is, each of these satellites revolves about its mother planet in a direction contrary to the principal motions of the solar system. It may be presumed that these satellites were captured after their planets were formed.

## 6.11 Man-made Satellites and Orbits

The subject of satellites is more inclusive now than ever before, for on October 4, 1957, the Russians sent the first man-made satellite, Sputnik I, into an orbit about the Earth. The question is often asked: "How does a satellite stay up there?" The answer to this also answers the questions: "Why doesn't the Earth fall into the sun?" and "Why does a rock whirled at the end of a string keep the string taut?"

As was stated in the discussion of Newton's second law (see page 115), if any object in motion has no force acting on it that object will move in a straight line and with a constant speed. If a force should now act on this object, then according to Newton's second law of motion the object must accelerate. The amount of acceleration depends not only on the force applied but also on the mass of the object. The direction of acceleration is the same as the direction in which the force is acting. If the force is applied in a direction that is perpendicular to the direction of travel, the direction of travel will change accordingly. If this force is constant and always acts perpendicular to the direction of travel, the object will move in a circular path or orbit (Figures 5-3 and 6-21).

Any force that causes an object traveling in a straight line to be diverted into a curved path is called a *centripetal force*. The centripetal force acting on the whirling rock is the tension in the string. The centripetal force that keeps the moon or artificial satellite in its orbit is supplied by the

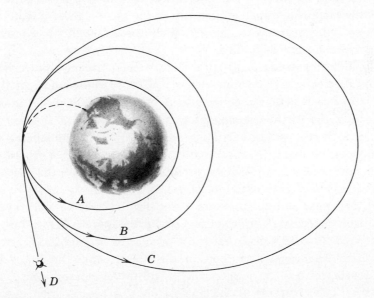

**Figure 6-21**

Three possible orbits for a satellite. The velocity with which the satellite was placed in the orbit helps determine the type of orbit.

mutual gravitational pull of the Earth and the satellite. This force of gravity is just strong enough to continually divert the satellite from its tendency to travel in a straight line, into an orbit that may be circular or elliptical.

The fact that the force of gravity is exactly enough to keep the satellite traveling in an orbit has consequences for any object that rides in the satellite. For example, suppose that a satellite were cut into two pieces, one half moving along just outside the original orbit and the other half moving just inside. The two satellites would travel along side by side for some time, neither one falling to the Earth any more than the other. Of course, eventually the outer satellite would lag behind, for its orbit is slightly larger and thus its velocity slightly less and its period slightly longer (Kepler's third law).

If, now, rather than cutting a satellite in two, we place one satellite inside the other, they would then have the same orbit and they would travel around together, neither one tending to fall toward the Earth any more than the other. When man rides inside a satellite, he is the inside satellite; thus he is in an orbit and feels weightless. The lack of apparent weight results from the fact that the entire force of gravity is used up in keeping the satellite and man in an orbit. In the same manner a man diving off a high diving board is weightless, since the entire force of gravity is used to accelerate him as he falls.

The motions of artificial satellites, since they are not powered, are described by Kepler's three laws just as natural satellites and planets are. The length of each period depends on the mean radius of its orbit as well as the mass of the Earth. The mass of an artificial satellite is so small when compared with that of the Earth that its orbit may be considered to be independent of its mass. The mean radius and the shape of the orbit depend on the height to which the final-stage rocket finally propels the spacecraft, the final speed, and its direction of travel.

If the final speed and direction are just right, the satellite will travel in a circular orbit concentric with the Earth (Orbit *B*, Figure 6-21). If, however, the final velocity is less than that for a circular orbit, and its direction of travel is parallel to the local horizon, the altitude at which the final velocity was achieved will become apogee, and it will travel closer to the Earth to establish perigee (Orbit *A*). If the final velocity is greater than that for a circular orbit, it will establish this altitude as perigee and travel farther from the Earth to apogee (Orbit *C*). The velocity for a circular orbit whose

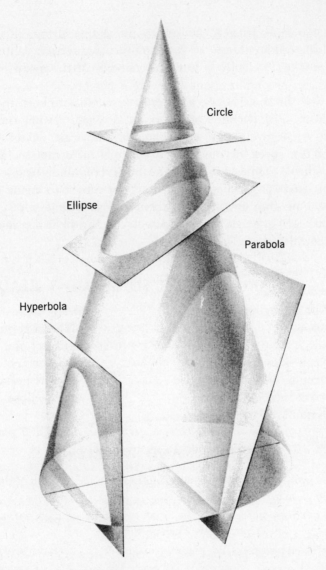

Circle

Ellipse

Parabola

Hyperbola

**Figure 6-22**

The intersection of a plane and a cone is called a *conic section*. If that plane is perpendicular to the axis of the cone the conic section is a *circle*; if that plane is oblique to the axis of the cone, yet cuts every element of the cone, the intersection is an *ellipse*. If that plane is parallel to one element, the intersection is a *parabola*; if the plane is parallel to the axis of the cone the intersection is a *hyperbola*.

radius is 4,400 miles (that is, 400 miles above the surface of the Earth) is about 18,000 miles per hour or about 5 miles per second. With orbits of increasing radii the velocity is less; the velocity of the moon in its orbit is 0.66 miles per second or about 2,400 miles per hour.

If, however, the speed of the spacecraft just equals the escape velocity of the Earth, its orbit (Orbit *D* of Figure 6-21) would assume the shape of a *parabola* (see Figure 6-22). The shape of the orbit would become a *hyperbola* if the speed of the craft exceeded the escape velocity. The ellipse is a closed figure, but both the hyperbola and the parabola are open curves, that is, the spacecraft would not return to Earth. It would, however, assume an elliptical orbit about the sun, unless, of course, its speed should reach or exceed the escape velocity for the sun at this distance from that central body.

## BASIC VOCABULARY FOR SUBSEQUENT READING

| | |
|---|---|
| Annular eclipse | Perigee |
| Apogee | Sidereal period |
| Lunar eclipse | Solar eclipse |
| Neap tides | Spring tides |
| Occultation | Synodic period |
| Partial eclipse | Total eclipse |
| Penumbra | Umbra |

## QUESTIONS AND PROBLEMS

1. Draw the Earth–moon system to scale and indicate the location of the center of mass of the system.

2. After the next new moon follow the moon from night to night through at least one complete cycle of phases (weather permitting).

3. Describe the appearance and possible origin of three features of the lunar landscape.

4. Explain the relationship between solar eclipses, lunar eclipses, and the lunar nodes.

## FOR FURTHER READING

Alter, D., *Pictorial Guide to the Moon*, Thomas Y. Crowell Co., New York, 1963.
Baldwin, R. B., *A Fundamental Survey of the Moon*, McGraw-Hill Book Co., 1965 (paperback).

Glasstone, S., *Sourcebook on the Space Sciences*, Van Nostrand Co., Princeton, N.J., 1965, Chapter 9.

Whipple, F. L., *Earth, Moon and Planets*, Harvard University Press, Cambridge, Mass., 1968.

Cannon, P. J., "Lunar Landslides," *Sky and Telescope*, p. 215 (Oct. 1970).

Faller, J. E., and E. J. Wampler, "The Lunar Laser Reflector," *Scientific American*, p. 38 (March 1970).

Greenacre, J. A., "A Recent Observation of Lunar Color Phenomenon," *Sky and Telescope*, p. 316 (Dec. 1963).

Hess, W., et al, "The Exploration of the Moon," *Scientific American*, p. 54 (Oct. 1969).

Hillenbrand, R., "The First Men on the Moon," *Sky and Telescope*, p. 144 (Sept. 1969).

Hillenbrand, R., "The Apollo 12 Explorers on the Moon," *Sky and Telescope*, p. 95 (Feb. 1970).

Kuiper, G. P., "The Lunar and Planetary Laboratory," *Sky and Telescope*, part I, p. 4 (Jan. 1964); part II, p. 88 (Feb. 1964).

Saari, J. M., and R. W. Shorthill, "Hot Spots on the Moon," *Sky and Telescope*, p. 327 (June 1966).

Shoemaker, E. M., "The Geology of the Moon," *Scientific American*, p. 38 (Dec. 1964).

Wildey, R. L., "Measuring the Shape of the Moon," *Sky and Telescope*, p. 147 (March 1966).

"The Apollo 11 Experiments," *Sky and Telescope*, p. 149 (Sept. 1969).

"Findings from a Sample of Lunar Material," *Sky and Telescope*, p. 144 (March 1970).

"Oldest Moon Rock," *Sky and Telescope*, p. 3, July 1970.

"Results of Apollo 11 Research," *Sky and Telescope*, p. 226 (April 1970).

"Standing Waves on the Moon," *Sky and Telescope*, p. 166 (Sept. 1969).

CHAPTER
7
CHAPTER

# MINOR PLANETS, METEORITES, AND COMETS

Man leaves no stone unturned in his eternal search for system and law. This search has led from the Pythagorean numbers games to the bases of modern mathematics; and numbers games are still being played, with results that are sometimes as fruitful.

## 7.1 Numbers and Planetary Distances

Take, for instance, the mean distances in astronomical units of the planets from the sun. Beginning with Mercury, we find that these give, naturally enough, a sequence of increasing numbers:

$$0.387, 0.723, 1.00, 1.52, 5.20, 9.54, 19.2, 30.1, 39.5$$

An early attempt to impose logic and system upon this sequence is known as *Bode's law*. No law was ever less legitimate; it is merely a scheme for systematically duplicating an approximate progression of planetary distances from the sun. If we start with 0.0, 0.3, 0.6, etc., doubling each number until we reach 38.4, we obtain the following sequence simply by adding 0.4 to each of the resulting numbers:

$$0.4, 0.7, 1.0, 1.6, 2.8, 5.2, 10.0, 19.6, 38.8$$

This sequence roughly approximates the natural one, except that between the numbers 1.6 and 5.2, which represent the distances of Mars and Jupiter respectively, there is a 2.8. Also, the number 30.1, representing the distance of Neptune, is missing, and the number 38.8 differs considerably from the actual distance of Pluto.

All this is quite meaningless. It is a game of numbers.

It is possible, however, to derive a sequence that is more descriptive of the physical facts. Let us divide the mean radius of each planet's orbit, starting with Venus, by that of the planet just inside its orbit. The result indicates how many times farther from the sun a given planet is than its neighbor toward the sun. For example, 0.723 divided by 0.387 equals 1.9, which tells us that Venus is 1.9 times as far from the sun as Mercury. Doing this for all the planets, from the nearest to the farthest, gives us the sequence:

$$1.9, 1.4, 1.5, 3.4, 1.8, 2.0, 1.6, 1.3$$

## Chapter Opening Photo

The Lost City meteorite photographed by a meteor patrol camera. (Smithsonian Astrophysical Observatory)

All these numbers lie between 1.3 and 2.0 except for that single 3.4, obtained by dividing Jupiter's distance from the sun by that of Mars. These two planets, then, are farther apart with respect to their distance from the sun than are any other neighboring planets.

## 7.2 The Minor Planets (Asteroids)

If Bode's law is not taken seriously today, it did gain support in 1781 when the discovery of Uranus appeared to verify it; the predicted 19.2 is not far from the observed 19.6. On January 1, 1800, Bode's law acquired still another friend. The Italian astronomer Piazzi discovered a very small planet, later named *Ceres*, whose distance from the sun is the expected 2.8. In March, 1802, a second small planet, *Pallas*, was discovered at about the same distance from the sun—and the race was on. Since that time, thousands of very small *minor planets* (sometimes called *asteroids*—but they are "star-like" only when seen from great distances) have been discovered, most of which revolve about the sun in that gap between Mars and Jupiter. Each minor planet is nothing but a chunk of rock revolving about the sun. These chunks of rock, however, should contain an invaluable record of historical events in the solar system.

The distance of each minor planet from the sun can be found by computing its orbit from observed positions. Most minor planets have orbits that lie between the planets Mars and Jupiter. Some, however, come inside Mars' orbit and others inside the orbits of Venus and even Mercury. As a group, then, these very populous members of our solar system travel over a considerable range. Their orbits are sometimes quite eccentric, that is, elliptical and considerably distorted from a circle. Two of these orbits are shown in Figure 7-1.

Minor planets are discovered most frequently by accident. A photograph of a field of stars is usually a time exposure because the stars are faint; and since a minor planet moves with respect to the stars, it will leave a blurred streak instead of a point as do the stars (Figure 7-2). (The telescope is usually driven and guided to follow the stars.) If the minor planet leaves a long streak it is in all likelihood relatively close to the Earth and is therefore of interest. It is then further observed in order to obtain positions that are required for the calculation of its orbit. If the minor planet leaves a short streak, however, it is of little interest as it is likely to be far from the Earth and one of the many that may be forgotten.

It may seem "unscientific" that the astronomer should disregard any

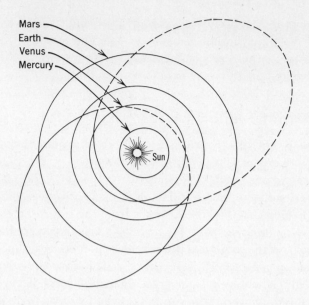

Mars
Earth
Venus
Mercury
Sun

**Figure 7-1**

The orbits of two minor planets that come inside the orbit of Mars. The dotted portions of the orbits are south of the plane of the ecliptic.

evidence at all, but it is time-consuming to make observations and compute orbits. There are so many minor planets, so few facilities, and so many pressing unknowns in the universe that he is forced to deal only with those unknowns which he decides are most important.

When the periods of a large number of minor planets are arranged according to their length, we may note an interesting phenomenon (Figure 7-3): there are periods that seem to be avoided. The most conspicuous are those of 6.0, 4.8, and 4.0 years which happen to be $\frac{1}{2}$, $\frac{2}{5}$, and $\frac{1}{3}$ of the 12-year period of revolution of Jupiter. Other avoided periods are $\frac{1}{4}$, $\frac{1}{5}$, and $\frac{3}{5}$ of the period of Jupiter.

This is not a coincidence but a direct result of the perturbing effect massive Jupiter has on these small objects. The missing periods can be translated into mean distances from the sun by Kepler's third law; they represent actual gaps in space that are populated by fewer minor planets than other regions in the "minor planet belt." These gaps are similar to the gaps between Saturn's rings which are caused by the perturbing effect of Saturn's innermost satellite.

Although minor planets are very small, a few are large enough to have been seen as disks in large telescopes. By measuring the apparent size of the disk of a minor planet, and knowing its distance at the time it was photographed, it is possible to calculate the minor planet's actual diameter. Ceres is the largest with a diameter of 480 miles; Pallas is next with a diameter of 300 miles; then Vesta, 240 miles; and Juno, 120 miles. Since their diameters have been measured, their albedoes can be determined; as might be expected they are quite small. The smallest albedo, 0.06 for Ceres, is about the same as that of the moon. A few exhibit higher albedoes, for example, Juno 0.12 and Vesta 0.26; this indicates that these must have a surface that is different from, and more reflective than, that of the moon and Mercury.

It is not uncommon to find a minor planet whose brightness varies. Since every minor planet is seen by reflected sunlight, the natural assumption is that it must rotate on its axis. Consequently, it presents to the Earth different faces which may have different reflecting qualities; or, more probably, the minor planet itself may be irregular rather than spherical

**Figure 7-2**

The motion of a minor planet causes it to appear as a streak on a time exposure of the stars. (Courtesy of C. Wirtanen, Lick Observatory)

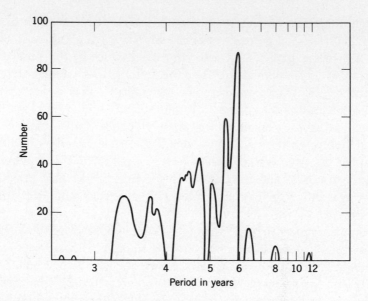

**Figure 7-3**

A distribution of minor planets according to their periods indicates that certain periods are rare.

in shape and by rotating may present now a larger and now a smaller cross section to the Earth. From its variations in brightness, it has been determined that Vesta rotates with a period of 5 hours and 20.5 minutes.

Because of their small size and large number, the minor planets are considered as a separate group in the membership of the solar system. There are only a few that are larger than 100 miles in diameter, but their number increases into the thousands as those of smaller and smaller diameters are counted. The very large number of small minor planets raises questions about their origin.

## 7.3 Meteorites

Compared to the nine major planets the minor planets are very small, but they are far from being the smallest objects that revolve about the sun. *Meteorites* are not only smaller, but since they collide with or fall on the Earth they have become the first extraterrestrial objects to be touched by man and examined in his laboratory.

**a. Seen as Meteors.** Meteorites travel in orbits about the sun at velocities of 10 to 30 miles per second. When one collides with the Earth it must first penetrate our protective atmosphere. During this high-speed penetration, the air in front of the meteorite does not have time to move to the side, so the meteorite acts like a piston, compressing the air in front of it. The heat resulting from this compression causes the meteorite to melt and burn. That is, the energy, both kinetic and potential, which the meteorite had with respect to the Earth is converted into heat. Small meteorites never survive this burning; they filter down to the surface of the Earth as small particles of ash or dust. From the length of time it takes an average meteorite to burn—just a fraction of a second—one can estimate their size to be not much larger than an "o" on this page.

As the meteorite streaks and burns through the upper atmosphere, it may be visible from the earth as a "shooting star" or *meteor*. The term meteorite is reserved for the object itself; the term meteor, for historical reasons, has come to designate the flash of light in the nighttime sky. The term meteor derives from the fact that before the very early part of the nineteenth century meteoric streaks were thought to be a strictly atmospheric phenomenon.

A patient observer away from city lights may observe 5 to 10 meteors every hour. If he continues watching all night long he will find that the frequency of meteors increases as the night progresses. Not only will the number increase each hour but after midnight they will become, on the average, more brilliant.

We can understand the increase in meteor activity in the early morning hours if we recall how the Earth revolves and rotates. It rotates in the same direction in which it revolves, so the leading side of the Earth is the morning side (Figure 7-4). That portion of the Earth in dawn, is, therefore, "out in front" of the Earth as it travels in its orbit about the sun. As a consequence it is able to pick up more meteors; it runs into them. Thus the early morning meteors have a higher velocity when they strike our atmosphere. The brightness of a meteor depends to a large extent upon its relative velocity, so these meteors are on the average more brilliant than those seen in the evening. An evening meteor overtakes the Earth and therefore has a low relative velocity.

Most meteors are rather faint, but sometimes a very bright one rivals the brightness of the full moon and even casts shadows. Such a bright meteor is generally called a *fireball*. Since the difference between a meteor

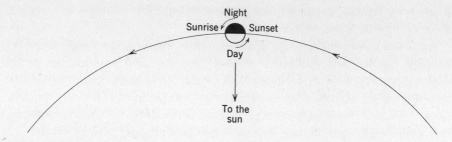

**Figure 7-4**

The sunrise portion of the Earth leads the way as the Earth revolves about the sun.

and a fireball is dependent on brightness, it is difficult to make a definite distinction between the two. In his book *Between the Planets,*[*] Watson calls attention to a very quaint way of distinguishing a meteor from a fireball. He says that, "One astronomer adopted a very practical definition: 'A fireball is a meteor sufficiently bright to make people report it.'" Some fireballs may be so bright as to be seen over the areas of several states.

At best meteors are difficult to study, for they give no warning when or where they will be seen, and they last only a fraction of a second once they strike our atmosphere. Modern observational techniques, however, have permitted a great deal of information to be gathered about these elusive objects.

**b. Techniques of Observation.** The first study of meteors was made in 1798 by two German students who realized that each had seen the same meteor even though they had been separated by several miles on the Earth. In following up this observation with others, they learned how to determine the height of the meteors by the method of parallax (Figure 7-5). They knew the distance AB and the angles formed at the points A and B, so they needed only to draw a similar triangle to scale in order to estimate the height of the meteor. Or they could make a more accurate determination by using trigonometry. Their results indicated that meteors appear at heights of about 50 miles.

Current studies are being directed to determine the motion of meteors with the hope of learning not only the height at which they appear in the atmosphere, but also their speed and direction of travel. These results enable the astronomer to calculate the orbit of the meteor before it was

*Fletcher G. Watson, *Between the Planets*, Cambridge, Harvard University Press, 1956.

intercepted by the Earth. In this way, it is hoped to identify the source of meteors with other astronomical objects.

Their velocities may range from about 6 miles per second to about 50 miles per second. The faster meteors become visible at heights of up to 75 miles and generally disappear at a height of about 60 miles; the slower ones appear at about 50 miles and disappear at about 30 or 40 miles. Since the faster meteors are more subject to heating by the atmosphere, they make their appearance higher where the atmosphere is thinner.

Current studies are being made to determine the motion of meteorites with the hope of being able to determine their orbit before colliding with the Earth, and, if possible, of recovering them after their fall. The first success in this program, carried out by the Smithsonian Astrophysical Observatory, was the recovery of a 22-pound meteorite near Lost City, Oklahoma, on January 9, 1970 (Figure 7-6a).

The photograph that opens this chapter helps tell the story. Two of many cameras used in the study photographed the meteor streak. Each camera made a time exposure of the skies (note that the stars leave streaks) and each camera had a special shutter that interrupted the light entering the

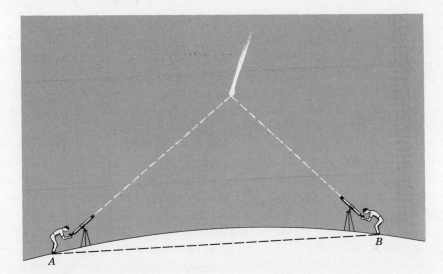

**Figure 7-5**

The height of the meteor can be determined if it is observed either visually (without a telescope) or photographically (with a telescope) from two well-separated positions on the Earth.

**Figure 7-6 (a)**

The Lost City meteorite, a chondrite, that was recovered after having been photographed during its fall (see photograph that opens this chapter). (Smithsonian Astrophysical Observatory).

camera 20 times each second. By counting the interruptions in the meteor trail, it is possible to determine how long the meteorite was burning.

From its speed and direction of fall, it was possible to determine not only the orbit of this meteorite before it hit the Earth, but where it was located after it fell. Luckily it fell right along side a road, and so was easily recovered only six days after it fell even with a snowstorm causing some delays.

When the Lost City meteorite struck the Earth's atmosphere its speed was 8 miles per second; at a height of 15 miles, its speed had decreased to 6 miles per second. Its trail ended at a height of only 12 miles when its speed was 2 miles per second. This was not a fast meteorite, so it did not create much of a crater where it fell.

Meteors may help us understand the upper atmosphere but meteorites are of greater interest to astronomers as extraterrestrial objects. Not only are they tangible evidence that the chemical elements on the Earth are no different from those in the solar system in general (which is certainly to be expected), but also they have been helpful in determining the solar system's age. They are, incidentally, a bundle of clues that need to be untangled to help solve the mystery of the origin of the solar system.

c. **Three Groups of Meteorites.** The meteorites can be classified into three general groups: *iron–nickel, stony, and stony–iron.* (The iron–nickel meteorites average about 90% iron and 8% nickel (although these percentages vary considerably) with the balance consisting of other elements. The stony meteorites have quite a different composition; their main chemicals are 36% oxygen, 24% iron, 18% silicon, 14% magnesium (again, these are averages and there is considerable variation), and a smattering of other elements. The stony–irons form a group that bridges the gap between the other two main groups.

More iron–nickel meteorites have been found than all others put together. But iron meteorites are easily distinguished from terrestrial rocks, even by a farmer tilling the soil, or a hiker or hunter. Iron meteorites have been found which, according to radioactive studies (see Chapter 8), have been on the Earth for as long as 800,000 years. Many of the stony meteorites would have weathered away in that time interval. Of those observed to fall and then recovered, only a little more than 40 have been iron meteorites; more than 600 stony meteorites have been seen to fall and then collected! Nevertheless, all of the big meteorites are significantly iron–nickel and they far outweigh all of the stony meteorites.

The reason for the existence of three main types of meteorites is not really known, but they have fed speculation on a possible relationship between the minor planets, the meteorites, and the "missing planet" between Mars and Jupiter. It is reasoned that perhaps there were small planets in this region while the solar system was forming. If after those planets had formed, and if their interiors became hot enough, then the heavier elements such as iron and nickel would have settled to the center. It is then supposed that several pairs of these minor planets collided each breaking into many smaller meteorite pieces, giving rise to the thousands of small minor planets now observed. Because of the resulting explosion some of the debris would have been spread out over a fairly large region of the solar system; and some of this debris would certainly have landed on Mars, the Earth, and the moon.

The cores of those minor planets would have broken into what we now find as the iron–nickel meteorites, the outer parts would have given rise to the present stony meteorites. There is no other reasonable explanation for the separation of the heavier iron and nickel from the rest of the rocky material except that it occurred inside of a much larger body, such as a planet, with a hot interior.

**Figure 7-6 (b)**

Crystals in an iron–nickel meteorite found in Willow Creek, Wyoming.

However, in order for the interiors of those minor planets to become hot enough for the settling of the heavier elements, they would have to have been fairly large planets. Recall that the moon apparently does not have much of an iron core. But perhaps the heating processes in those minor planets was different from the heating processes in the moon.

There is other evidence that leads us to suspect that the meteorites were formed inside a much larger body. When meteorites are examined in the laboratory it is found that they contain crystals. This in itself is nothing startling, but the crystals are large (Figure 7-6b). If a rock contains large crystals the geologist knows that the rock cooled and the crystals formed very slowly; if they are small he knows that the rock cooled and the crystals formed more rapidly. Since the crystals in meteorites are large they may once have been a part of a much larger body that took some time to cool from a molten state.

Detailed studies of the crystals by J. A. Wood of the Enrico Fermi Institute for Nuclear Studies have permitted good estimates of the rate of cooling of the iron meteorites. One group of meteorites studies apparently cooled at the rate of 1°C per 100,000 years; the other group cooled 10 times more slowly. This evidence indicates that at least two parent bodies are involved: the faster-cooling meteorites came from a body about 60 to 120 miles in diameter, the slower-cooling meteorites from a body about 160 to 200 miles in diameter. These are not very large bodies, and it is wondered whether they are large enough for the settling of heavier elements to the center. It can only be concluded that the origin of the three forms of meteorites remains a mystery, but one that is yielding to improved techniques of observation.

**d. Tektites.** There are suggestions, on the other hand, that meteorites come from the moon. Certainly, if a large meteorite or comet struck the surface of the moon, some of the material would be ejected at velocities exceeding the escape velocity of the moon. Curious objects called *tektites* have been particularly singled out as lunar in origin. Tektites are found in strangely limited areas of the world: in Australia, Java, the Philippines and on the Indochina peninsula. Quite different tektites are found in Czechoslovakia, Africa, Texas, etc. The tektites are unique in their glasslike composition and more notably in their shapes resulting from heating apparently upon entering our atmosphere at velocities less than that which burn meteorites.

It is significant that one of the many shapes of tektites has been duplicated in the laboratory by using pieces of tektite glass mounted in a high-speed wind tunnel (Figure 7-7). But more than that, by a thorough study of surfaces and flow lines, the experimenters, Dean R. Chapman and Howard K. Larson of the Ames Research Center, California, have been able to establish the primary origin of tektites in an atmosphere with a pressure less than 0.0002 times the atmospheric pressure at the surface of the Earth. This observation would eliminate terrestrial volcanoes as the source of tektites. Furthermore, these studies indicate that their velocity upon entering the Earth's atmosphere must have been just slightly higher than 7 miles per second, the escape velocity at the surface of the Earth. There seems little question that tektites indeed originated outside the Earth's atmosphere and there is reasonable (but not conclusive) evidence that the ones that struck Australia resulted from the collision and resulting explosion that caused the crater Tycho on the moon. This evidence comes from a consideration of their probable orbital motion in the Earth–moon gravitational field after impact explosion, as well as from their distribution on the Earth.

**e. Chondrites.** Another class of meteorites has proved every bit as interesting as the iron–nickel meteorites. The stony meteorites have

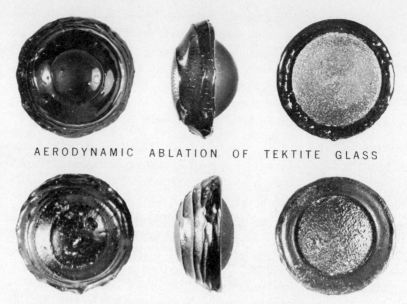

AERODYNAMIC ABLATION OF TEKTITE GLASS

AUSTRALIAN TEKTITES

**Figure 7-7**

Three different pieces of tektite glass which have been placed in a wind tunnel and subjected with an arc jet to simulate entry into the atmosphere. These are compared with three natural tektites found in Australia. (Courtesy Dean R. Chapman, NASA)

been divided into two subclasses: the *achondrites* and the *chondrites*. The chondrites contain little spherical bodies only about 0.1 cm in diameter, called *chondrules*. Study of these chondrites has opened up new avenues of approach to the age-old problem of how the solar system came into being. Achondrites do not have chondrules.

The study of chondrites has proceeded along lines similar to studies of the iron–nickel meteorites. Both objects, after all, are basically crystalline in structure. As such they are the subject of the geochemist. The geochemist studies the growth of crystals, the chemical reactions which lead to the particular formations, and conditions of temperature and pressure which must prevail before a particular crystal or chemical can form. With this background, he can learn a great deal about the history of the different kinds of meteorites. Studies of chondrites have, it appears, led to some enigmas.

The chondrules are held in a matrix composed to a large extent of two minerals found in terrestrial rocks: olivine and pyroxene (Figure 7-8a). The matrix also contains other minerals, including glass; the matrix is often rather friable and as such crumbles easily. But the principal distinction between the matrix and the chondrules is their iron content.

This distinction is particularly evident in a class known as *carbonaceous chondrites*, because of the black, loosely held matrix (Figure 7-8b). The matrix contains almost no free iron; nearly all of the iron has been combined chemically to form oxides and other compounds. The iron in the chondrules, however, is almost entirely free; less than 1% occurs as iron oxide.

Furthermore, it is found that the chemical composition of the chondrules is almost identical with that of the sun. This cannot be said of any rock on the Earth. During the Earth's history, rocks have been heated and melted so that chemicals have become separated. This has not happened to the chondrules in the carbonaceous chondrites.

Radioactive studies lead to the conclusion that the chondrules were formed in the very earliest stages of the solar system—in the original gases from which the planets were made. Their content of free iron further indicates that they were formed during the passage of a shock wave set up by the forming and partially unstable sun, a shock wave which was transmitted through the primordial gases. It is presumed that during the shock wave some of the material would condense into liquid droplets and harden into the chondrules after the shock wave passed. Material not condensing into liquid droplets would later form the dust from which the matrix of the chondrite is formed. The chemical action of the primordial gas following the passage of the shock wave could have oxidized the iron in the fine dust particles, but not so in the larger—and thus protected—chondrules.

Some of the chondrules have suffered recrystallization. In these, as expected, some of the iron has become oxidized. It is supposed that chondrules which have been recrystallized became part of a larger body which was heated. This heating would account for the recrystallization.

It seems to be generally accepted that the round little chondrules embedded in the matrix of the chondrites were once molten. Although much study remains to be done on these very fascinating objects, their once molten condition, their chemical similarity to the sun, their free iron, make them unique in meteoritic studies. It is anticipated that they hold still more clues to the history of the solar system.

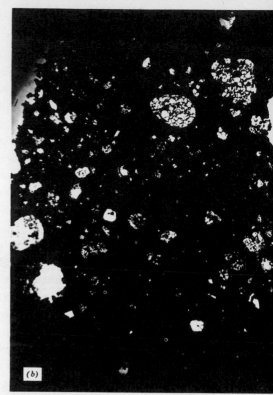

## Figure 7-8

(*a*) Chondrules embedded in the surface of the Bjurböle (Finland) chondrite meteorite. (*b*) A thin section of a carbonaceous chondrite taken from the Murray meteorite which fell in Kentucky. (Courtesy Brian Mason, U.S. National Museum)

Whether chondrules are solidified droplets from the primordial gases of our solar system, and whether there is any connection between the minor planets and the meteorites is not known with certainty. But the entire story of the solar system cannot be learned in one day or one step. Speculation gives direction to observational research. If a speculation proves incorrect, progress has been made even though a new direction must be taken. Man has never seemed to be at a loss for new ideas and new directions, at least not in the past 400 years.

## 7.4 Meteorite Craters

When a large meteorite falls, it leaves a crater on the surface of the Earth. Some of these craters are large, although none is as large as many of those on the moon. The largest that is definitely known to be of meteoritic origin is the Barringer crater near Winslow, Arizona (Figure 7-9), which is about 4,150 ft in diameter. The Chubb crater in Quebec, however, is nearly 2 miles in diameter and unlike any of the lakes in the surrounding area. These lakes were formed by glaciers and consequently are long and narrow. The lake in the crater, however, is quite circular. The crater is assumed to be meteoritic in origin, although, unlike the Barringer crater, no meteoritic fragments have been found around it. This may be because the area around the Chubb crater is covered with boulders which make it quite difficult to find fragments, whereas the area around the Barringer crater is flat and sandy.

The biggest fall in recorded history occurred in Siberia on June 30, 1908. The explosion resulting from this fall was big enough to make any man-made explosion seem small in comparison. At least ten craters were formed; presumably the meteor broke into smaller fragments when it hit the atmosphere. The trees outside the region of craters, to a distance of nearly 20 miles, were all blown down, with their tops pointing away from the explosion. Windows were broken for a distance of 50 miles, and an engineer about 450 miles away stopped his train in fright. The pressure wave traveled through the atmosphere as far as England, where it was recorded on barographs (instruments that measure atmospheric pressure). The shock waves traveled through the surface of the Earth and were recorded on seismographs (instruments that measure earthquakes) in Europe. Had the meteorite struck only 4 hours, 47 minutes later, the Earth would have turned enough so that it would have struck the city now called Leningrad!

Recent studies have shown that this explosion was too powerful to be accounted for by craters of the size found in the region of impact. It has therefore been speculated that it was caused by a comet and not by a meteorite. The gases in the comet, when heated in the Earth's atmosphere, could account for the magnitude of the explosion, for a gas such as methane would react chemically with the atmosphere.

That there are not more meteoritic craters on the Earth has caused some concern. Even with the geological changes, it seems that if the moon is so scarred and if Mars has been pelted by meteorites, then the Earth ought to have more such craters. Increased searching has revealed a number of *astroblemes*, fossil remnants of meteoritic craters. These scars of ancient collisions between the Earth and meteorites have been found on each of the great continents of the world. A number of them have been found in

**Figure 7-9**
The Barringer Meteor Crater, Winslow, Arizona. (Great Meteor Crater photograph)

the Hudson Bay area of Canada in some of the oldest rock formations in the world.

## 7.5 Meteor Showers

On most nights during the year we see only *sporadic meteors*, but every now and then we are fortunate enough to witness a *meteor shower*. On these occasions there may be hundreds of meteors streaking across the sky in a single hour. Like eclipses, these showers frightened more superstitious peoples into thinking that the end of the world was imminent; but the world is still here with us, or we with it.

A photograph of meteors that constitute a shower makes it appear as though they originated in the same part of the sky like spokes from the hub of a wheel. This is the result of perspective. The point from which the trails radiate is called the *radiant*. The radiant effect can exist only if these meteors have been traveling in the same direction in space before they collide with the Earth's atmosphere.

To find the actual speed and direction of travel of meteors in space we observe their speed and direction of fall; from these we can subtract the velocity of the Earth and the effect of the Earth's gravitational field. It has been found that the particles forming each meteor shower traveled around the sun in slightly elongated ellipses with periods ranging from 1.6 to 415 years for different groups. Such a group of particles is called a *stream*.

The Earth travels through one such stream about August 12 of each year, and since the radiant of this shower is in the constellation Perseus these meteors are called the Perseids. Their rate of fall averages about 50 meteors per hour. On October 9, 1933, there appeared a meteor shower (the Draconids) in which some 350 meteors were seen each minute at the peak of the shower. Again on October 9, 1946, the same meteor swarm was intercepted and another brilliant shower was observed. There are more than 25 meteor showers each year.

It is interesting to calculate how many meteors strike the Earth's atmosphere each day. This may be done by considering that any one observer will see only a small fraction (about 1/100,000) of the entire upper atmosphere. Since an observer sees about 10 meteors per hour on a night when no shower is in progress, the entire Earth is bombarded by about 24 million meteors each day. This figure is astonishing, to be sure, but even so it includes only those meteors that are visible. When we include the invisible ones the estimate runs to about 8 billion in one day. The number of invisible meteors can be roughly estimated by finding how many visible meteors of each size strike the atmosphere. The number increases regularly as the size decreases, and from this we can extrapolate to estimate the number of invisible meteors.

It might seem that so much meteoritic material filtering through the atmosphere would add a considerable amount of mass to the Earth. But each meteor is so small that the increase in mass is insignificant when compared with the Earth's mass. The increase becomes more understandable if expressed in weight; it has been estimated that as much as 5 million tons of extraterrestrial material reach the surface of the Earth each year. This is a sizable amount of material, but the calculated weight of the Earth is $7 \times 10^{21}$ tons. At the present rate of collection, it would take roughly $10^{15}$ years to collect an amount of material equal to the mass of the Earth.

## 7.6 Comets

Only the comets have made a greater impression on mankind than the meteor showers and solar eclipses. To nearly every comet visible to the naked eye, in all but the most recent recorded history, man has been able to lay the blame for some disaster, whether it be war, massacres, the black plague, assassinations, floods, or earthquakes. Comets have been looked upon as omens of terror, blood, evil. Powerful men (who happened to live when a comet was seen in the sky) have been associated with them. The comet of 43 B.C. was supposed to have been the soul of Julius Caesar transported to heaven. This belief was encouraged by Augustus for political reasons. A comet was associated with the coming of William the Conqueror when it appeared in April of 1066. The comet of 1811–1812, of course, was called "Napoleon's comet."

It is true that comets are a spectacular sight when seen in the sky; little wonder that people who did not know what they were associated them with the evils that plagued man! When we learn more about comets, however, we find that they are of little danger to us; they do not strike the Earth very often, although one may have struck it in 1908!

The first fruitful scientific work on a comet was done by Tycho Brahe, the astronomer who supplied Kepler with such a splendid list of planetary positions. Brahe compared the positions of the comet of 1577 as seen from Prague and from his island in the Baltic Sea, and by the method of parallax determined that it was farther away than the moon. Therefore it could not be an atmospheric phenomenon, as had been thought up to that time.

**a. Orbits of Comets.** It remained for Halley, a contemporary of Newton, to predict the return in 1758 of a comet he had observed in 1682. Halley based his prediction on the orbit he had derived and on the fact that comets had appeared in the past at times that agreed with that orbit. He cited the comets of 1607, 1531, 1456, and 1305 as evidence that these were but passages of a single comet with a period of 75 or 76 years. On Christmas Day of 1758 Halley's comet did indeed appear in the skies, although he himself did not live to see his prediction confirmed.

A *periodic comet* revolves about the sun in an elliptical orbit which is usually quite eccentric. Halley's comet, for instance, comes within about 0.6 A.U. of the sun at perihelion and goes out as far as 35 A.U. at aphelion. Since all comets move in accordance with Kepler's three laws, they must travel faster at perihelion than at aphelion.* We are able to see comets with very eccentric orbits for only a small part of their journey because when they travel far from the sun they become too faint to be seen with our telescopes. Since a comet in a very eccentric orbit is observable only during a small part of its period, such an orbit is exceedingly difficult to compute. Our example, Halley's comet, has an orbit that brings it inside of Venus' orbit and takes it out beyond Neptune's.

Many of the periodic comets have orbits less eccentric than that of Halley's. Encke's comet, for example, has a perihelion distance of about 0.3 A.U., but an aphelion distance of only 4.1 A.U. Its period of 3.3 years is the shortest known. The orbit of comet Schwassmann–Wachmann lies between the orbits of Jupiter and Saturn and the comet is therefore observable every year.

Jupiter has such a strong gravitational influence on the comets that it has managed to "capture" a large number of them. There are about 60 comets whose periods are less than 100 years; these are termed the *short-period comets*. Some of them have aphelia that are close to Jupiter's orbit, comprising what is called "Jupiter's family of comets."

Although the planes of the orbits of all the planets are inclined by only small angles from one another, and their motions are all direct, the same is not true of comets. Halley's comet, for example, revolves in retrograde motion, and a number of comets have orbits that are nearly perpendicular to the ecliptic. The planes of near-parabolic orbits are, more often than not, tilted by a large angle from the ecliptic, whereas the orbits of periodic comets conform more to the average plane of the solar system. The orbits of comets, then, are more erratic than those of minor planets, as those of minor planets are more erratic than the orbits of planets.

**b. The Anatomy of a Comet.** The aspects of comets are nearly as varied as their orbits. Among individual comets there are strong similarities and yet many variations. When viewed through a telescope, a comet will be seen to have a central *nucleus* which is quite bright and small. Surrounding the nucleus is a fainter and more nebulous structure called the *coma*; and this together with the nucleus composes the *head* of the comet.

---

*As a rule of thumb, the velocity of a comet is comparable to the velocity of a planet at the same distance from the sun. Actually, since a comet's orbit is generally more elliptical than that of a planet, its velocity will be somewhat greater near perihelion and less near aphelion than that of a planet at the same distance.

Extending from the head is the *tail* of the comet, although unlike most tails in nature it is sometimes in front of the comet. The tail (and not all comets have tails) points away from the sun, so as the comet recedes from the sun the tail will lead the way. This seems contrary to what we might expect, but we must remember that in the space between the planets there is little that would cause the tail to trail behind as the resistance of the air in our atmosphere causes the smoke from a locomotive to trail.

As the comet approaches the sun the coma and tail grow larger. (Their growth is shown schematically in Figure 7-10 and pictorially in Figure 7-11). The obvious cause of this effect is the sun.

The streams of particles (electrons, protons, and nuclei of other light atoms) comprising the solar wind are the cause of both the coma and the tail of a comet. Gases are given off by the warming nucleus of the comet, are caused to emit their own light, and are driven away from the sun all by the solar wind. But studies of Comet Ikeya-Seki in 1965 indicate that only a thin outer shell of the comet is heated during a close approach to the sun. As the comet whirls about the sun, the gases in the tail are farther from the sun; they therefore lag behind, because each particle of gas assumes an orbit of its own.

Cometary tails will often show knots of material that progress outward from the comet, as with Comet Morehouse in October 1908 (Figure 7-12). The timing and velocity with which these knots move indicate that comets are subjected to sudden increases in the solar ejection of particles. These sudden increases in the solar wind are apparently caused by solar activity related to the sunspots and flares.

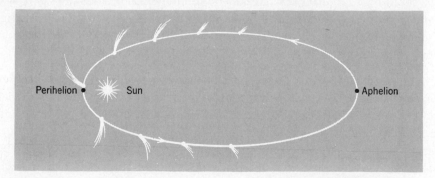

Figure 7-10
The comet's tail always points almost directly away from the sun.

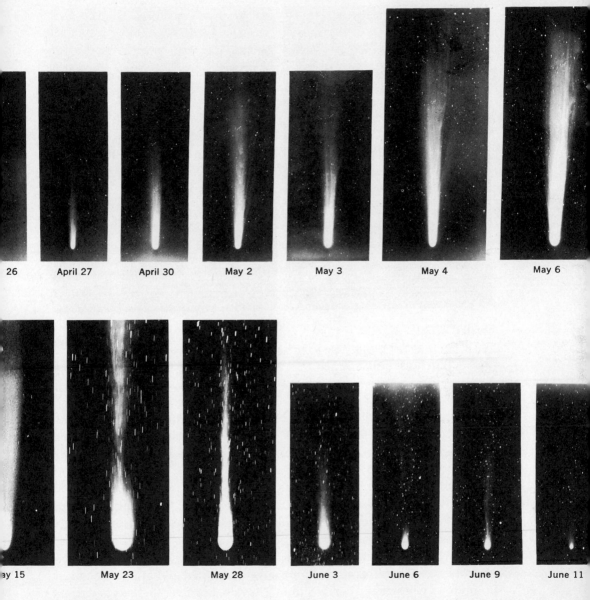

**Figure 7-11**

As Halley's comet approached the sun in 1910, its coma and tail grew in size.
As the comet receded from the sun both the tail and the coma grew smaller.
(Photographs from the Hale Observatories)

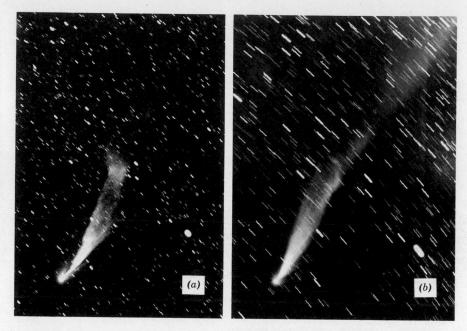

**Figure 7-12**

Comet Morehouse showing changes within 3 hours on September 30, 1908.
(Yerkes Observatory photograph)

The size of the coma and tail is rather staggering. The tail of Halley's comet reached a maximum length of about 94 million miles, a little more than the distance of the Earth from the sun. The comet of 1843 had a tail that was nearly twice as long as this. The head of a comet (including the coma) may reach a diameter that rivals that of Jupiter; in fact, this is about the average size. The smallest coma observed was about 10,000 miles in diameter and the largest was that of Holmes's comet in 1892 with a diameter of 1,400,000 miles, nearly twice the diameter of the sun! However, the dimensions of any comet are not constant. The tail and coma are at their largest when the comet is near perihelion, but both shrink at great distances from the sun. They therefore do not give a complete picture of the comet, for both vary in size and composition. They do, however, help us determine the composition and structure of the nucleus.

If we analyze the light from a comet with a spectrograph we are able to learn much more about it. The nucleus gives a continuous spectrum which is apparently reflected sunlight. The coma and tail yield an emission spectrum composed of both the bright lines caused by atoms and bright *bands* from molecules. (Molecules, because they are several atoms together, emit a great many lines in groups which are called bands.) These molecules absorb some of the energy of the sunlight and reradiate it. Many of these molecules detected in comets, however, are not the ones we are likely to find on the Earth or any other planet. Instead, we find such strange molecules as $C_2$ (carbon), CN (cyanogen), NH (nitrogen hydride), CH (methylidyne), and OH (hydroxyl). In addition, there are a few common molecules, such as CO (carbon monoxide) and $N_2$ (nitrogen). Some of these have become ionized, that is, they have lost one electron (for example, $CH^+$, $CO^+$, $OH^+$, and $N_2{}^+$).

The strange molecules are incomplete and are called free radicals. If any one of them were placed in the Earth's atmosphere it would quickly grab one or more atoms and by doing so become a stable molecule. Under very low pressure, however, atoms are too far apart for an incomplete molecule to capture free atoms. The fact that these strange molecules are found in the gases composing the coma and tail of a comet, then, leads to the conclusion that these gases are in what would be considered a vacuum by terrestrial standards.

Since the gases that leave the comet by way of the tail never return, and since their supply is not inexhaustible, the comet becomes less spectacular after repeated trips near the sun. The bright "naked eye" comets that appear unexpectedly from time to time have apparently made only a few trips around the sun, whereas the short-period comets that spend more of their time close to the sun have very short tails or none at all.

**c. Nucleus of Comets.** It is of interest to consider the nucleus of the comet, since both coma and tail originate in this small and less conspicuous part. From the nucleus we obtain a spectrum that is reflected sunlight, and we therefore assume that it is denser than the gaseous portions. Beyond this we are left with little to go on. It was thought that when Halley's comet passed directly between the Earth and the sun in 1910 its silhouette would be visible as it transited the sun. But nothing was seen!

The comet's tail was to sweep past the Earth; this, it was hoped, would enable astronomers to make observations of any reaction between the tail and the Earth's upper atmosphere. However, the Earth's passage through the tail occurred on May 21, 1910, when the moon was unfortunately nearly full; if anything unusual happened it was washed out by the bright moonlight.

It had been announced to the newspapers that the tail consisted of poisonous gases, an announcement, which when interpreted by the newspapers and the public, led to a minor form of mass hysteria and fright for fear that the world would come to an end. It not only initiated the sale of nostrums reputed to confer immunity from the noxious effects of the comet's tail but is thought to have been the cause of a number of suicides. But the tail of the comet, which terrified people then as in centuries past, could be neither seen nor detected when the Earth was quite in the midst of it.

Most of a comet's mass obviously resides in the nucleus, but this mass is very difficult to determine. The first indication of a comet's mass was obtained by Dr. Elizabeth Roemer at the University of Arizona. It was noticed in 1957 that Comet Wirtanen, 1956c (the c indicates that it was the third comet discovered in 1956) had split into two parts (Figure 7-13). By continued observation over 2 years, Dr. Roemer was able to determine the velocity with which the two nuclei separated. Making the assumption that this velocity was the velocity of escape of the nuclei, she was able to estimate that the total mass of the comet was about $10^{14}$ kilograms. On the Earth this would weigh about $10^{15}$ pounds. A rock with that weight on Earth would have a diameter of about 2 miles. A ball of ice and snow would have a diameter closer to $3\frac{1}{2}$ miles.

We do have one clue to the structure of a comet's nucleus. In a preceding section it was mentioned that meteors sometimes appear in showers, and that these showers result from myriads of tiny meteorites called *micrometeorites* traveling in a common orbit around the sun. When a careful study is made of these orbits it is found that nearly all agree quite closely with the orbit of a particular comet. For example, the shower that occurs about August 12, the Perseids, follows the orbit of comet 1862 III; the Leonids, which form a shower about November 14–18, have the same orbit as Comet 1866 I; the May Aquarids are associated with Halley's comet; and the Draconids of October 9 are closely associated with the comet Giacobini–Zinner. In fact, on October 9, 1946, the Earth missed the latter comet by only 140,000 miles; that is, we passed through its orbit only 15 days after the comet passed the same point in space.

It therefore appears fairly obvious that meteor showers result from the Earth's attracting and sweeping up debris that was left in the orbit by a

**Figure 7-13**
Comet Wirtanen, 1956c, showed a double nucleus in 1957 and 1958. This photograph was taken on April 27, 1958, at the U.S. Naval Observatory, Flagstaff, Arizona. (Official U.S. Navy photograph)

comet. The comet, then, must be composed of these tiny particles and the nucleus could be a swarm of micrometeorites trailing its stragglers like a swarm of bees.

But the gases that form the coma and tail could not originate from micrometeorites; the comet is too cold to vaporize them. It is therefore postulated that the micrometeorites of the comet's nucleus are held together by a spongy mixture of ice, frozen methane, ammonia, and perhaps carbon dioxide and $C_2N_2$. As the comet nears the sun some of this frozen mixture begins to vaporize, yielding the coma and the tail. The coma is composed of the gases that have evaporated from the nucleus, and the tail is caused by the action on these gases of the pressure of radiation and atomic particles from the sun.

Hundreds of meteorites have been collected, but until recently no micrometeorites have been collected; they are too small to burn in the upper atmosphere. They are simply slowed by the air, and drift down as undistinguished dust. A group of astronomers headed by C. L. Hemenway at the Dudley Observatory in Albany, New York, have collected micrometeorites by the use of rocket and balloon flights. Collecting panels of nitrocellulose or of aluminum foil coated with mineral oil were sealed in dust-free boxes. Every effort was taken to make sure that the panels contained only material collected above the atmosphere: 100 miles for the rockets and 20 miles for the balloons. At the proper height the sealed boxes were first opened and then closed by a radio signal. The micrometeorites collected are very small, on the order of $10^{-3}$ cm. Some of them are strange jagged shapes, others are spherical, still others are fluffy. Balloon flights made during a meteor shower have collected up to 80 times the number of micrometeorites collected when no shower was in progress. There seems to be every reason to believe that these particles were once part of a comet!

It is becoming clear that the sporadic meteors and the meteorites should be distinguished from the meteor showers and the micrometeorites. The sporadic meteors and meteorites apparently come from the minor planets; some of them may result from the catastrophic collisions of larger ones, but some may result from collisions of lesser minor planets. The micrometeorites, on the other hand, are apparently cometary debris and have nothing to do with the minor planets.

**d. Famous Comets.**  A discussion of some of the more notable comets will help to emphasize and clarify some of the features already mentioned. The most famous, Halley's comet, has been mentioned already, but there are a few interesting facts concerning it that remain to be told.

Since Halley's comet is so bright, it has been seen by many peoples in recorded history. Only one passage since 240 B.C. has not been found in historical or archeological records, many of which were made by the ancient Chinese astronomers. It has made twenty-nine passages close to the sun and it will again appear as a spectacular sight in the sky in 1985 or 1986. There are indications that at one time it was more spectacular than in its more recent visits. But it is dangerous to take literally the descriptions of comets given by ancient peoples; they were too awed and frightened by the sight to speak objectively of the appearance.

Two unexpected comets which appeared in 1957 became bright enough to be seen by the naked eye and to make the year an unusual one. The

**Figure 7-14**

Comet Arend–Roland showed a long "spike" pointing toward the sun when the Earth passed through the plane of its orbit on April 25, 1957. (Uppsala Observatory photograph)

first was Comet Arend–Roland (1956*h*) which was discovered on November 18, 1956, at the Uccle Observatory, Belgium, by Sylvain Arend and Georges Roland. It was followed until February when it went behind the sun. Observations made it possible to predict that when the comet came out from behind the sun it would be bright enough to be seen without the help of the telescope. By April 23, 1957, many people had seen it in the northwestern sky shortly after sunset, and it remained visible for some weeks although it grew fainter and fainter as it receded from the sun and the Earth.

This comet was unusual not only because it was so bright but also because of its "spike," which appeared to be an "antitail" pointing toward the sun. There is no known reason why a jet of gas should point toward the sun, since the solar wind pushes all the gases away from the sun, and gravity does not play favorites. The "antitail" looked somewhat like a fan when first seen on April 22–23, but on April 25 (Figure 7-14) it appeared as the "spike," long and narrow. This feature is thought not to be actually a spike pointing toward the sun but merely sunlight reflected by the debris that was left behind in the comet's orbit. According to this view, when the Earth passed through the plane of the comet's orbit on April 25, the debris was seen edgewise; it thus appeared brighter and narrower than at other times.

We might think that one naked-eye comet would be enough for one year, but no; shortly after Comet Arend–Roland faded from view there appeared Comet Mrkos, technically known as 1957d. Apparently first seen on July 29 by Sukehiro Kuragano in Japan, it was again noted on July 31 by an airline pilot, Peter Cherbak, as he was flying over Nebraska. Antonin Mrkos of Czechoslovakia sighted the comet on August 2; since he was the first to report it officially it was given his name. Comet Mrkos had an unusual tail in that it was broken into two main parts: one part seemed to be straight but to exhibit occasional kinks as though it were wagging; the other arched away from the straighter tail (Figure 7-15). The formation of more than one tail, as in Comet Mrkos, is apparently caused by the fact that molecules may or may not be ionized. Molecules with no electric charge will move differently through the magnetic field of the solar system than molecules with an electric charge.

The fabulous Great Comet of 1965, Comet Ikeya–Seki, was discovered on September 18, 1965, by two Japanese amateurs K. Ikeya and T. Seki (Figure 7-16). This exceptional comet had a perihelion distance of only 290,000 miles from the surface of the sun and a perihelion velocity of 300 miles per second! That the comet survived such a close approach to the sun (recall that the moon is 240,000 miles from the Earth) indicates that really splendid insulation properties operate near the surface of the comet. It was expected that the tidal forces from the sun would affect the comet in some manner; on November 4, 1965, just 2 weeks after perihelion passage, its nucleus was observed to be split into two pieces. There were even reports of a third nucleus. The fact that cometary nuclei split into smaller parts is direct evidence that they are not solid in the same sense that a minor planet is.

## 7.7 Interplanetary Dust

The material that leaves comets may in part account for the myriads of tiny particles that pervade the solar system. These particles become visible just after sunset or before sunrise as a faint glow extending up into the sky from a point on the horizon just above the sun. The faint glow follows the ecliptic and is called the *zodiacal light*, for it is seen along the zodiac (the band of twelve constellations along the ecliptic). The ecliptic is the plane of the Earth's orbit and is also nearly coincident to the plane of the entire solar system; consequently as we look along the ecliptic we see more of the particles reflecting the sun's light. (It thus resembles the spike of Comet Arend–Roland.)

Observations tell us that the zodiacal light is not redder than the sun's light, which indicates that the particles reflecting the sun's light are larger than minute dust particles. Perhaps we see the micrometeorites reflecting the sunlight.

The zodiacal light is partially polarized; this indicates that electrons reflecting sunlight contribute some of the total amount of light seen. There is every reason to believe that the zodiacal light is part of the sun's vastly extended atmosphere. The sun's outer atmosphere is called the

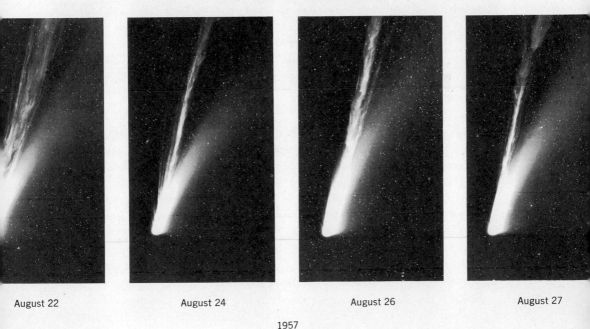

August 22        August 24        August 26        August 27

1957

**Figure 7-15**

Four views of Comet Mrkos taken with the 48-in. Schmidt telescope on Mount Palomar. (Photographs from the Hale Observatories)

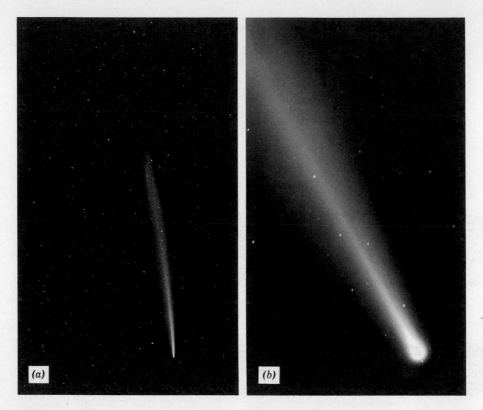

**Figure 7-16**

(*a*) Comet Ikeya–Seki as photographed with 35 mm camera. (Courtesy Elizabeth Roemer). (*b*) The head of comet Ikeya–Seki taken with the 40-in. reflector at the U.S. Naval Observatory, Flagstaff, Arizona at nearly the same time that (*a*) was taken. (U.S. Navy photograph)

corona and it is best seen during a total solar eclipse. The brightness of the corona decreases with increasing distance from the sun, and this can be expressed mathematically. The brightness of the zodiacal light also decreases with increasing distance from the sun, and by the same mathematical relationship used to describe the corona!

Around midnight there can be seen in the sky directly opposite the sun a glow similar to, though fainter than, the zodiacal light; it is called *Gegenschein* (German for counterglow). This very faint light may be either the sun's light reflecting from micrometeorites, or the material that is temporarily trapped by the Earth's magnetic tail.

In order to see either the zodiacal light or Gegenschein the observer must look into a sky that does not receive light from cities, towns, or even the moon. Zodiacal light is fainter than the milky Way, and Gegenschein is so faint that some people cannot see it under the best of conditions.

We see that the entire composition of the solar system includes many different types of bodies: the sun, a hot gaseous sphere much larger than any planet; the Jovian planets (Jupiter, Saturn, Uranus, and Neptune), composed principally of hydrogen and helium, the second largest bodies in the system; the terrestrial planets (Mercury, Venus, Earth, and Mars), which are rocky, metallic, and consequently denser than the Jovian planets; the satellites that revolve about the planets; the minor planets, which lie mainly between the planets Jupiter and Mars; the comets which may be frozen gases impregnated with small particles; and the tiny particles and gases that pervade the entire solar system.

## BASIC VOCABULARY FOR SUBSEQUENT READING

| | |
|---|---|
| Achondrite | Meteorite |
| Chondrite | Micrometeorite |
| Chondrule | Minor planet |
| Comet | Particle pressure |
| Gegenschein | Zodiacal light |
| Meteor | |

## QUESTIONS AND PROBLEMS

1. Describe how the height of a meteor can be determined.
2. Explain one observation which indicates that meteors are not much larger than a grain of sand.
3. What evidence indicates that meteorites may be small minor planets swept up by the Earth?
4. What evidence indicates that many meteorites may have belonged to larger bodies?
5. Describe the appearance of a comet as it approaches the sun from beyond the orbit of Neptune, passes close to the sun, and returns to aphelion again.
6. What information leads astronomers to suspect that meteor showers are related to comets?

## FOR FURTHER READING

Glasstone, S., *Sourcebook on the Space Sciences*, Chapter 7, Van Nostrand Co., Princeton, N.J., 1965.

Mason, B., *Meteorites*, John Wiley and Sons, New York, 1962.

Nininger, H. H., *Out of the Sky*, University of Denver Press, Denver, Colorado, 1952.

Page, T., and L. W. Page, ed., *Neighbors of the Earth*, The Macmillan Co., New York, 1965.

Wood, J. A., *Meteorites and the Origin of Planets*, Chapters 1, 2, 3, and 4, McGraw-Hill Co., paperback, New York, 1968.

Anders, E., "Diamonds in Meteorites," *Scientific American*, p. 26 (Oct. 1965).

Barringer, D. M., "The Meteorite Search," *Natural History*, p. 56 (May 1964).

Fireman, E. L., "Freshly Fallen Meteorites from Portugal and Mexico," *Sky and Telescope*, p. 272 (May 1969).

Florensky, K. P., "Did a Comet Collide with the Earth in 1908?" *Sky and Telescope*, p. 268 (Nov. 1963).

Futrell, D., "Some Notes on Tektites," *Sky and Telescope*, p. 272 (May 1967).

Hartmann, W. K., "Craters—A Tale of Three Planets," *Natural History*, p. 58 (May 1968).

Marsden, B. G., "The Great Comet of 1965," *Sky and Telescope*, p. 332 (Dec. 1965).

Mason, B., "The Unearthly Visitors," *Natural History*, p. 18 (April 1961).

Matthews, M. S., "The Asteroid Conference in Tuscon," *Sky and Telescope*, p. 22 (July 1971).

McCrosky, R. E., "The Lost City Meteorite Fall," *Sky and Telescope*, p. 154 (March 1970).

O'Keefe, J. A., "Tektites and Impact Fragments from the Moon," *Scientific American*, p. 50 (Feb. 1964).

Wood, J. A., "Chondrites and Chondrules," *Scientific American*, p. 65 (Oct. 1963).

"Bosomtwe: An African Meteorite Crater?" *Sky and Telescope*, p. 15 (July 1965).

"The Great Comet of 1970," *Sky and Telescope*, p. 351 (June 1970).

"Micrometeorite Collecting with High Altitude Rockets," *Sky and Telescope*, p. 84 (Feb. 1962).

"Origin of Some Iron Meteorites," *Sky and Telescope*, p. 282 (May 1965).

"Shower Meteorites Collected by Balloon," *Sky and Telescope*, p. 276 (Nov. 1965).

"Two More Ancient Canadian Meteorite Craters," *Sky and Telescope*, p. 198 (Oct. 1963).

# THE AGE AND ORIGIN OF THE SOLAR SYSTEM

Man's speculations and researches into his origins have led him to theories of evolution, both for himself and for the entire universe. The concept of evolution implies a beginning and leads to such questions as: "When and how did the Earth come into being? From where came the moon and the sun, and the comets, meteors, stars?" Peoples of all ages and races have asked these questions and have answered them to the best of their abilities and according to their beliefs. But none of the answers given in past ages satisfy the requirements of modern scientific thought and observation. Neither, for that matter, has any claim to complete accuracy yet been made for our modern description of the beginnings and evolution of the universe as a whole or of the solar system in particular. But the search alone has its fascinations, stimulated as it is by the hope that ultimately the story of the universe may unfold from a mass of data and a few germinative ideas.

## 8.1 Age of the Solar System

How old is the solar system? The question is basic to any theory of origins, but how is it to be answered? Age reveals itself in many different ways. The age of a tree is determined by the number of rings in its trunk, that of a horse by its teeth. If there is some analogous sign or evidence in the solar system that will help us determine its age we may expect to find it first at home, on the Earth.

**a. Techniques of Observation.**   To obtain evidence of the Earth's age we must first find some material or object that shows signs of change over a long period of time. Then to determine this period, three factors must be considered:

1. The material's present condition.
2. Its assumed condition when formed.
3. Its rate of change.

Rain water, for instance, is nearly pure (that is, fresh); the oceans are salty; and the Great Salt Lake and Dead Sea are even saltier. The rain falls

### Chapter Opening Photo

The Lagoon nebula (M8) with dark "globules" of gas and dust that could perhaps form new stars with planets about them. (Photograph from the Hale Observatories)

**Star Plate 1**
The Earth from outer space (NASA photograph)

**Star Plate 2**
The Crab nebula (*above*)
(Photograph from the Hale
Observatories)

**Star Plate 3**
The Rosette nebula (*right*)
(Photograph from the Hale
Observatories)

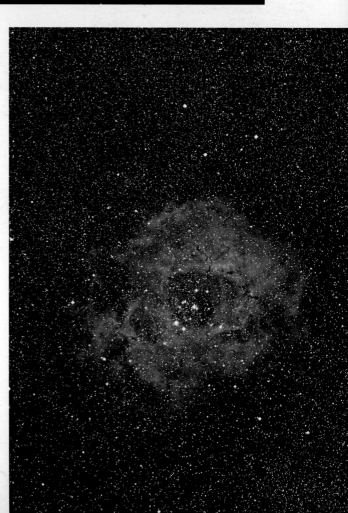

**Star Plate 4**
The Horsehead nebula (Photograph from the Hale Observatories)

**Star Plate 5**
The Andromeda Galaxy (Photograph from the Hale Observatories)

on the land and carries salts from it into the ocean. The ocean's water evaporates, leaving the salt behind, to form clouds that cause rain to wash again down the mountain slopes and deposit more salt into the oceans. The cycle goes on; how long has it been doing so?

The ocean's present condition is one of water with dissolved salts. It is reasonable to assume that its condition when formed was one of fresh water, as fresh as river water. Its rate of change from a fresh to a saline condition may be deduced from the amount of salt carried into the oceans each year by the present rivers. If we divide the number of tons of salt in the oceans at the present time by the number of tons deposited each year, we obtain the length of time this process has been going on. From this calculation it has been estimated that the oceans are about 300 million years old.

But the geologists warn us to be careful because the Earth that we now see is *not* typical of its past history. There are many more rugged mountains now than in former geological epochs. This means that the rivers erode the Earth's surface more rapidly and thus carry more salt than usual into the oceans. Consequently, we postulated too rapid a rate of salt cumulation and 300 million years turns out to be too short a time interval. When we take into consideration the slower rates of cumulation before the present era, we arrive at an age for the oceans that is closer to a few billion years.

This method, approximate though it may be, does enable us to be certain of one thing: the oceans are not infinitely old; they did begin sometime. If salt had been deposited in them for an infinite length of time they would have become completely saturated and could have held no more salt, for they are finite in size and therefore the amount of salt required to saturate them is also finite. If the oceans had become saturated with salt any additional salt would have settled to the bottom. If this had happened we would have to investigate the amount of sediment on the ocean floors to arrive at an estimate of the length of time that sedimentation has been going on.

The age of the oceans, however, need not be the same as the age of the Earth. Indeed, we expect that they would be younger, for the assumption has been made (and not without good reason) that the surface of the Earth was originally too hot for any water to collect or even for any rain to fall. We must look for some other change that will give us a clue to the age of the Earth itself.

One branch of physics (nuclear physics) concerns itself partly with radioactivity. A radioactive element is one whose atoms are unstable and therefore break up or *decay* into other atoms. It has been found that the rate of radioactive decay is constant and quite independent of any pressure or temperature that man has imposed on any radioactive substance. Some of these elements appear naturally in the rocks of the Earth, and the fact that they decay at a known rate gives us a clue to the age of the rocks.

Uranium 238 (the 238 is the atomic mass of this isotope of uranium) is one of these natural radioactive elements. It decays into another radioactive element, thorium 234, which decays into yet another radioactive element, protactinium 234, and so on through a number of radioactive elements until lead 206 is reached. Another naturally occurring radioactive element, thorium 232, decays into lead 208, another stable isotope of lead. Neither lead 206 nor lead 208 decays into any other element. Furthermore, lead is a solid, and consequently accumulates in the rock that contains the uranium 238, just as salt accumulates in the oceans.

The rate of decay of a radioactive element is generally measured by the *half-life* of that element. The half-life of any radioactive element is the time it takes for half of the original atoms to decay. For example, the half-life of the uranium 238–lead 206 process is 4.5 billion years, which means that no matter how many atoms of uranium 238 we start with, in 4.5 billion years half as many atoms of lead 206 will have been formed. The half-life of the thorium 232 to lead 208 series is $1.3 \times 10^{10}$ years.

**b. Ages of the Earth and Meteorites.**   To find the age of a particular rock we first determine its present condition by measuring the amount of lead and uranium in it. We can simplify this example by assuming that the rock contained no lead when it was formed; therefore all the lead it now contains comes from the radioactive decay of uranium. We can then obtain the age of the rock, that is, the length of time since it solidified, by applying the known rate at which lead has been formed through the decay of uranium.

This sounds very easy. Unfortunately, however, we were not justified in assuming that the rock started without lead; if there was uranium in the primordial rock there may certainly have been some lead. In fact, we are confronted with the initial problem of determining the amount of primordial lead.

The composition of the rock when it solidified may be determined by investigating several different radioactive substances. For example, there are two uranium isotopes each with its own decay scheme, and three isotopes of lead: uranium 238 decays into lead 206, uranium 235 decays into lead 207. Lead 204 seems to have been part of the primordial material. By comparing the ratios of the amounts of these various isotopes a good estimation of a rock's original composition can be made and consequently an age of the rock can be determined.

The maximum age for terrestrial rocks is between 3.2 and 3.5 billion years, although not all rocks are of the same age. After all, some volcanic mountains are still in the process of formation. Since the terrestrial rocks that have been studied have different ages, we cannot be sure that the oldest one has yet been found, or that we will ever find part of the original crust of the Earth.

The ages of meteorites, on the other hand, are very consistent and yield an age of $4.6 \times 10^9$ (billion) years. Since these were probably formed early in the life of the solar system, the age of the system is taken as $4.6 \times 10^9$ years.

The age of the solar system, important though it may be, is only part of the evolutionary story; the question "how?" still faces us. Here we shall discuss only the most important of many theories of the solar system's origin.

## 8.2 Origin of the Solar System

**a. Features of the Solar System.** A complete description of the origin of the solar system must be able to explain the main features of the system as it now exists, unless we assume (though it is very unlikely) that there has been some strong outside disturbing force since the origin. In general we could say that the solar system is fairly flat with almost everything spinning in the same direction. The following are the specific features which such a description must explain:

1. The motions (revolution and rotation) of nearly all the bodies that revolve about the sun are in the same direction, except those of Uranus and Venus, some of the satellites, and many of the comets.
2. The planes of revolution of these bodies are within a small angle of one another. The exceptions are the satellites of Uranus and most of the bright comets.

3. The planes of the equators of the planets as well as that of the sun are nearly parallel to the planes of revolution of the planets. Uranus is again an exception.
4. The densities of the four terrestrial planets are considerably greater than those of the four Jovian planets (Table 8-1).

### Table 8-1 ● Densities of the Planets in Terms of Water

| Terrestrial planets | | Jovian planets | |
|---|---|---|---|
| Mercury | 5.5 | Jupiter | 1.3 |
| Venus | 5.2 | Saturn | 0.7 |
| Earth | 5.5 | Uranus | 1.6 |
| Mars | 4.0 | Neptune | 2.3 |

**b. Early Hypotheses.**  One of the early hypotheses (that of the Frenchman Buffon) was based on the belief that a comet is a huge mass of material that, if it were to strike the sun, would surely cause debris to fly out into space. The Earth and planets would be formed of such debris.

We now know that a comet's mass is too small for this to be possible. The hypothesis, however, was revised; instead of colliding with a comet, the sun might have collided with another star. To be sure, if two stars were to collide, things would happen on a grand scale. But the stars are immensely distant from each other, and the possibility that a collision might occur between them is so remote that the idea may be dismissed on these grounds alone.

If a collision is out of the question, what about the close approach of another star? Two stars might come so close to one another that tidal actions set up in each star might draw out a cigar-shaped blob of gas from the sun (or both stars) which might eventually condense into the planets and satellites. But the probability of this happening is also extremely remote; it is estimated that during the lifetime of the Earth only ten planetary systems could have formed by this process in our Milky Way galaxy of 200 billion stars.

The astronomer, faced by billions of stars, cannot fall heir to the belief that the Earth is something special, the planetary system unique. If the universe is to be investigated objectively at all, he cannot accept the belief that all the billions of stars exist solely for the benefit of mankind. He assumes that there are many other stars circled by planets that support life not too unlike our own. And for this reason he is very critical of hypotheses, like those of stellar collision or close approach, whose acceptance would imply the uniqueness or near-uniqueness of our solar system.

Furthermore it has been shown that even if two stars were to collide or suffer a close approach, the gases spewed out into space would not condense into planets; the density would be too low and the motions would only lead to further dispersal of the gas. Therefore, astronomers have sought to explain the formation of the solar system as a self-contained event with no outside influences.

**c. Nebular Hypothesis.**   The hypotheses which satisfy the restrictions of self-containment have one basic similarity: they all base their arguments on the supposition that the sun originated in a great nebulous cloud of gas and dust. *Gaseous nebulae*, composed primarily of hydrogen gas, are very prevalent in our Milky Way galaxy. It is believed possible that a portion of such a nebula might become localized to the extent that it separates from the main body of gas by becoming a local "whirlpool" or *eddy*. Photographs of some bright nebulae that reveal dark "globules" of gas between them and the Earth add weight to this argument (see the frontispiece of this chapter).

Now, if the gas in such an eddy has a density that is high enough it will be able to contract under its own gravitational influence. Any object or system that is rotating *and* at the same time contracting will have to rotate more rapidly still as a consequence of the law of *conservation of angular momentum*.

The angular momentum of a contracting blob of gas is difficult to determine. However, the angular momentum of each particle in that gas is simply the product of the particle's mass $m$, the particle's velocity $v$, and the radius $r$ of its orbit about the center of rotation ($AM = mvr$). If no outside forces act on the particle, this product, $mvr$, will remain constant.

If it is assumed that the particle's mass $m$ does not change, then as the radius $r$ of its orbit decreases (because the entire blob of gas is contracting), its velocity $v$ will increase. The phenomenon may be illustrated by the example of a spinning ice skater. As she starts to spin her arms are outstretched; then as she brings her arms in close to her body she spins more rapidly.

If the radius of the orbit decreased and the velocity did not increase, the angular momentum would also have to decrease. But the law of conservation of angular momentum denies this possibility—if no outside forces act on a rotating system, its angular momentum must remain constant.

As a result of the more rapid rotation of a contracting eddy in a nebula, then, a large disk of gas would form around a central nucleus in a plane perpendicular to the axis of rotation. The result would appear much like a slightly out-of-focus picture of Saturn and its rings. At this stage the sun is a new star which has a very large and flat disk of gas and dust around its midriff. The flatness of this disk would account for the fact that the planetary orbits lie nearly in the same plane.

1. *According to Kant–Laplace.* From this point on, various hypotheses concerning the origin of the solar system begin to diverge. The first to postulate a gaseous disk were the German philosopher Kant and the French mathematician Laplace. Once the disk had formed, Laplace suggested, it would in turn contract. In so doing it would periodically leave smaller rings or "zones of vapors" which would become detached from the main disk by ceasing to contract. Each of these detached rings would then coalesce into a planet by forming a little eddy of its own; and the rotation of this eddy would in turn form smaller gaseous rings from which the new planet's satellites would form. Laplace cited Saturn's rings as "existing proofs of the primitive extension of the atmosphere of Saturn, and its successive condensations."

Ingenious as this explanation is (it partially anticipated contemporary hypotheses), it fails to account for one important aspect of the solar system: the very unequal distribution of angular momentum between the planets and the sun. Although the sun is much more massive than the rest of the solar system (it contains 99.9% of the total mass), it comprises only 2% of the angular momentum! Physically this means that the sun rotates quite slowly while the tiny planets with 98% of the system's angular momentum revolve at great speeds and at great distances from the center.

The validity of the theory of Kant and Laplace can be tested by using it, together with additional knowledge gained since their time, to predict a period of rotation for the sun. Not only have astronomers been able to estimate the size of the gaseous cloud before it contracted, but they have also been able to measure the speed of rotation of gases in observed nebulae. Using this assumed size, the observed speed of rotation, and the law of conservation of angular momentum, the Kant–Laplace theory predicts that the period of rotation of the sun should be about $\frac{1}{2}$ day; the observed period of rotation is about 26 days. Such a large discrepancy between theory and observation is not acceptable; the Kant–Laplace theory must omit some important aspect of the process of formation. Even though it has proved a fertile guide to the thinking of later astronomers, it must be discarded.

2. *The Accretion Hypothesis.* Given the nebular disk about the sun, the next job is to figure out the process by which the planets formed. There are two methods under general consideration. The first is formation by *accretion*, that is, the building up of a large body by the collection of many smaller bodies. It is clear from the impact craters on Mars, the moon, and the Earth that each of these bodies collected many smaller bodies during the first billion years or so of the solar system's existence. Some of those collected by the moon to form the maria, and the Orientale Basin must have increased the moon's mass significantly. To this extent, then, the moon was formed by accretion.

The main difficulty with the accretion theory is that it is not easy to imagine how it all started. How did all those little tiny particles in the original nebular disk get together to form a few very large planets?

Apparently the first solid objects in the nebular disk were the chondrules found in the chondrite meteorites. These were somehow gathered into a matrix to form larger bodies, which may in turn have collected to form perhaps minor planets, and eventually the planets. But the mechanism that permitted the first small particles to collect together is not at all clear.

The chondrules and the dust particles in the nebular disk were each very small particles; their mutual gravitational forces were not strong enough to effectively attract other particles.

Electrical forces, however, are much stronger than gravitational forces. Those particles were bombarded with high energy photons, with electrons and protons in the solar wind, and with cosmic rays. It is conceivable, therefore, that they could have become electrically charged. The Apollo 12 astronauts found many tiny particles clinging to their suits and equipment. These particles, it is presumed, were charged electrically.

Two objects with opposite electric charge attract each other, and an object with an electric charge (either positive or negative) will attract another particle with no charge. Try picking up bits of dust and paper with a plastic ruler rubbed with a paper towel. Certainly electrical forces would be strong enough to enable particles in the nebular disk to not only attract each other but to stick together once they collided.

The next question to ask is why are the densities of the two major groups of planets so different?

The temperature of the nebular disk and the solar wind could have played a part in separating out the chemical elements to leave the rocky terrestrial planets in the center. Surely the inner part of the nebular disk was hotter than the outer regions, so it is reasonable to suppose that the rocky materials would have solidified there first. Certainly the gases methane, ammonia, water vapor, and carbon dioxide would not have solidified in the hotter parts of the nebular disk. The very lightest elements, hydrogen and helium, were probably blown away from the inner part of the solar system by the solar wind.

However, if the solar wind was energetic enough to blow the gases away, it was also energetic enough to ionize those gases. The inner part of the nebular disk would then consist of a rotating ionized gas. Electrically charged particles in motion set up a magnetic field, and it is the reaction between the magnetic field of the rotating ionized nebular disk and the magnetic field of the sun that is generally considered to have caused the sun to slow down. Angular momentum was transferred from the sun to the disk, and if a good part of that gas was blown back out into space, some of the sun's angular momentum went with it. Just how far along the planets were in their process of formation when the hydrogen and helium were blown away is not clear.

In the accretion hypothesis, which has been developed to a large extent by H. C. Urey of the University of California and F. L. Whipple of Harvard University, many bodies with a diameter of several hundred miles could have formed which in turn collected to form the terrestrial planets and the moon. In fact, Urey suggests that the moon was originally in an orbit about the sun. It was then somehow captured by the Earth. The fact that the plane of the moon's orbit is more nearly parallel to the plane of the solar system than the plane of the Earth's equator is a strong argument in favor of the idea that the Earth captured the moon. The planes of the orbits of the Galilean satellites are nearly parallel with the plane of Jupiter's equator, indicating that they were formed along with Jupiter. The outer satellites of Jupiter, however, have orbits that are considerably inclined from the plane of Jupiter's equator and in fact some of them revolve in retrograde motion. It is, therefore, generally accepted that Jupiter captured its outer satellites.

The Jovian planets are composed primarily of hydrogen and helium. It can be surmised, then, that these gases, along with some rocky material perhaps, were collected together with the help of methane, ammonia, water, and carbon dioxide. Apparently, the nebular disk in this region had enough material to permit the formation of giant planets, whose gravitational fields soon became strong enough to gather in gases, and thus each planet grew bigger. The bigger each one grew, the faster it grew until the nebular disk became depleted with gases.

Beyond the region of the Jovian planets it is postulated that only smaller bodies could form, largely of the frozen compounds methane, ammonia, water, and carbon dioxide. These smaller bodies each revolve about the sun in its own orbit, which normally keeps it beyond the planets Neptune and Pluto. According to this hypothesis, however, one of these small bodies of frozen compounds is occasionally perturbed in its orbit so that it comes into the inner part of our solar system. It then becomes heated by the sun and forms one of the spectacular comets that surprise astronomers.

3. *The Contraction Hypothesis.* A second hypothesis based on the idea of a nebular disk about the newly forming sun has been proposed by G. P. Kuiper, of the Stewart Observatory of the University of Arizona. He bases his idea on the planet's beginnings not on accretion but on the fact that two opposing forces act on any eddy (rotating parcel) of gas in a nebular disk: (1) the force of the gas's own gravity, which tends to pull the eddy of gas together, and (2) the gravity of the sun, which exerts tidal forces in the eddy of gas that try to pull it apart (see Figure 8-1). If the density of the gas is high enough (higher than a critical value Kuiper called the Roche density), its gravitational field will gain control, and the material composing the eddy will contract into a *protoplanet*, a newly forming planet, with a central nucleus surrounded by a disk of gas much as Laplace envisioned.

The tidal forces from the sun cause each protoplanet to become elongated with the long axis toward the sun, and consequently to rotate in the same

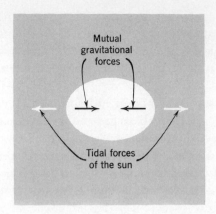

**Figure 8-1**

Tidal forces of the sun.

length of time and in the same direction that it revolves about the sun (Figure 8-2). But as the protoplanet contracts under its own gravitational influence it rotates more rapidly, since its angular momentum must remain constant so long as no outside forces are acting upon it. Thus we see that the planets should rotate in the same direction in which they revolve. Similarly, the satellites form from the nebular disk about each protoplanet and thus revolve and rotate in the same direction that the planets rotate.

At first, this explanation of satellite formation does not appear to account for the rings of Saturn. However, E. Roche has shown that the tidal forces existing within a distance of 2.4 times the radius of a planet (the so-called Roche limit) are so great as to prevent the formation of a satellite. The concept of the Roche limit may then help explain the existence of Saturn's rings: the material originally composing the nebular disk surrounding Saturn which was within that limit has remained without coalescing into a satellite. We may recall that two centuries earlier Laplace also thought of Saturn's rings as evidence of a preexisting nebular disk.

Kuiper suggests that the entire process of formation can be discussed in four stages.

*The first stage* is the collapse of the original nebula into a *proto-sun* with a nebular disk rotating about it. The proto-sun in this stage would still be dark.

*The second stage* includes the formation of the protoplanets in the nebular disk. The disk would break up into two parts: the inner part from which the four terrestrial protoplanets form, and the outer part from which the four Jovian protoplanets form. In the region between these two parts the density of the nebular disk would not exceed the Roche density which is necessary for contraction to take place, and consequently the tidal forces imposed by the sun would prevent the formation of any one large body. It is conceivable that several bodies smaller than the planets would form, which upon colliding would break up and become the minor planets. The nebular disk beyond the Jovian protoplanets would also have a low density which would result in the formation of comets.

*The third stage* of formation brings the sun from darkness to brightness. This intense radiation would ionize gases in the nebular disk relatively close to the sun. The interaction between this rotating, ionized disk and

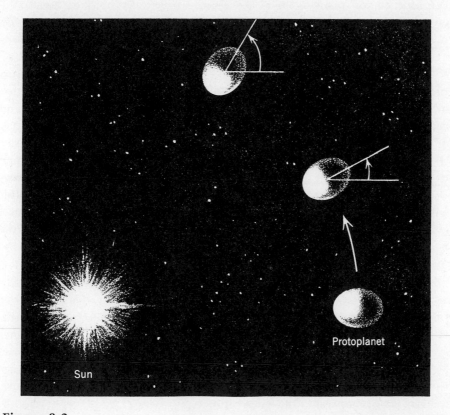

**Figure 8-2**

The tidal bulge in a protoplanet must by its very nature line up with the sun and thus the protoplanet is given an initial rotation in the same direction that it revolves.

the sun's magnetic field would quite adequately account for the sun's loss of angular momentum and its consequential longer period of rotation.

*The fourth stage* in the formation process is a further reaction between the nebular disk and the radiation from the newly brightened sun. Just as comets' tails are forced away from the sun by radiation and particle pressure, so were most of the gases in the nebular disk. Only those gases that are tightly held by the gravitational field of a protoplanet would remain in the vicinity of the sun; the rest of the material (actually the vast majority of it) would be ejected from the solar system back into interstellar space. Those planets closest to the sun and thus subject to more intense solar radiation and particle pressure would lose more material than would the four Jovian planets. Since the ejected material would be composed largely of lighter substances (mostly hydrogen) this would account for the greater density of the terrestrial planets.

The main difficulty with the contraction hypothesis is that it is hard to imagine how the terrestrial planets formed while the gas was being blown outward by the solar wind. In order for planets to form, the density of the gas in the nebular disk must exceed the Roche density, and this requires that hydrogen, which amounted to three-quarters of the mass of the nebular disk, be present during the collapse process. On this basis, then, our planet should have a lot of hydrogen in it; but it doesn't, neither does Mercury, Venus, the moon, nor Mars. If hydrogen and helium were present and became trapped inside the terrestrial planets, how could the solar wind blow them away?

It is possible, of course, that the terrestrial planets formed by the accretion process, and the Jovian planets formed by the contraction process. This is supported to some extent by observation. Witness the extensive craters on the moon and Mars. And recall that the rings of Saturn and the inner satellites of both Jupiter and Saturn support the idea that these planets formed from a collapse of a dense portion of the nebular disk. It is perfectly possible that the terrestrial planets formed under conditions and by processes that were different from those that formed the Jovian planets. The gap between Mars and Jupiter could then be explained by the fact that because of the density, solar tidal forces, and temperature neither process operated in that region.

Just how the planets and other members of the solar system formed is clearly the subject of conjecture right now. But new studies have led to a new approach, an approach that carries with it the hope of achieving our goal of understanding. This new approach is the ability to make significant measurements of time, and of crystal structure and content. It is a result of cooperation between the astronomer, the physicist, and the crystallographer.

## 8.3 The Formation of Life

Once the Earth was created and its oceans formed, there remained one vitally important development—the formation of life. At this point, however, the astronomer serves only as a consultant to the biologist; he can tell the biologist what the conditions might have been on that barren primordial Earth, and he can tell the biologist what compounds were part of the gases and dust from which the Earth was formed.

It would not be unreasonable to assume that the Earth's atmosphere was initially similar to that of Mars and Venus, that is, it probably was composed of a great deal of carbon dioxide. Other compounds that are found astronomically are water, methane, ammonia, and formaldehyde. All of these have been discovered with radio telescopes to exist in small regions of gaseous clouds in the Milky Way galaxy.

Once the surface of the planet had cooled down to the temperature where water could exist as a liquid, the condensation and evaporation of water started, that is, rains began to fall.

During the early stages of the Earth, the sun shone brightly on the barren Earth, and that sunlight supplied energy for chemical reactions. During its earliest stages, the Earth may have had more radioactivity. If some of those initial radioactive elements had short half-lifes, they would have decayed completely by now. But in the process they would have heated the interior of the Earth to a higher temperature than now. This could have resulted in more volcanic activity, with the more rapid release of water trapped beneath the Earth's surface.

Volcanic eruptions frequently give rise to lightning, and lightning, like the sunlight, is a supply of energy which could have caused the methane, ammonia, water vapor, and carbon dioxide to react with one another. S. L. Miller, a colleague of Harold Urey, has passed mixtures of these gases repeatedly over a spark and then made a chemical analysis of the results. He found that the spark caused those gases to react and form more complex molecules. The molecules he found were several amino acids as well as other organic compounds of biological interest. Ultraviolet radiation has also caused those gases in modern laboratories to combine and form amino acids.

Amino acids can combine to form more complex molecules, proteins, polypeptides, sugars, and finally deoxyribonucleic acid, the DNA molecule.

The exact processes by which a DNA molecule can be formed from the rudimentary methane, ammonia, carbon dioxide, and water, is not clear, but many of the steps have been worked out and performed in the biological laboratory. The process is not a direct one, nor is it likely to happen by the process of passing simple gases over a spark. However, if the oceans formed one billion years after the Earth, then according to our best estimates those oceans were without life for another billion years or so. During that time, the chemical reactions continued, the oceans were converted into what has been called a broth: lifeless water teeming with complex organic compounds. Given enough time, it is expected that the DNA molecule first formed in this broth and the DNA molecule holds the key to life. It is the DNA molecule that contains the information which determines what an individual will be, a bug or a man or a mouse. It also determines that individual's characteristics, the color of his eyes, etc.

It is the vast amount of time that makes possible the development of the DNA molecule. If you flip a coin once, it will have a 50–50 chance of turning up heads. However, the chance that a head will appear, if it is flipped 10 times, is 999 out of 1,000; almost a certainty—at least certainly a good bet. If an event is unlikely to occur in one reaction, it will be nearly a certainty to occur in 10,000 or 10,000,000 reactions. Given a billion years, the DNA molecule was certain to appear in that primordial broth that nurtured the life it gave rise to. However, we don't expect to repeat that identical process in our laboratories.

Once life appeared in the seas, it multiplied and developed, and as it did it began to alter the seas and the atmosphere that gave rise to it. It is felt that about one billion years ago, free oxygen began to accumulate in our atmosphere. Life altered the atmosphere and the face of the Earth, and we are still doing it.

## BASIC VOCABULARY FOR SUBSEQUENT READING

Accretion

Protoplanet

Radioactive decay

Radioactive half-life

## QUESTIONS AND PROBLEMS

1. As an example of the methods of age determination, 3 cubic cm of water are poured into a large graduate cylinder every 15 seconds (analogous to the accumu-

lation of lead from the radioactive decay of uranium). After some time it is determined that the graduate cylinder contains 57 cubic cm of water.

(a) Assuming that the graduate cylinder was empty before the first water was poured in, determine how long the process has been going on.

(b) If the graduate cylinder had *not* been empty and you cannot determine how much water was in it to begin with, can you make a statement as to the maximum or minimum duration of time this process has been going on?

2. Explain the technique used in determining the age of the solar system.

3. What observational evidence is there in support of the accretion hypothesis of the origin of the solar system?

4. What observational evidence is there in support of the contraction hypothesis of the origin of the solar system?

5. A rock is whirled on the end of a string 3 ft long. Its velocity is such that it makes one revolution in $\frac{1}{2}$ second. The string is then shortened to 1 ft while the rock continues to whirl.

(a) What is the speed of the rock when the string is 3 ft long?

(b) According to the principle of conservation of angular momentum, what is the speed of the rock after the string has been shortened to 1 ft?

(c) What is the period of revolution of the rock with the 1-ft string?

## FOR FURTHER READING

Faul, H., *Ages of Rocks, Planets, and Stars*, McGraw-Hill paperback, 1966.

Gamow, G., *Biography of the Earth*, The Viking Press, New York, 1959.

Glasstone, S., *The Book of Mars*, National Aeronautics and Space Administration, Washington, D.C., 1968, Chapters 8, 9, 10, and 11.

Hurley, P. M., *How Old is the Earth?* Doubleday Anchor, New York, 1959.

Jastrow, R., *Red Giants and White Dwarfs*, Harper and Row Publishers, New York, 1967.

Page, T., and L. W. Page, ed., *The Origin of the Solar System*, The Macmillan Co., New York, 1966.

Shklovskii, I. S., and C. Sagan, *Intelligent Life in the Universe*, Holden-Day, Inc., San Francisco, Calif., 1966.

Struve, O., and V. Zebergs, *Astronomy of the 20th Century*, Crowell, Collier and Macmillan, New York, 1962, Chapter IX.

Whipple, F. L., *Earth, Moon, and Planets*, 3rd edition, Harvard University Press, Cambridge, Mass., 1968, Chapter 14.

Wood, J. A., *Meteorites and the Origin of Planets*, McGraw-Hill paperback, 1968.

Fox, S. W., and R. J. McCauley, "Could Life Originate Now?" *Natural History*, p. 26 (Aug.–Sept. 1968).

"Oldest Solids," *Scientific American*, p. 72 (March 1963).

# CHAPTER 9

# THE SUN

CORONAL
POLAR RAYS

QUIET
PROMINENCE

SUNSPOT
GROUP

SOLAR
FLARE

ACTIVE
PROMINENCE

FILAMENTS

PLAGE

At the center of the solar system resides its largest and most important member, an average star which we call the sun. The fact that it is relatively close makes possible fairly detailed observations that can tell us something about the other more distant stars. But of more immediate importance to us is the sun's influence on the Earth. Not only is it our source of heat and light, but it has other effects on the Earth that, although less noticeable, are interesting none the less.

## 9.1 General Features

The sun is a huge ball of gas* 864,000 miles in diameter, with a mass 333,000 times that of the Earth. But since its volume is 1,300,000 times that of the Earth, its density, 1.41 times that of water, is less than the Earth's density of 5.5, about the same as that of the Jovian planets. The force of gravity operating on its surface is 28 times that on the surface of the Earth. Therefore, if a man weighing 180 lb on the Earth could stand on its surface, he would weigh 5,040 lb. Such a large force of gravity also means that the escape velocity on the sun is very high, about 386 miles per second as compared with 7 miles per second for the Earth.

The portion of the sun that we see in the sky, its visible surface, is called the *photosphere* (sphere of light). This surface is gaseous rather than solid like the surface of the Earth. The photosphere is visible because it is the sun's outermost region, where the pressure is great enough and the temperature high enough to emit a continuous spectrum. The photosphere is therefore opaque and hides the solar interior from our view. There is no abrupt change as there is between the Earth's surface and its atmosphere, but there is a more gradual change in density and temperature.

Above the photosphere there are two clearly differentiated regions of gas. Immediately above the photosphere and extending for an estimated 5,000 to 10,000 miles is the *chromosphere*. The chromosphere can be seen during a total solar eclipse just seconds before and after totality. During

---

*This gas is quite unlike any found on Earth because its temperature, pressure, and density are so much higher than gases in our atmosphere. Yet the gases in the sun obey the gas laws met with on Earth.

### Chapter Opening Photo

Composite of two photographs of the sun, one of the disk in hydrogen light and the other of the limb with the disk masked out by an artificial eclipse. (Lockheed Aircraft Corporation)

these fleeting moments, the moon covers the more intense photosphere and permits observation of the reddish chromosphere. In fact its reddish color is the origin of its name which means the "colored sphere." The gases of the chromosphere cause the absorption lines which show up in the solar spectrum by selectively absorbing light emitted by the photosphere. Accordingly, the chromosphere must be at a lower pressure and temperature than the photosphere which emits the background continuous spectrum.

From the chromosphere the *corona* extends for millions and millions of miles as a very tenuous gas. There is good evidence to conclude that many of the planets, including the Earth, actually revolve within the limits of the corona. The corona is best seen during a total solar eclipse, although both the chromosphere and the inner corona can be seen by optical means which simulate a solar eclipse.

## 9.2 Solar Radiation

One of the first questions that must be asked about the sun concerns the light and heat that it radiates into space. This amount of energy has been determined by measuring its heating effect at the Earth's surface and making allowance for the energy absorbed by the Earth's atmosphere. When this absorption is taken into account, we find that the Earth receives 2.0 calories per $cm^2$ each minute.* But at the distance of the Earth from the sun this 1 $cm^2$ represents a very small fraction of the total energy output of the sun. To find the total output, we must determine how many square centimeters there are on the surface of a sphere with a radius equal to the radius of the Earth's orbit. We can then multiply the amount of energy passing through 1 $cm^2$ by the total number of square centimeters on that sphere. The surface area of this large sphere is $2.9 \times 10^{27}$ $cm^2$. Consequently, the total output of radiation passing through this sphere must be $2.0 \times 2.9 \times 10^{27} = 5.8 \times 10^{27}$ calories per minute! This must be the total output of energy from the sun, which when translated into power rating is $5 \times 10^{33}$ horsepower!

**a. The Radiation Curve.**    Not only is the total amount of energy radiated by the sun important to the astronomer, but the amount of energy radiated

---

*The calorie, a unit of heat, is the amount of heat required to raise the temperature of 1 cubic cm of water 1°C. The calorie referred to in dietetics is actually the kilocalorie, or 1,000 calories.

at each wavelength is of vital interest. If the amount of energy radiated by the sun, for example, is measured in small adjacent regions of the electromagnetic spectrum, the *radiation curve* for the sun can be obtained (solid curve in Figure 9-1a). The peak of the solar radiation curve is observed to be at 4750 Å.

Some stars appear bluer in the sky than the sun, for example, Rigel in the constellation Orion. The peak of Rigel's radiation curve is observed to be at a wavelength shorter than 4750 Å. Other stars, such as Antares in Scorpio, are distinctly red in color and the peaks of their radiation curves appear at a wavelength longer than that for the sun.

**b. The Radiation Laws.** Differences in the radiation curves urged astronomers in the late 19th century to find an explanation. Three successful explanations were forthcoming, two being partial explanations and

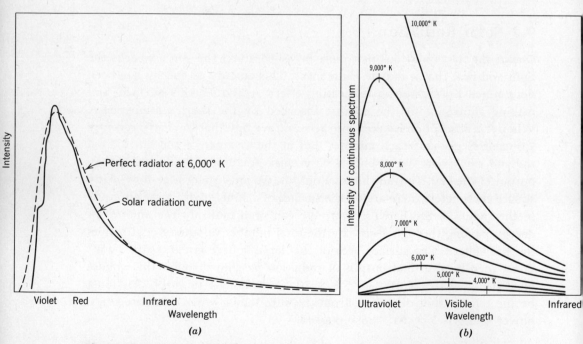

**Figure 9-1**

(a) The solar radiation curve is the solid-line curve; it nearly matches the radiation curve for a blackbody radiator with a temperature of 6000°K. (b) Radiation curves for sources of various temperatures.

the third, the most thorough one. One of the important aspects of these explanations, called the *radiation laws*, is that each permits the astronomer to determine the temperature of any source emitting radiation.

The first of the radiation laws, *Wien's law*, identifies the temperature of the source by determining the wavelength of the peak of the radiation curve, $\lambda_{peak}$.

As different stars of *increasing* temperature are studied, the peak of their radiation curves appear at ever *shorter* wavelengths (Figure 9-1b). Mathematically, it can be stated simply as:

$$T = \frac{2.9 \times 10^7 \text{ Å}}{\lambda_{peak}} \text{ K}°$$

where the temperature is given in degrees Kelvin.* Hence, if the wavelength of the peak of the radiation for a particular star is observed to be $\lambda_{peak} = 7250$ Å, the temperature of the surface of that star must be

$$T = \frac{2.9 \times 10^7 \text{ Å}}{7.25 \times 10^3 \text{ Å}} °K$$

or

$$T = 4,000°K$$

Using Wien's law, the temperature of the sun's surface is determined to be 6,100°K.

Wien's law describes the observation that blue stars are hot stars, some with temperatures higher than 20,000°K; red stars are cool stars with temperatures as low as 2,500°K. Some objects must be so cool that all of their radiated energy is in the radio and microwave region of the spectrum. These objects are not visible to the eye, so must be detected with radio telescopes.

The second radiation law, the *Stefan—Boltzmann law*, relates the amount of energy radiated to the temperature and surface area of the radiating object. For example, of two stars the same size, the hotter one will radiate more energy. This will be observed as a difference in the area under the

*All of the countries in the world, except the United States and a few others, use a temperature scale called Celsius (centigrade) designed so that water freezes at 0°C and boils at 100°C. Another scale, with divisions the same size as the Celsius scale, is fundamentally important to science. It is called the Absolute or Kelvin scale. Zero on the Kelvin scale (−273°C or −460°F) is absolute zero, the temperature at which the energy of atoms is at a minimum. A temperature of 6,000°K is equal to about 11,000°F.

radiation curve, the hotter star exhibiting a larger area (see Figure 9-1*b*).
Mathematically the law is stated as:

$$E = \sigma A T^4$$

where $E$ is the total amount of energy radiated each second. The proportionality is changed into an equation by the insertion of a constant of proportionality $\sigma$ (Greek small sigma). The value of $\sigma$ is $5.67 \times 10^{-5}$ if $T$ is in degrees Kelvin and the area is expressed in square centimeters. For a given temperature, the greater the surface area, the more energy radiated. For a given surface area, the higher the temperature, the more energy radiated. But we know the surface area of the sun, and we have just calculated the amount of energy it radiates; from these we can calculate the sun's surface temperature as 5,770°K.

The third of the three radiation laws, *Planck's law*, is complete in the sense that it describes the *shape* of the radiation curve. From the shape of the curve, after all, both the wavelength of the peak of the curve and the area under the curve can be obtained. Planck's law is mathematically so complex, however, that the other two laws are still used because of their simplicity.

Planck's law describes what is called a "perfect radiator." To define a perfect radiator we must, strangely enough, first describe a perfect absorber of radiation, because a perfect absorber is also a perfect radiator. The closest man-made thing to a perfect absorber is a cavity in a piece of iron. If the cavity has a larger diameter than the opening, nearly all of the radiation that enters the hole will be absorbed. Conversely, if the piece of iron is heated to incandescence, the hole will glow more brightly than the piece of iron.

A cavity will radiate energy as described by Planck's law (see Figure 9-1*a*). Such radiation is called either *black-body radiation* or *cavity radiation*. Most stars are nearly perfect radiators, but not quite, so their surface temperatures as determined by each of the three radiation laws are not quite equal. The surface of the sun is taken to be 6,000°K.

## 9.3 The Solar Spectrum

The studies of the solar spectrum now exceed the bounds of the optical spectrograph. The sun is being studied in nearly all wavelengths of the electromagnetic spectrum. Both spectrographic studies and direct photographs of the sun are being made in the gamma ray, X-ray, and ultraviolet region by rockets and satellites which operate above our shielding atmosphere. Optical studies (including the near-ultraviolet and infrared) can be made with ground-based telescopes. Radio astronomers are carrying on an intense study of the solar radiation in the long wavelengths of the electromagnetic spectrum.

Spectral studies in the optical region reveal some 26,000 spectral lines! Each of these lines must result from a particular element or chemical compound. Of the more than 100 elements on the Earth, only 67 have been observed in the sun. The others may be exceedingly rare or have spectral lines in regions of the electromagnetic spectrum not well studied yet.

It is not easy to identify spectral lines; temperature and pressure conditions on the sun are too different from those prevailing on the Earth. We can identify many lines such as those of hydrogen, iron, ionized calcium, helium, and others. But even for some of these elements, such as iron, a spectrum obtained in a terrestrial laboratory will not contain all the iron lines that are found in the solar spectrum.

**a. Line Intensity and Electron Transitions.** The identification of spectral lines is facilitated by a consideration of their *intensity*. The intensity of an emission line is, of course, the amount of light in that line, whereas the intensity of an absorption line is the amount of light subtracted from the continuous background spectrum. The intensity of a spectral line resulting from a given type of atom in a given state of excitation is a direct indication of the number of atoms in that state. Thus, if many atoms of the same kind experience the same transition, the resulting line will be more intense than if relatively few atoms experience that transition. For example, as more atoms of iron become ionized the resulting lines become more intense, whereas the lines resulting from neutral atoms of iron become less intense.

The factors which determine the extent of ionization are temperature and pressure. The higher the temperature, the greater the extent of ionization; the lower the pressure, the greater the extent of ionization. In a gas at very low pressure, an atom once ionized persists as an ion for a long time since there are no electrons near enough to capture. At high pressures it would quickly pick up a stray electron.

Theoretical studies which indicate how the spectral lines of each element depend on temperature and pressure permit us to better understand the conditions on the sun by studying the intensities of the lines in the solar spectrum.

A striking example of the identification of lines unseen in terrestrial sources presents itself in the so-called "coronal" lines. These lines in the spectrum of the solar corona remained unidentified for years before a theoretical study showed that they result from transitions in iron atoms that have been ionized no less than thirteen times. Since an atom of iron, when electrically neutral, contains 26 electrons, some of the atoms in the solar corona have lost half of their electrons. Such extensive ionization indicates not only a very high temperature but also a very low pressure. Similar studies identified helium in the sun before it was known to exist on the Earth. The name (from *helios*) gives due credit to the locale of its discovery.

**b. The Chromosphere and the Flash Spectrum.** We have seen that the solar absorption spectrum (not including the coronal spectrum which is seen at the time of total solar eclipse) results from the gases in the chromosphere. Studies of this spectrum indicate that the bottom of the chromosphere has a temperature of about 4,500°K and a pressure only about 0.1 that of the Earth's atmosphere at its surface. But gases at this temperature and pressure ought to emit a bright-line spectrum rather than cause an absorption-line spectrum. The chromosphere causes an absorption-line spectrum because it is normally seen in front of the photosphere, which is hotter (6,000°K) and therefore emits more radiant energy at each wavelength.

During a total solar eclipse, however, when the moon covers the rest of the sun, the chromosphere can be seen projected against a dark sky; it then emits a bright-line spectrum. This bright-line spectrum occurs within less than a minute before and after totality, when the visible portion of the chromosphere is shaped like a very thin crescent. If the light from this crescent is allowed to pass through a spectrograph that has no slit, it produces an effect similar to that produced by a slit, except that the lines appear not as lines but as thin crescents (Figure 9-2). Such a spectrogram is called a *flash spectrum* because the appearance of the chromosphere just before and after totality is a sudden flash.

The flash spectrum is very interesting because it gives us the opportunity to study the distribution of elements in the solar atmosphere. There are a few lines that form longer crescents than others. These are the lines of

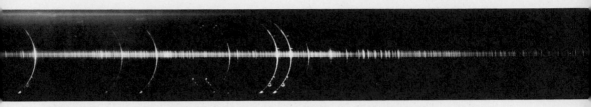

**Figure 9-2**

Flash spectrum of total solar eclipse of August 31, 1932, in Fryeburg, Maine. (Lick Observatory photograph)

calcium that have been ionized once and of hydrogen. Since their crescents are longer, these gases must reach higher into the atmosphere than the other gases. The visible hydrogen spectrum is composed of one very bright line in the red (*H*α) and other lines of the Balmer series in the blue and violet. The two bright lines of ionized calcium (called H and K, following Fraunhofer's designation) are in the violet. The combination of these lines causes the reddish color of the chromosphere during a total solar eclipse.

## 9.4 The Sun's Interior

The sun's interior is of utmost importance in understanding the more obvious solar features; it is the furnace that keeps the sun "burning." The interior, however, must be more than a furnace, for the sun is too hot to burn—fire as we know it cannot exist there because the sun is too hot for the common chemical reactions to take place. The source of light and heat in the sun is the energy derived from nuclear reactions. These nuclear reactions take place deep inside the core of the sun where the temperature is hot enough (about 15,000,000°K) to force atoms together so fast that instead of rebounding some of them combine to form new atomic nuclei.

**a. Converting Hydrogen to Helium.** There are two principal nuclear reactions that could occur in the sun's core—the *proton–proton reaction* and the *carbon cycle*. The proton–proton reaction is not only simpler but also is the most likely explanation for the sun's source of energy, since the temperature of the sun's core is only moderately hot as stellar interiors go. At 15,000,000°K, two protons (nuclei of hydrogen atoms) may combine, if they collide, to form a deuterium nucleus in accordance with the following equation:

$$_1^1H + _1^1H \rightarrow _1^2H + _1^0e$$

The H is the chemical symbol for hydrogen. The number in front of and below each letter gives the *atomic charge* (the number of protons in the nucleus). The number in front of and above the chemical symbol represents the *atomic mass number* (the number of protons and neutrons in the nucleus). The $_1^0e$ is a *positron* that takes away the remaining positive charge and some energy. It does not have a long life, however, for it quickly combines with a nearby electron to form two gamma-ray photons. The $_1^2H$ is an isotope of hydrogen called *deuterium*. It behaves chemically like ordinary hydrogen, but has twice the mass. The deuterium nucleus will, upon collision, combine with another proton to form an isotope of helium whose mass is three:

$$_1^2H + _1^1H \rightarrow _2^3He + \text{gamma ray}$$

One helium-3 nucleus may react with another to form helium-4 along with two protons:

$$_2^3He + _2^3He \rightarrow _2^4He + 2_1^1H$$

A total of six protons go into this reaction, and two protons come out plus one helium-4 nucleus, with the net result that four hydrogen atoms are converted into one helium atom.

The net result of the carbon cycle is the same as that for the proton–proton reaction: four hydrogen atoms are converted into one helium atom. The carbon acts only to permit this reaction to occur; there is no net change in the carbon. The details of the reaction are not important here, only the net result.

**b. Converting Mass to Energy.** The energy from both these sets of reactions results from the fact that four protons have a total but uncombined mass of 4.0291, whereas one helium nucleus has a mass of 4.0015. Thus, 0.0276 mass units are lost in each reaction. But according to the law of conservation of matter and energy, mass cannot be destroyed; any loss of mass must be converted into energy according to the equation formulated by Einstein, $E = mc^2$, where $m$ represents the amount of mass converted to energy $E$, and $c$ is the velocity of light. We previously calculated the amount of energy radiated by the sun to be $5.8 \times 10^{27}$ calories each minute. This is equal to $4.0 \times 10^{33}$ ergs per second (more convenient units) which can be substituted into the equation as $E$. The velocity of light is $c = 3 \times 10^{10}$ cm per second. By solving the equation for the mass $m$, we find that the sun loses $4.5 \times 10^{12}$ grams of mass each *second*. On the Earth, that amount of matter would weigh nearly $5 \times 10^6$ tons!

But can the sun continue losing mass at this rate for $4.6 \times 10^9$ years and still have some left? At this rate the sun would lose $6.5 \times 10^{28}$ grams in that time span. But the mass of the sun is $2.0 \times 10^{33}$ grams, so the sun would lose only 0.03% of its present mass! By way of comparison, if a 125-lb lady on a diet were to lose 0.03% of her weight, she would lose the insignificant amount of $\frac{1}{2}$ ounce!

The rate of both these reactions depends not only on the temperature but also on how closely the atoms are packed together. From theoretical studies, it seems very probable that the core of the sun is under a pressure that is equal to about $3 \times 10^{12}$ lb per $in^2$ ($2 \times 10^{11}$ atmospheres). The resulting density, about 110 times that of water, means that 1 cubic foot of material from the sun's core would weigh about 6,850 lb if placed on the Earth!

The energy that is generated deep in the interior of the sun must make its way to the surface in order to radiate out into space; if the energy could not get out, the sun would get hotter and hotter and consequently would expand. Once the energy leaves the core of the sun, it travels outward through the *radiative zone* in the form of X-rays and gamma rays until it reaches a region called the *convective envelope* (Figure 9-3) which is quite opaque to X-rays. The energy must travel in the form of convective currents similar to the air above a hot road in summertime, but on a much grander scale. Because such a vast amount of energy strikes the convective envelope from below, these currents become turbulent; the result is the turmoil of the photosphere.

## 9.5 Photospheric Activity

a. Granules.    The photosphere's mottled appearance indicates regions of different temperature called *granules* (Figure 9-4). The bright granules are several hundred miles in diameter, last for only a few minutes, and are hotter than the darker regions. They result from a parcel of gas which is heated from below, then rises and appears bright. In rising, the gas expands, radiates energy, cools, and sinks back into the lower levels to

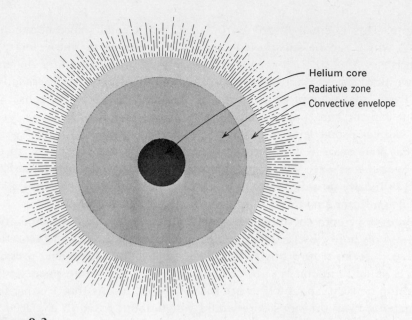

Helium core
Radiative zone
Convective envelope

**Figure 9-3**

A cross section of the sun.

become part of the darker portion of the photosphere. There the gases are again heated and rise once more as a bright granule.

The turbulent nature of the photosphere is revealed not only indirectly by these rapid changes in the granulation but spectroscopically as well, for spectral lines formed in a turbulent gas will be broader than lines formed in a quiet gas. Since atoms in a turbulent gas move both up and down, their motions cause simultaneous Doppler shifts to the red and blue. The spectral lines are therefore broadened in both directions. This is called *turbulent broadening*. Because it is evident in addition to and can be differentiated from pressure broadening (see p. 61), astronomers have learned that the gases in the chromosphere must move with velocities of about 1 mile per second. Such turbulence results from energy transported up from the sun's interior.

With the very high dispersion of the big solar telescopes, the turbulent broadening of the solar spectral lines can be resolved into irregularities in the lines (Figure 9-5). Immediately adjacent portions of the three spectral lines are shifted in different directions. The direction in which each portion is shifted, depends upon where in the lower chromosphere that

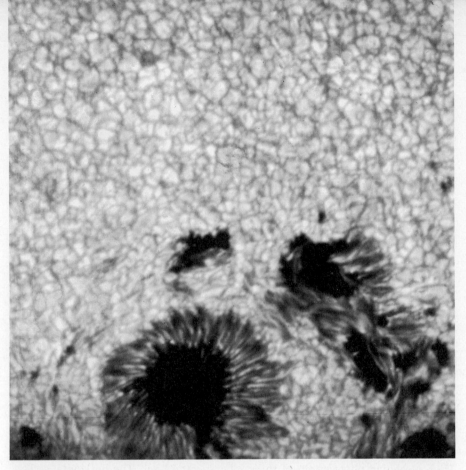

**Figure 9-4**

A sunspot group and solar granulation as photographed from a balloon at an altitude of 80,000 ft on August 17, 1959. This photograph was taken as part of Project Stratosphere of Princeton University, sponsored jointly by the Office of Naval Research and the National Science Foundation. (Courtesy of M. Schwarzschild)

portion of the line is formed. If it is formed in a rising column of gas, that portion of the line is shifted to the violet; if it is formed in a sinking parcel of gas, it is shifted toward the red. The same portion of each of the three lines is shifted in the same direction, for they were each formed in the same parcel of gas.

**b. The Sunspots.**    In our everyday life we consider the sun's outpouring of radiant energy very nearly constant; nevertheless it does manifest slight cyclical variations over a period of about 11 years. The most obvious evidence for this is the spotty appearance of the photosphere. Dark regions, called *sunspots*, appear from time to time. Some of the sunspots are small

**Figure 9-5**

High resolution photographs of three solar spectral lines. The horizontal solid black line represents a black spot on the sun's surface, but is actually in the spectrograph. Actual locations on the sun's surface can be recognized by the horizontal white and grey streaks. The white streaks result from hot gases rising to the surface. The three lines and wavelengths are (left to right): iron, 5,187.9 Å; ionized titanium, 5,188.7 Å; calcium, 5,188.9 Å. (Sacramento Peak Observatories, Air Force Cambridge Research Laboratory)

and normally do not last very long. Others are very large by terrestrial standards and may last as long as several months. Some spots are single and travel with the rotating sun as a lone spot. It is more common, however, for the spots to appear in groups (Figure 9-6). Each group is usually divided into two main parts—the preceding spot (preceding in the sense of solar rotation) and the one following. Spots will last any where from a few days or weeks to more than a month. The following spot is apt to break up and disappear first.

The center of a sunspot, being dark, is called the *umbra*. It is surrounded by the *penumbra*, which is brighter than the umbra but still not as bright as the photosphere. The spot is dark when compared with the photosphere which is indicative of a lower temperature; this fact is verified by spectrograms made of sunspots. In these spectrograms we find lines that are typical of a gas at 4,000°K. But a sunspot is not black in any sense of the word; if the rest of the sun were to be masked off we could see a large

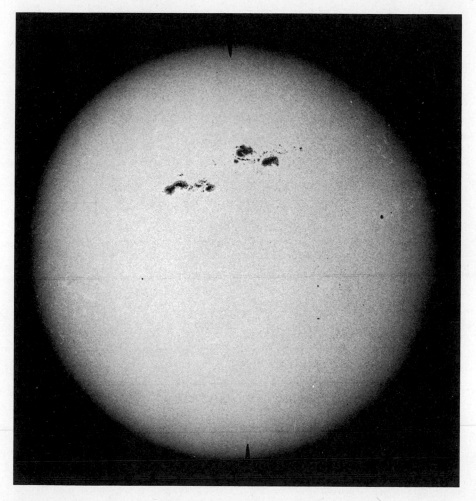

**Figure 9-6**

A large group of sunspots photographed on July 31, 1949. (Photograph from the Hale Observatories)

bright spot with the naked eye. It is dark only when compared with the photosphere, because its temperature is about 2,000°K less.

**c. Sunspots and Solar Rotation.**   As seen from the Earth, the sunspots move across the disk of the sun in a regular manner. This motion was first observed in 1610 by Galileo who suggested that the sun rotates with a period of about 26 days. It isn't all that simple, however. Studies more detailed than those made by Galileo indicate that the period of rotation of the sun's surface depends on latitude. For example, if a spot appears near the equator, its period of rotation will be about 25.1 days; if it appears at 15° north or south latitude, its period of rotation will be about 25.5 days; and if it appears at 30° latitude, its period of rotation will be about 26.5 days.

In order to verify this finding we can obtain the period of rotation by placing the limb of an image of the sun on the slit of the spectrograph and determine the Doppler shift. Spots are rare at distances of more than 35° north or south of the equator, and consequently the spectrographic method proves more useful because it can be used at much greater latitudes. The periods of rotation determined spectrographically, however, are not in complete accord with the periods determined from sunspots. The spectrograph tells us that the equator rotates in 24.6 days; 15° latitude in 25.4 days; 30° in 26.4 days; 60° in 31.0 days; and 75° latitude in 33.0 days. That a discrepancy should exist may be accounted for by the fact that whereas the spots yield a period of rotation of the photosphere, the spectrographic determination yields a period of rotation of the lower chromosphere where the absorption lines originate.

**d. The Sunspot Cycle.**   Since about 1750 the number of sunspots appearing each day has been recorded. By 1851 enough observations had been compiled to recognize that the number of spots varies over the years. Figure 9-7 shows that this variation called the *sunspot cycle* is not regular; neither all the maximum years nor all the minimum years have the same number of spots. In fact, even the period of 11 years is not the same. In 1917, for example, the maximum year occurred with a sunspot number of 104; 11 years later, in 1928, there was another maximum year with a sunspot number of 78; only 9 years later, 1937, with a sunspot number of 114, was another maximum year.

The number of spots is not the only variation, however, for the latitude at which spots occur changes as the cycle progresses. A cycle begins

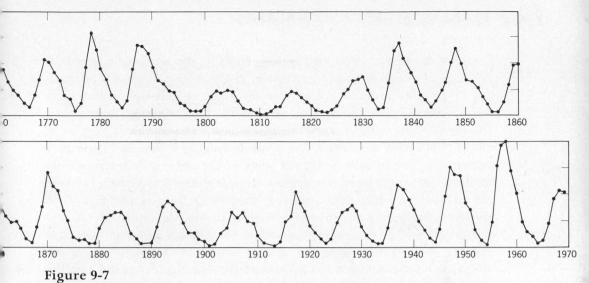

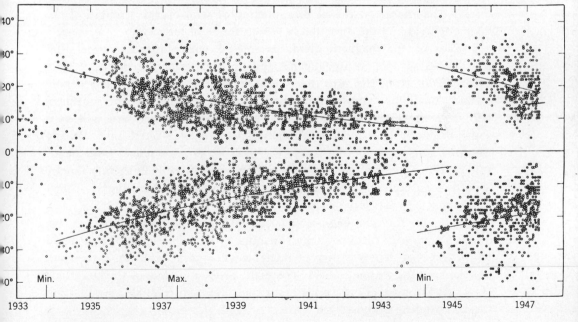

**Figure 9-7**

The sunspot cycle. The variations in the number of sunspots since 1760. The records before 1860 are not very reliable for they were not kept systematically.

**Figure 9-8**

The distribution of sunspots during the 11-year cycle. At the beginning of each cycle the sunspots form at some distance from the equator. As the cycle progresses they form ever nearer the equator. (Courtesy of Giorgio Abetti, *The Sun*, The Macmillan Co., 1957)

during a minimum, at which time the spots in the new cycle appear at latitudes of about 30° north and south. A given spot maintains approximately the same latitude, but as the cycle progresses the succeeding spots appear ever closer to the equator, until by the time minimum again occurs the spots in that cycle appear at about 10° north and south of the equator. Often two cycles overlap; a few spots belonging to the old cycle may appear near the equator, while the spots of the next cycle are appearing at about 30° from the equator. Spots are rarely seen more than 35° from the equator, and only occasionally do they reach the equator. Figure 9-8 shows a distribution of sunspots according to latitude and years.

During the 11-year cycle, then, both the number of spots and their latitude vary. But there is yet another variation, magnetic in nature. Associated with each spot is a magnetic field that behaves as if a long bar magnet had been inserted into the spot with one end sticking out. The evidence for this is spectrographic: when a spectral line is formed in a strong magnetic field it will be split into several components (the Zeeman effect). The characteristics of this splitting depend on the strength of the magnetic field and the direction in which the light leaves it.

Studies of the magnetic fields associated with sunspots reveal that if a preceding spot in a group has the north pole of the magnetic field pointing out from the sun, the following spot will have the south pole sticking out. The spots in the other hemisphere (northern or southern) will have the opposite magnetic polarity; the preceding spot will be a south magnetic pole and the following spot will be a north pole. During the next sunspot cycle, however, the polarity is reversed in both hemispheres. Any two succeeding cycles, then, are different in their magnetic aspects; consequently the cycle does not repeat itself in every respect until about 22 years have elapsed.

The reason for this magnetic variation is quite unknown; in fact, we do not know with certainty why sunspots have magnetic fields associated with them at all. But it is suspected that the magnetic field causes the spot, not the spot the magnetic field. The main problem is one of refrigeration! How can the magnetic field maintain the temperature 2,000° cooler than the surrounding photosphere?

**e. The Photosphere and the Spectroheliogram.**   We could learn more about the photosphere in general and sunspots in particular if we knew where the various spectral lines originated in the sunspots and in the

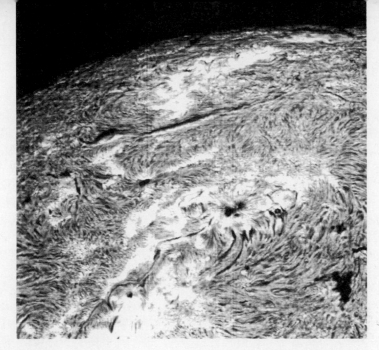

**Figure 9-9**

(*a*) The sun photographed on May 4, 1958, in the light of the Hα line of hydrogen (*above*). (Photograph from the Hale Observatories.) (*b*) Detail of two closely spaced sunspot groups is seen in this photograph taken on October 26, 1969, in the Hα light of hydrogen (*below*). The bright spots are hydrogen in emission and thus hotter regions. The darker (cooler) filaments follow the magnetic field. (Big Bear Solar Observatory)

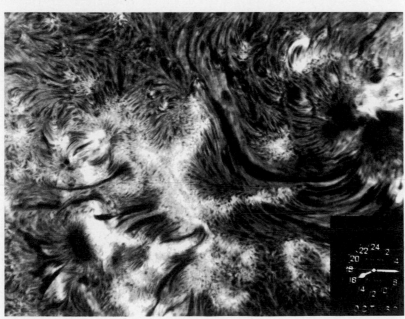

chromosphere. Spectral lines are, after all, indicative of the conditions under which they form. Since some of the lines in the solar spectrum are rather broad and quite intense, it is possible to take a picture of the sun by the light of any one of these broad lines alone. (It should be recalled that even absorption lines are not black but contain some light that has been absorbed and re-emitted by the gas.) Special filters, both mechanical and optical, have been devised that eliminate all the solar spectrum except that spectral line of special interest. The resulting photograph, called a *spectroheliogram*, reveals the distribution of that particular gas whose spectral line has been allowed to pass through the filter. The three principal lines used in studying the sun's surface are *H*α of hydrogen and the H and K lines of ionized calcium, since these lines are sufficiently intense to yield a good picture, and broad enough to be isolated by filters.

Figure 9-9 shows two photographs of the sun taken in the light of hydrogen alone. The brighter spots are regions of higher temperature. Such detail is not obtainable with photographs taken with the light of the entire visible spectrum. The differences in the distribution of hydrogen and calcium on the sun are shown in Figure 9-10. The top photograph was taken in hydrogen light, the bottom in the light of ionized calcium.

**f. The Plage.**   Each brighter region is called a *plage* and is evidence of some disturbance on the sun. Sunspots are located in plages (Figure 9-10).

When the sun is photographed by means of hydrogen light, the mottled appearance of photographs made by calcium light becomes less evident, and we see striations or streaks as in Figure 9-10a. Some of the streaks are bright, indicating a higher temperature, and nearly all of them appear to follow magnetic lines of force as iron filings do about a bar magnet. One other feature that is often visible in spectroheliograms of hydrogen light is the long *dark filaments*. These are cooler gases above the surface that absorb light.

**g. Solar Flares.**   Another form of photospheric activity is the *solar flare*, a very hot and explosive region. A flare normally forms in a sunspot group, generally along the line dividing the opposite magnetic polarities (Figure 9-11). One of the first signs is a local increase in brightness and the emission of short bursts of radio energy. After 15 to 30 minutes, the explosive phase begins, and for 15 minutes or more the flare emits short-wavelength X-rays, ultraviolet radiation, visible light, and radio energy. Gases are ejected from the region at velocities in excess of 600 miles per

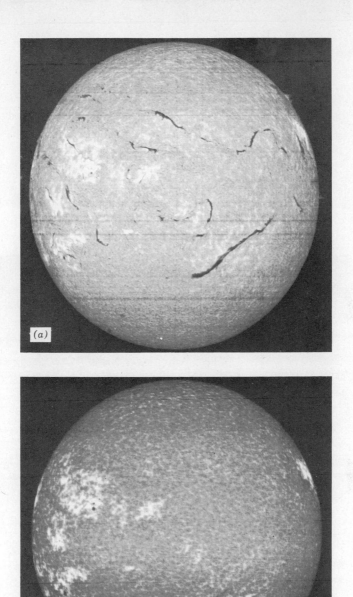

**Figure 9-10**

The sun photographed in (a) hydrogen Hα light and (b) ionized calcium light. The plages (bright areas) and hydrogen dark filaments are clearly seen. (Sacramento Peak Observatory, Air Force Cambridge Research Laboratories)

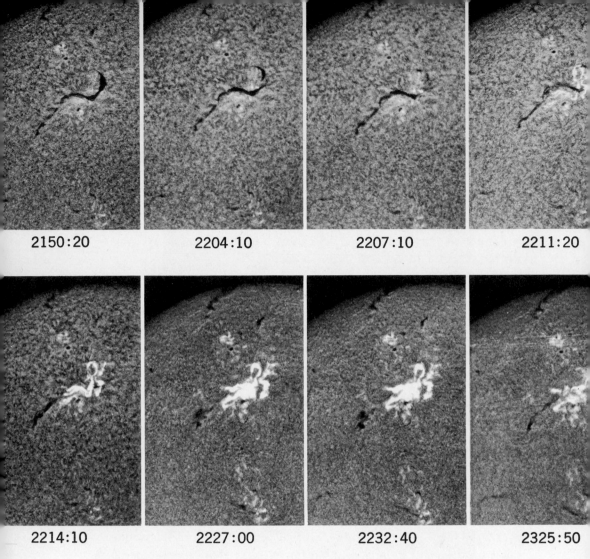

| | | | |
|---|---|---|---|
| 2150:20 | 2204:10 | 2207:10 | 2211:20 |
| 2214:10 | 2227:00 | 2232:40 | 2325:50 |

**Figure 9-11**

The eruption of a solar flare as seen in hydrogen light; total time lapse is about 1 hour 35 minutes. (Lockheed Aircraft Corporation)

second. Shock waves are propagated over the solar surface at velocities of 1,000 miles per second. Vast amounts of energy (about $10^{33}$ ergs) are emitted during a flare. Clearly, the Earth does not escape the effects of such an explosion. The gases, composed mostly of ionized hydrogen, that is, protons and electrons, reach the Earth in a very short time. Some of the most energetic particles travel at nearly the speed of light and reach the Earth in just over 8 minutes. Other less energetic particles take longer. Some of the very energetic protons reach the Earth in an hour or so, indicating that they do not travel in straight lines. The path of these particles is presumably deviated by the sun's magnetic field. Upon arriving at the Earth, the low-energy particles are deviated by the Earth's magnetic field and enter the atmosphere in the polar regions. These particles are related in a manner not well understood with the *aurora*—the northern lights and the southern lights. The more energetic particles come right on through the magnetic field and strike the surface of the Earth. They have been known to play havoc with electrical equipment, such as causing circuit breakers for towns and cities to go out.

Ultraviolet and X-ray radiation disrupt the ionized layers in the Earth's atmosphere, which by reflecting radio broadcasting waves ordinarily enable radio operators in New York to receive messages from London, Sydney, Caracas. The ionized layers themselves are a result of the sun's usual ultraviolet radiation, but the intense radiation associated with solar flares has a disrupting effect and causes a corresponding fading of long-distance radio communications. Since we rely so heavily on long-distance radio communication (especially for ships and airplanes), there is a very practical purpose in observing the sun continuously: we may predict disruptions in the radio service and permit advance arrangement with other means of communication such as a communication satellite.

However, not all solar flares result in the corresponding terrestrial activity; nor does such terrestrial activity occur only when visible markings are in evidence on the photosphere. Magnetic disturbances often occur at intervals of 27 days, some of them when no corresponding visible markings appear on the sun. Since this 27-day interval between some of the magnetic storms is too frequent to be accounted for by mere chance, and since it is the synodic rotation period of the sun at 10° solar latitude, it is suspected that some regions of the sun remain disturbed for months at a time, even though they may not be visible.

**h. Limb Darkening.** The sun's photosphere, then, is mottled by granulation, marked with spots, and occasionally blemished by a flare. But even aside from these irregularities the photosphere does not present a uniformly bright disk when photographed. Its brightness decreases from the center of the disk to the limb. This is called *limb darkening* (Figure 9-6). We cannot see into the sun as deeply at the limb as at the center of the disk, for at the limb light must travel through a thicker layer of gas. Consequently the visible regions of the limb are nearly 190 miles higher in altitude and consequently cooler than the center of the disk. It is this lower temperature that accounts for the decrease in brightness. By taking successive spectrograms of small regions of the sun from the center of the disk out to the limb, we actually obtain spectrograms of ever-higher levels above the photosphere and therefore learn more about the distribution of temperature and pressure on the sun.

## 9.6 Chromospheric Activity

We have seen that above the photosphere lies the chromosphere extending some 5,000 to 10,000 miles. The thickness of the chromosphere varies somewhat with the sunspot cycle. During a sunspot minimum its thickness at the poles is about 10% greater than at the equator, and during the maximum the thickness is the same all around the sun.

**a. Temperatures.** The range of temperature in the chromosphere is considerable. At a height of only about 400 miles above the photosphere, the temperature reaches a minimum value of 4,200°K. At an altitude of 2,500 miles, it is the same as that of the photosphere, 6,000°K. But at its rather indefinite top, where it merges with the lower corona, its temperature is 50,000°K.

Like the gases beneath it, the chromosphere is turbulent. When seen during a total solar eclipse or with a coronagraph (to be discussed) it presents a very uneven outline that has many small (in solar terms) irregularities called *spicules* projecting upward (Figure 9-12). These are several hundred miles in diameter and go as high as 10,000 miles. Their lifetime varies from 2 to about 12 minutes, with an average of about 3 to 5 minutes. Thus they have about the same diameter and lifetime as a bright granule. That the spicules are somehow related to the granules on the photosphere is difficult to prove but tempting to assume.

**b. The Prominences.** It is quite obvious that gases do extend up from the photosphere (Figure 9-13). These projections, called *prominences,* were first seen during a total solar eclipse. Since prominences are such an obvious sign of solar activity they are very important in the study of the sun, but to photograph them only when the sun is shielded by the moon not only demands great patience on the part of the astronomer but used to mean that he could not study them systematically.

Several types of instruments have been built to alleviate this difficulty. One of these, the *corongraph,* was devised to produce an artificial eclipse

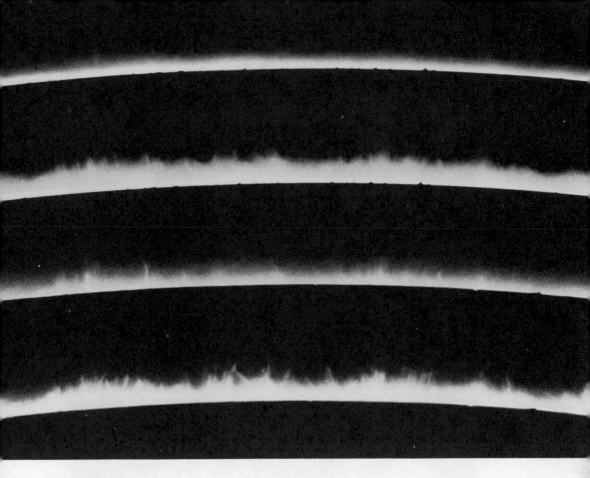

**Figure 9-12**

Spicules photographed with a 15-in. coronagraph. (Sacramento Peak Observatory, Air Force Cambridge Research Laboratories)

during which the prominences and the inner part of the corona can be studied systematically. Since the sun's light is so bright that reflections from dust and air cause blurring, coronagraphs must be installed and used at very high altitudes where the air is thin and clear. The most famous have been placed in the Rocky Mountains at the High Altitude Observatory at Climax, Colorado, on the Pic du Midi high in the French Pyrénées, and on Sacramento Peak near Sunspot, New Mexico.

Since the coronagraph is able to simulate a solar eclipse whenever the sun is visible, it is possible to take time-lapse movies in which each frame is taken at a specified time interval (for example, 1 per minute). When such a movie is run through a motion picture projector at 16 frames each second the motions of the prominences are speeded up. In this way it is possible

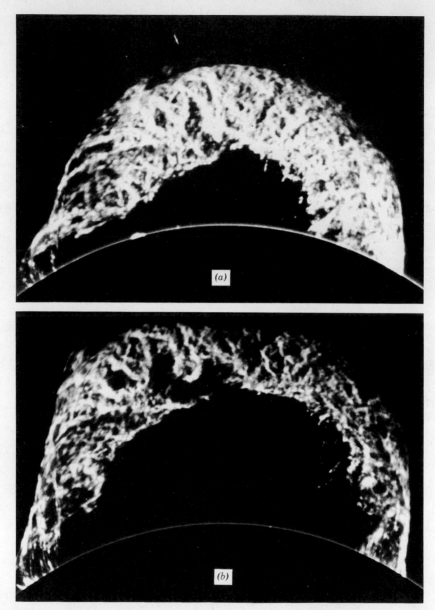

**Figure 9-13**

The great arch prominence of June 4, 1946. The two exposures were taken only 15 minutes apart. The diameter of the sun, roughly 864,000 miles, gives some idea of how rapidly this arch left the sun. (Sacramento Peak Observatory, Air Force Cambridge Research Laboratories)

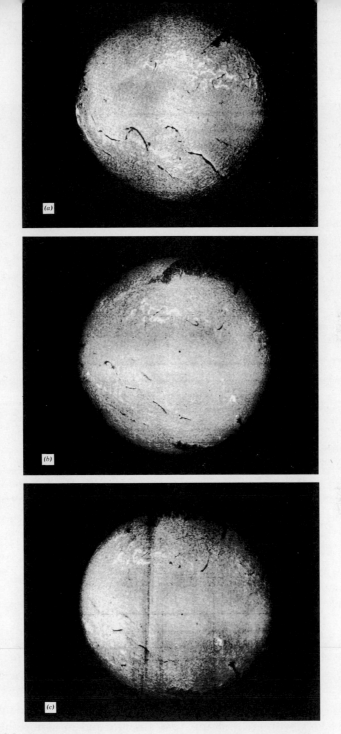

## Figure 9-14

The great arch prominence was seen as a dark hydrogen filament on (a) May 30, (b) June 2, and (c) June 3, 1946. (Observatoire de Paris, Meudon)

to determine not only how fast the gas in the prominences moves but also along what path it moves.

By this means we find that there are many varieties of prominences: some look like geysers, some take the shape of loops, some appear as arches, and still others hang above the sun quiescently like stringy clouds.

The *arch prominence* is apparently a dark filament seen from the side. The most spectacular arch prominence yet photographed occurred on June 4, 1946 (Figure 9-13). The first of the time-lapse photographs was taken shortly after sunrise when the prominence had already risen to a height of 250,000 miles. The prominence moved upward from the sun at velocities in excess of 100 miles per second, so that it was a matter of only $\frac{1}{2}$ hour before the prominence had extended beyond the limits of the coronagraph. This prominence was actually seen as a dark filament during the few days before it erupted into this fantastic upheaval (Figure 9-14).

A *surge prominence* (Figure 9-15a) may burst forth from the photosphere like a huge column of gas, stand with its top perhaps 30,000 miles above the photosphere, then fade and fall back into the sun along the same path. Such surges apparently result from some sort of explosion under the photosphere.

*Loop prominences* (Figure 9-15b) are associated with sunspots and are peculiar in that the material of both branches of the loop flows downward. Thus, gases seem to enter the loop at the top, divide, and flow into the associated sunspot along the two halves of the loop. Ionized gases falling into the sun are apparently constrained into the loop by the magnetic field about the sunspot.

Falling gases may also take the form of *coronal rain*, a flow of gases into the sun along definite paths that do not appear as loops. It is possible, however, that coronal rain may sometimes be loops seen edgewise. Often coronal rain will flow from different directions into a common region of the sun, as if the sun were "sucking" the gases in like a vacuum cleaner.

A *quiescent prominence* (Figure 9-15c) is also a dark filament, seen from the side. It may remain for months before it breaks up by either vanishing or exploding, like the great arch prominence of June 4, 1946. It seems rather amazing that on the turbulent sun quiescent prominences can remain so long with so little apparent change, and in fact, although their form may not vary, time-lapse movies reveal that the material composing them continually changes. Here, as in loop prominences and coronal rain,

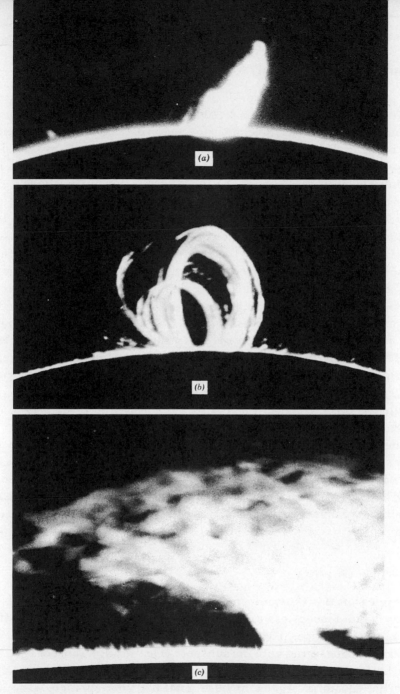

**Figure 9-15**

(*a*) Surge prominence. (High Altitude Observatory of the University of Colorado).
(*b*) Loop prominence, and (*c*) Quiescent prominence. (Sacramento Peak Observatory, Air Force Cambridge Research Laboratory)

we find that gases reveal their presence by falling into the sun, becoming incandescent only when traveling in the region defined by the prominences.

It is not known precisely why gases should emit light only in a restricted region, but there is reason for thinking that this region is one whose temperature is lower than that of the inner corona which surrounds it. It is possible for a gas to be so hot that it becomes completely ionized (this occurs most readily with hydrogen). Hence, its atoms retain no electrons that may make transitions, and consequently the gas cannot emit visible light! When a gas is too cold, therefore, it does not emit visible light because there is not enough energy to excite the electrons, and when it is too hot it does not emit visible light because the electrons have become so excited that they have escaped the atoms completely. Between these two extremes the gas emits visible light. We have seen that the corona is hot enough to ionize iron thirteen times; so it is hot enough to completely ionize hydrogen, the main constituent of the prominences. For this reason we suppose that gases falling through the corona must fall into a cooler region before they can emit visible light. Why these regions are cooler remains to be determined, but the study of magnetic fields may lead us to the answer.

The fact that gases *fall into* the sun to produce many of the prominences may seem puzzling. But we see only the rains on the Earth; we do not *see* water evaporating from the oceans and lakes. Perhaps gases, too hot to be seen, leave the turbulent photosphere in a like manner to go thousands or millions of miles above the sun and then fall back down to produce the ''rain.''

## 9.7 The Corona

**a. Appearance.**  Outward from the chromosphere there extends the pearly white tenuous gas called the *corona*, which is seen so beautifully during a solar eclipse (Figure 9-16). Its height above the photosphere, as well as its shape, varies with the sunspot cycle. During a sunspot maximum it is fairly regular, but during a minimum it extends out much farther from the equatorial regions in huge petal-like configurations. The inner part of the corona has a brushlike appearance which seems shaped by magnetic lines of force near the polar region. Since the corona extends millions of miles into space, it may be associated with the particles (discussed on p. 197) which cause the zodiacal light.

**Figure 9-16**

A specially designed filter was used for the March 1970 eclipse to bring out details in the bright inner corona as well as the much fainter outer corona. (Courtesy Gordon Newkirk, Jr., High Altitude Observatory)

**b. Temperature and Pressure.** It is the emission spectrum of the corona that contains the lines of iron ionized 13 times. We have noted (p. 228) that such ionization in the corona indicates not only a very high temperature of perhaps 1,000,000°K, but also a low pressure as well. The emission lines, like the absorption lines, are broadened by the rapid motion of the atoms that compose the very hot gases of the corona. This type of broadening is called *Doppler broadening* and is slightly different from turbulent broadening, in that the motions involved are not those of gaseous masses, as with turbulence, but are those of randomly moving particles.

In order for the corona to maintain a temperature of roughly 1,000,000°K it must receive energy from the sun. Since it is quite transparent to electro-magnetic radiations, it does not seem likely that it can be heated by this mechanism. It has been suggested that the turbulence of the photosphere sets up shock waves which are transported into the corona. The pressure of the corona is so low, however, that shock waves cannot continue as such; they die out and the energy they contain is transformed into thermal energy.

c. **The Corona and the Solar Wind.** The fact that the temperature of the corona increases with increasing altitude has led to some rather startling observations. An ionized gas at low pressure is a good conductor of heat. The conduction of heat outward from the sun results in the ex-pansion of the outer corona. At about 6,000,000 miles (about 14 solar radii) from the solar surface, the corona is expanding at a speed in excess of 100 miles per second—faster than the speed of sound in the same gas. The outer corona is turned into a supersonic wind; the outer corona becomes the *solar wind* which whizzes past the Earth at a velocity roughly 250 miles per second. The solar wind continues out beyond the orbit of Jupiter, but just how far remains to be seen by future research. The existence of the solar wind has been verified by measurements made by spacecraft on their way to the moon, Venus, and Mars. The solar wind at the Earth's distance from the sun is composed of 10 protons per square cm, or perhaps more. A comet is truly a "solar wind-sock."

## 9.8 The Sun and the Earth

The sun has been and is now being studied by all available methods and with all the imagination and ingenuity that can be brought to bear on it. Very important to us here on the Earth is its 11-year cycle of activity with its sunspots and flares, prominence activity, varying coronal shape and size, and their corresponding effect on our planet. It has been found, for example, that the rate of growth of trees, which depends on the weather, varies in an 11-year cycle which corresponds to the cycle of the sun. In fact, a study and understanding of this growth as seen in tree rings has been used to date pieces of wood found by archeologists in their diggings.

The effect of solar activity on radio broadcasting reception has already been mentioned. Its effect on the weather of the Earth has led meteorologists, too, to become interested not only in the 11-year cycle but also in solar flares; there is reason to believe that those powerful eruptions affect our weather to an as yet undetermined degree. Furthermore, since large flares emit cosmic rays in such deadly amounts, they will also concern the space traveler who may find it healthier to venture out beyond our shielding atmosphere only during a sunspot minimum, when solar flares are quite rare.

The astronomer, however, studies the sun primarily because it is a star, the only one whose surface we can see and study in some detail. And stars, after all, are the astronomer's first calling, for they are the building stones of the universe.

## BASIC VOCABULARY FOR SUBSEQUENT READING

| | |
|---|---|
| Chromosphere | Plage |
| Convective envelope | Prominence |
| Corona | Solar flare |
| Dark filament | Solar wind |
| Flash spectrum | Spectroheliogram |
| Granules | Spicules |
| Limb darkening | Sunspot |
| Photosphere | |

## QUESTIONS AND PROBLEMS

1. What would be the horsepower rating of the sun if it were to:
   (a) double its diameter but maintain the same temperature?
   (b) double its temperature but maintain the same diameter?
   (c) double both its diameter and its temperature?

2. Explain by a diagram how the flash spectrum can be used to determine the heights of various elements in the solar atmosphere, as well as the heights of elements in various degrees of excitation.

3. Explain why a dark hydrogen filament appears bright when seen at the sun's limb projected against the dark sky.

4. List the similarities and differences between the photosphere, the chromosphere, and the corona.

5. Under what conditions can the chromosphere produce a bright-line spectrum?

6. How much energy is emitted per unit area each second by:
   (a) a toaster filament, $T = 1,000°K$
   (b) a furnace, $T = 2,000°K$
   (c) a star, $T = 4,000°K$
   (d) a hot star, $T = 16,000°K$
7. Find the wavelength (in Ångstroms) of the peak of the radiation curve for each of the four objects in question 6.

## FOR FURTHER READING

Bray, R. J., and R. E. Loughheed, *Sunspots*, John Wiley and Sons, New York, 1965.

Glasstone, S., *Sourcebook on the Space Sciences*, Van Nostrand Co., Princeton, N.J., 1965, Chapter 6.

Menzel, D. H., *Our Sun*, Harvard University Press, Cambridge, Mass., 1959.

Struve, O., and V. Zebergs, *Astronomy of the 20th Century*, Crowell Collier and Macmillan, New York, 1962, Chapter VII.

Babcock, H. W., "The Magnetism of the Sun," *Scientific American*, p. 52 (Feb. 1960).

Bahcall, J. N., "Neutrinos from the Sun," *Scientific American*, p. 29 (July 1969).

Carroll, G. A., "The Star Telescope of Lockheed Solar Observatory," *Sky and Telescope*, p. 10 (July 1970).

Chapman, S., "The Earth in the Sun's Atmosphere," *Scientific American*, p. 64 (Oct. 1959).

Dunn, R. B., "Sacramento Peak's New Solar Telescope," *Sky and Telescope*, p. 368 (Dec. 1969).

Evans, J. W., "Motion in the Solar Atmosphere," *Sky and Telescope*, p. 321 (June 1963).

Goldberg, L., "Ultraviolet Astronomy," *Scientific American*, p. 92 (June 1969).

Malitson, H. H., "The Solar Energy Spectrum," *Sky and Telescope*, p. 162 (March 1965).

Minnaert, M., "Solar Symposium at Sydney," *Sky and Telescope*, p. 151 (March 1964).

Ney, E. P., "Balloon Observations During July Eclipse," *Sky and Telescope*, p. 251 (Nov. 1963).

Parker, E. N., "The Solar Wind," *Scientific American*, p. 66 (April 1964).

Pneuman, G. W., "The Chromosphere–Corona Transition Region," *Sky and Telescope*, p. 148 (March 1970).

Schwarzschild, M., and B. Schwarzschild, "Balloon Astronomy," *Scientific American*, p. 52 (May 1959).

Zirin, H., "The Big Bear Solar Observatory," *Sky and Telescope*, p. 215 (April 1970).

"OSO 4 Ultraviolet Solar Observations," *Sky and Telescope*, p. 362 (Dec. 1967).

"Round-the-Clock Solar Movies," *Sky and Telescope*, p. 296 (May 1969).

"Solar Flare Symposium," *Sky and Telescope*, p. 89 (Feb. 1965).

"The Moon's Wake in the Solar Wind," *Sky and Telescope*, p. 93 (Feb. 1968).

"The Temperature Minimum in the Sun," *Sky and Telescope*, p. 362 (June 1969).

CHAPTER 10 CHAPTER

# STABLE STARS

Studying the stars is like putting together a jigsaw puzzle for which we do not have all the pieces. What is more, the pieces that are available to us do not simply lie around waiting to be picked up and fitted into place; they are often hard to find and when found their shapes are sometimes indistinct. The pieces at hand are the observed physical characteristics of the stars. When fitted together they leave some gaps in our picture. We then deduce what ought to fill those gaps so that what we see may make sense. The astronomer hopes thus to obtain knowledge of the stars' present state, of their origin and past, and of their future. He wants to understand how stars are born, what stages they undergo, and how they end.

The prodigious amount of energy that a star emits each second is evidence that it is not a dead and static body, but that it undergoes an evolutionary process. The basic assumption of science, that of causality, tells us that any emission of energy from a star must have a source which is consumed during the star's lifetime. As this source of energy is consumed the star must change.

## 10.1 General Features

In order to understand and predict stellar changes we must be able to describe a star. Such a description should include a knowledge of each of the following:

1. Distance from the sun or, more exactly, the position of the star in our galaxy.
2. Space velocity
3. Size
4. Temperature
5. Luminosity
6. Mass
7. Density
8. Chemical composition
9. Age

### Chapter Opening Photo

Star clouds in the Sagittarius region of the Milky Way. The dark lanes are intervening gases that obscure the stars behind. (Photograph from the Hale Observatories)

## 10.2 Stellar Distances

The distance of a star from the sun is not only one of the more difficult characteristics to determine but also one of the more important. All the life processes of the star are determined by means of the amount and types of energy radiated, but the amount of energy a star radiates out into space cannot be known until its distance is known. Consequently, there has been a great deal of work done on the distance determination of stars, some of which has proved successful and some of which at one time was thought valid but was later proved incorrect. Hence, there is still much thought and work applied to this problem.

**a. Heliocentric Parallax.**   The first method used to determine a star's distance from the sun is the most fundamental, for it is an actual measurement of distance after the manner of the surveyor. It is the method of heliocentric parallax discussed on p. 117. If the nearer stars appear to oscillate back and forth in front of the background stars as the result of the Earth's revolution about the sun, we need only measure the *angle of parallax p* (Figure 10-1) through which a star appears to move. This angle when considered trigonometrically with the radius of the Earth's orbit as a base line yields the distance $oS$.

The need for an astronomical unit of measure to fit stellar distances becomes apparent when we consider that the nearest star is something like 24 trillion ($24 \times 10^{12}$) miles from the sun. The method of heliocentric parallax has given rise to a unit of distance called the *parsec*. One parsec is equal to about $19 \times 10^{12}$ miles and is the distance at which a star must be in order to have a *parallax of 1 second* of arc. The closest star must have the largest angle of parallax; it is called Proxima Centauri and its angle of parallax is 0.76 second of arc. Because all of the angles of parallax are

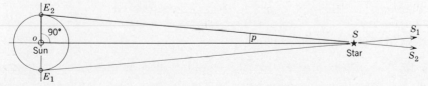

**Figure 10-1**

The angle of parallax is actually the angle subtended by the radius of the Earth's orbit as seen from a nearby star.

very small, the distance $d$ of the star in parsecs is defined as being equal to the reciprocal of the angle of parallax $p$ measured in seconds of arc:

$$d = \frac{1}{p}$$

Proxima Centauri must be $1/0.76 = 1.3$ parsecs away. Another unit of measurement employed in stellar distances is the *light year*, or the distance light travels in 1 year. Light travels 186,000 miles every second or about $6 \times 10^{12}$ miles in 1 year. A light year is, therefore, shorter than the parsec; 1 parsec equals about 3.26 light years. Proxima Centauri is about 4.2 light years from the sun.

**b. Limitations of Heliocentric Parallax.** As a method of determining distances, however, heliocentric parallax has inherent limitations. The farther the star is from the sun, the smaller the angle that must be measured. Since this angle must be measured by mechanical means (usually by the location of a star's image on a photographic plate) the measurement of the very small angles involved not only becomes difficult but the error in making the measurement soon becomes as large as the angle itself. Hence, only the distances of the nearer stars can be determined by heliocentric parallax. Distances for some 6,000 stars have been so determined, but only for the nearest 700 have they been determined with an uncertainty of less than 10%.

Astronomers also need to be able to determine distances for many of the billions of stars beyond the limit of heliocentric parallax. As we see in this and succeeding chapters there are many roundabout ways of determining stellar distances, all less accurate than the methods of heliocentric parallax applied to the nearest stars. Yet each contributes its bit to our knowledge of the size of our galaxy and ultimately of the size of our universe. One of these methods (see Chapter 15) is based on the space velocity of stars, the next characteristic to be discussed.

## 10.3 Stellar Velocities

We have seen (Chapter 2) that the Doppler shift of the spectrum can be used to determine the velocity of a star in the line of sight, that is, its *radial velocity*. This gives the velocity in only one direction; but certainly stars do not only travel directly away from or toward the sun. An airplane, for example, that is flying northwest may be due south of an observer at one moment; shortly thereafter it will not only be closer to the observer but to the west of due south. The observer could say, therefore, that the plane has two simultaneous velocities: one to the north, approaching him, and the other due west. These two velocities are called *components* of the actual velocity. When we observe the radial velocity of a star, we observe only one of the components of its motion, that one equivalent to the airplane approaching the observer. The other component, the one equivalent to the westward motion of the airplane, is called *tangential velocity* and is seen as the star moves perpendicular to the line of sight. This motion is revealed by the star's motion against the background stars, which are too far away to reveal any tangential motion of their own. When tangential velocity is expressed as an angular motion per unit of time, for example, seconds of arc per year, it is called *proper motion*.

Figure 10-2 shows the radial velocity component, the tangential velocity component, and the relative velocity of a star with respect to the sun. Figure 10-3 shows the proper motion of a star that has moved against the background stars during the time interval between the three photographs. The spectrum of this star reveals its radial velocity. Its velocity relative to the sun is obtained by the Pythagorean theorem from its radial and tangential velocities.

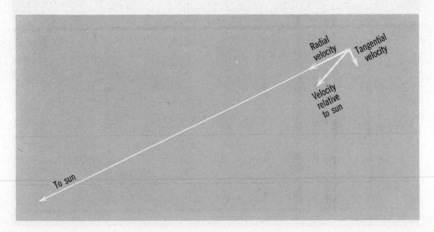

**Figure 10-2**
The velocity of a star with respect to the sun is broken up into two components— radial velocity and tangential velocity, for only these component velocities can be conveniently measured.

**Figure 10-3 (a)**

A composite of three photographs of Barnard's star (lower left). The three photographs have been superimposed so that the background stars (upper right) each show only one image. Motion to the north (down) is proper motion, motion east and west is the result of annual heliocentric parallax. (Courtesy Peter van de Kamp, Sproul Observatory)

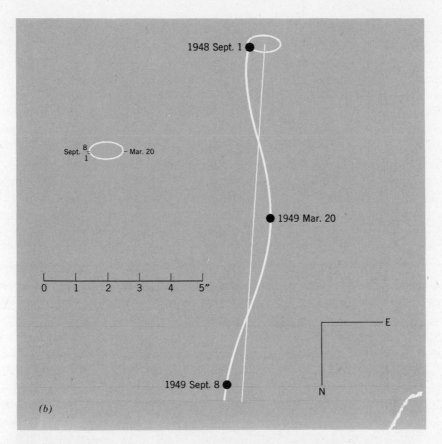

**Figure 10-3 (b)**

A drawing indicating the location of Barnard's star at the specific dates shown in Figure 10-3*a*. When proper motion is taken out, the star appears to move in the little ellipse shown to the left. This elliptical motion is heliocentric parallax.

The fact that stars do have proper motion interferes to a certain extent with the determination of their parallax, for in determining a star's position from year to year its proper motion must be untangled from its parallactic motion. Since parallactic motion is periodic and proper motion is cumulative, the separation of the two is not an impossible task. The proper motion of some 200,000 stars has already been determined.

Time is a factor in determining proper motion. The longer the time interval between photographs, the farther the foreground stars will have moved in front of the background stars. If one could wait 50 years between photographs, the proper motions of even quite distant stars could be determined by using the still more distant galaxies as a background against which a star's motion might be observed. Such a program is under way at the Lick Observatory. Beginning in 1947, the entire sky north of $-23°$ declination was photographed over a period of 7 years on 1,246 plates, each plate covering about 25 square degrees of the sky. The original plan was to rephotograph the sky in 50 years, but with the advent of new and better equipment for measuring stellar positions, and because of some eagerness on the part of the astronomers involved, it was decided to rephotograph the sky after an interval of only 20 years. The results of this program should tell us a great deal about stellar motions. It should also enable us to learn a great deal about the rotation of our galaxy, for the more distant galaxies do not, of course, participate in this motion.

*Relative velocity*, as used here, refers to the velocity of a star with respect to the sun. If the sun's motion with respect to the local group of stars is taken into account, that is, if it is subtracted from the relative velocity, we obtain the star's *peculiar velocity*—its velocity with respect to the local group of stars. The peculiar velocity of most stars in the neighborhood of the sun is about the same as that of the sun (about 12 miles per second) but there are certain stars, called *high velocity* stars, whose peculiar velocities are as great as 100 miles per second. As will be seen later, these are thought to be older than the sun itself. A star's velocity, as it revolves with the local group of stars about the center of our galaxy is called *rotational velocity*, since it is the star's motion as part of a rotating galaxy.

## 10.4 Stellar Sizes

The difficulties of studying parallax and proper motion are great because the stars are so very far apart in relation to their sizes and velocities. Although the velocity of a neighboring high-velocity star may be as much as 100 miles per second, its motion as seen from the Earth is very small indeed. Similarly, although some of the nearby stars are much larger than the sun, they appear only as points of light in our largest telescopes. To illustrate, let us imagine the sun shrunk to the size of an apple, and the nearest stars as well as the distances between each star shrunk in proportion. We would then have a group of apples, each about 1,500 miles apart, with the fastest of them moving at a rate of about $\frac{1}{2}$ in. per hour. To measure such distances and velocities is hard enough, but it is even more difficult to measure the sizes of the stars; an apple at 1,500 miles does not look very large. Nevertheless the astronomer has set himself the task of determining the sizes of the stars; how does he go about it?

The angular diameter of a star can be determined by the use of an interferometer (see p. 44). In 1925, two smaller mirrors were mounted on a 20-ft beam (it is called a *beam interferometer*) atop the 100-in. telescope on Mount Wilson. The angular diameters of 12 stars were measured by this instrument. Recently, however, an interferometer consisting of two separate telescopes each on a track has been installed at the Narrabri Observatory in Australia. Reports from work in progress indicate that the angular diameter of Sirius, for example, is 0.0062 seconds of arc. A 45-ft crater on the nearside of the moon appears about as big as Sirius does in our sky. At a distance of 2.7 parsecs (8.7 light years) Sirius must have a diameter of $1.52 \times 10^6$ miles, 1.76 times that of the sun.

The diameters of some of the largest stars are astounding. Mira, in the constellation Ceti, has a diameter 460 times that of the sun. Were Mira to replace the sun, its boundaries would include the orbit of the planet Mars! The star Betelgeuse is a pulsating star and has a diameter which varies from 700 to 1,000 times that of the sun! The star $\alpha$ Hercules has a diameter 800 times that of the sun.

The diameters of smaller stars cannot be measured with interferometers. They are simply too far away and too small. In some instances, however, double stars revolve about their common center of mass in such a way as to eclipse one another as seen from the Earth. If the motions of the stars are well known, their diameters can be determined by the length of the eclipse (see p. 364). A star's diameter can also be estimated by the Stefan–Boltzmann law if its temperature and luminosity are known. The smallest stars have diameters less than that of the Earth, diameters roughly $\frac{1}{500}$ times that of the sun. Consequently, the range in diameters is rather large; the biggest stars have diameters about 400,000 times the diameters of the smallest stars.

## 10.5 Stellar Temperatures

Stellar temperatures are fundamental in the study of the stars, for when a star's temperature is considered along with its size, we can find how much energy it radiates off into space and hence, for a stable star, how much energy is generated in its core. Fortunately, stellar temperatures are one of the more easily observable characteristics. The range of temperatures is less than that of diameters; the hottest known star, at about 100,000°K, is only about 70 times as hot as the coolest one (about 1,500°K). Since the temperature of the sun is about 6,000°K, it may be classified as one of the cooler stars.

The temperature of a star can be determined by using Wien's law. If the wavelength of the peak of the radiation curve can be determined, the temperature of the photosphere of the star can be calculated.

The temperature of the lower chromosphere, however, can be determined by a study of the spectral lines of the star's spectrum. Two main factors influence the electron transitions that cause the absorption lines in stellar spectra: the temperature and pressure of the gasses involved.

To illustrate this, let us first consider a gas at moderately low pressure whose temperature increases from a comparative coolness (let us say 3,000°K) to a high temperature. When the gas is cool only those transitions that require a little energy will occur and therefore we would expect most of the lines to appear in the red region of the spectrum. In a gas at this temperature it is even possible for some atoms to remain bound to each other, forming molecules. Consequently we would see some molecular bands, such as those of titanium oxide.

As the gas is heated the molecules begin to break up, and transitions occur that require more energy. Consequently the spectral lines (either bright or absorption, as the conditions of the gas dictate) toward the violet end of the spectrum begin to appear and become stronger. Furthermore, some of the atoms begin to lose their outer electrons and as ions produce spectral lines, such as ionized calcium. Finally, when the gas is hot more atoms have become ionized, and the many transitions which require that energy produce spectral lines deep into the ultraviolet.

If we keep this gas at high temperature and only moderately low pressure, the atoms and free electrons are relatively close together; consequently the ions will be able to capture electrons fairly easily and may not remain ions very long. If the temperature is maintained but the pressure reduced still more, the ions and free electrons will be farther apart and a given atom, if ionized, will remain ionized for a longer period of time. Therefore, of two gases at the same temperature, the one with the lower pressure will show more intense lines of ionized atoms.

That a hot star does have a very different spectrum from a cool star can be readily seen in Figure 10-4, where spectra are arranged according

| Type | Star |
|------|------|
| O6 | λ Cephei |
| B3 | η Aurigae |
| A0 | δ Cygni |
| F2 | β Cassiopeiae |
| G2 | η Pegasi |
| K5 | γ Draconis |
| M5 | α Herculis |

**Figure 10-4**

Principal types in the spectral sequence. (Photographs from the Hale Observatories)

to the temperature of the emitting star. On the left hand side of the figure, each spectrum has a letter which is used to designate that particular type. An O-type star is the hottest and the M-type is the coolest.

**a. The Spectral Sequence.**   The letters used to designate this sequence of spectra do not really make any sense: O B A F G K M R N S. It would almost appear as though the astronomer responsible for such a sequence of letters either did not know the alphabet or was simply not concerned with order. Fortunately, neither is the case. The history behind the formation of this sequence of letters helps explain the matter; it also shows that astronomers, like anyone else, have a tendency to become bound by an established system of nomenclature.

When the study of stellar spectra received the impetus of photography it became feasible to try to arrange the different spectra in some kind of order. Since hydrogen lines are present in so many of the spectra it was felt that these lines might be used as a criterion. Accordingly, the stars with the strongest hydrogen lines were called A-type stars, those with slightly weaker hydrogen lines B-type stars, etc. At the time this scale was established the reason for the differences in stellar spectra was not well understood, and only later did it become apparent that they represented differences in stellar temperatures. Further study revealed that an A-type star is not the hottest; a B-type star is hotter, and an O-type star is the hottest of all. So the order of the spectra was rearranged according to temperature, although the letters designating the types were retained except for a few that had to be discarded. Thus, we see that the spectral sequence is basically a temperature sequence.

Since the spectral sequence is so important in astronomy a scheme has been devised to help the beginning student remember it. If we let the letters in the sequence, O B A F G K M (R N S), be the first letters of each word in a well-chosen sentence we get the following: *Oh Be a Fine Girl Kiss Me (Right Now, Smack)!* It is a rare student who forgets this, after even a cursory glance, and therefore it has so far survived the vicissitudes of expression in the English language.

Actually the R-, N-, and S-type stars are a branch of the spectral sequence. These stars have chemical compositions slightly different from the rest and are relatively cool.

**b. Spectral Lines and Spectral Types.**   To explain the dependence of stellar spectra on temperature we can start with the coolest, the M-type star, and follow the changes as we move to the hotter stars. It should be recalled that in a gas of a given temperature the atoms are not all equally excited. Some atoms, because of collisions with other atoms, will have more energy, and some, for the same reason, will have less. The temperature is based on the *average* excitation of the atoms in a gas. This difference in excitation of the atoms in each star accounts for the fact that the spectral (temperature) sequence is a continuous one, free of any sharp breaks.

The M-type stars have a temperature of about 3,400°K, which is too low to excite electrons to very energetic levels. Thus the lines of the most easily excitable atoms (mostly those of the metals, which have many lines in their spectra) predominate the spectrum. Furthermore, the M-type stars are cool enough to allow some of the atoms in their chromospheres to combine and form molecules. The bands of these molecules, especially those of titanium oxide (TiO), are stronger in this type than in any of the other stars.

The K-type stars have a temperature of about 4,400°K. At this higher temperature some of the molecules have been broken up and therefore the molecular bands are not as strong as in the type M. Spectral lines resulting from more highly excited metallic atoms appear, and some metals become ionized at this temperature. In fact since calcium is so easily ionized, the H and K lines of ionized calcium are not only very intense in K-type spectra but are at their strongest here. The M-type star is not hot enough to ionize calcium to such an extent, whereas the hotter stars will begin to ionize calcium doubly or at least cause the outer remaining electron to make transitions that do not result in lines in the visible part of the spectrum.

The hydrogen lines of the K-type star, although weak, are stronger than in the M-type star. Since the first line of the Balmer series ($H\alpha$) is in the red region of the spectrum it may seem logical that it should appear strong in the cooler stars. It must be remembered, however, that this line (in absorption) is the result of a transition from the second to the third energy level and cannot appear until some of the hydrogen atoms have already been excited to the second energy level, which is a process requiring more energy than is generally available in the cooler stars.

The G-type stars have a temperature of about 5,500°K. As the temperature increases to this value, the hydrogen lines become more intense, for more of the hydrogen atoms are excited to the second energy level. The molecular bands essentially disappear except for bands of cyanogen (CN) and CH which, though weak, are still present. The H and K lines of calcium become slightly weaker than in the K-type star, since the outer remaining electron in the calcium atom is excited to levels higher than those that produce the H and K lines.

The F-type stars have a temperature of about 6,600°K. At this higher temperature the hydrogen lines increase in strength, the lines of most ionized metals are intensified, and the neutral metallic lines become weaker as more of the metallic atoms become ionized.

The A-type stars have a temperature of about 9,500°K. In this spectral type the hydrogen lines reach their maximum strength, and the lines of most of the singly ionized metals begin to weaken; the atoms become more highly ionized and produce lines in the far-ultraviolet region of the spectrum. The lines of neutral metals are quite absent from the A-type stars, since most of the metals have become ionized.

The B-type stars have a temperature of about 15,000°K, at which the hydrogen atoms start to become ionized and the lines of the Balmer series thus begin to decrease in strength. The lines of neutral helium appear, as well as those of ionized oxygen and carbon.

The O-type stars have a temperature so high that it is difficult to determine. The estimates run somewhere around 50,000°K or more. The lines that primarily distinguish the O-type spectrum are those of ionized helium, for they appear in no other spectral class.

The temperatures given for the various spectral types are really averages; each spectral class embraces a considerable range of temperatures. For example, the hottest A-type star has a temperature of about 11,000°K while the coolest is at about 8,000°K. This range in temperature gives rise to corresponding changes in the spectra. To account for these differences, each major spectral class is subdivided. There are stars, for example, that are classified as A0, others as A1, A2, A3, etc. An A3 star is cooler than an A2. Such subdivisions within each spectral type are recognized by differences in line intensities, the hydrogen lines being strongest in A0. The sun is a G2 star.

c. **Line Intensities.** Figure 10-5 shows how the line intensities of some of the more important lines or groups of lines vary throughout the spectral sequence. The lines of each element or group of elements reach a peak of intensity in some particular spectral type, that is, at a particular temperature. This temperature is related to the amount of energy needed to excite the atoms of the element in question to the level at which transitions producing these lines are most likely to occur. At this point, then, the lines will be at their maximum intensity. It should be noted once again that the intensity of lines from an ionized element or group increases as the lines for the neutral element or group weaken.

A peculiar terminology has arisen along with the spectral classes. Very often astronomers want to speak not of one particular spectral class but of a group of spectral classes or subclasses. It was once thought that the O-type stars were the youngest and that as stars grew older they cooled progressively into M-type stars. Therefore the stars classed as O, B, and A are loosely spoken of as *early-type* stars, and the K and M stars are referred to as *late-type* stars. An F0 or F1 star is sometimes called an early F star, but an F8 may be referred to as a late F star. This terminology is one of convenience only and has no strict definition or hidden meaning. It will be seen in Chapter 12 that an O or B star is indeed likely to be younger than a type M, although this need not be so.

Stars of very low temperature have recently been discovered and

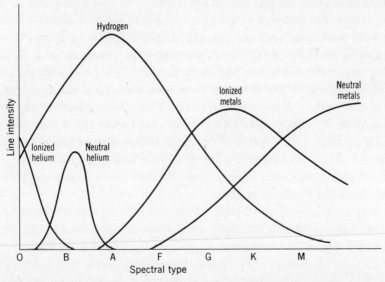

**Figure 10-5**

The intensity of a given stellar spectral line depends on the temperature of the star. Lines resulting from ionized atoms of a given chemical element will be more intense in a hotter star, and the lines of the neutral atoms will be more intense in a cooler star.

more are expected to be found as current researches continue at a number of observatories. These stars radiate mostly in the infrared region of the electromagnetic spectrum. Special equipment has been set up at Mount Wilson Observatory to survey the sky in search of *infrared stars*. Of those found so far, temperatures as low as 1,000°K have been measured. What sort of stars are these? Why are they so cool? These are questions being asked by astronomers.

## 10.6 Stellar Luminosities

The temperature of a star, according to the Stefan–Boltzmann law, determines how much energy is emitted per unit area of the star's surface. Of two stars the same size, the hotter will radiate more energy; of two stars the same temperature, the larger one will radiate more energy. The *luminosity* (intrinsic brightness) of a star, then, depends on two factors: its temperature and its size.

**a. Brightness and Apparent Magnitude.** A star's *brightness*, however, is simply its apparent brightness as seen in the sky and depends not only on its luminosity but also on the distance between it and the sun. If we had three stars of the same luminosity at distances of 1, 2, and 3 parsecs, their brightnesses would progressively decrease with increasing distance according to the *inverse square law*. That is, if the brightness of the first star were 1 (on an arbitary scale), the brightness of the second star would be $\frac{1}{4}$ which equals $(\frac{1}{2})^2$, and that of the third star would be $\frac{1}{9}$ which equals $(\frac{1}{3})^2$. If a star of the same luminosity as these three were 100 parsecs away, its brightness, on the same scale, would be $\frac{1}{10,000} = (\frac{1}{100})^2$. If the luminosity of any one of these were twice as great, its brightness would be twice as great.

The first astronomers concerned enough to make a record of the brightnesses of stars as they saw them were the Alexandrian Greeks. In the absence of a means for determining a star's distance from the sun they considered its brightness only. It was Hipparchus who first established a scale, called *apparent magnitude*, for comparing the brightnesses of stars. He did this in the course of compiling the first known star catalog, which contained the positions and brightnesses of just under 1,000 stars. The brightest stars he arbitrarily called of the *first magnitude*; the faintest stars visible to the naked eye were of the *sixth magnitude*, and all other stars were of intervening magnitudes according to their brightnesses. The system worked quite well, for it is fairly easy to estimate the brightness of a visible star on this scale. But with the advent of the telescope and of photography, the scale has had to be extended to ever fainter stars, until the *faintest* stars now photographed are of about the twenty-third magnitude.

Hipparchus misjudged the magnitudes of some of the brighter stars, however. When the magnitude scale was extended and expressed by a mathematical formula, it developed that the brighter stars are brighter than those of the first magnitude; indeed they are even brighter than those of zero magnitude. The only way to express these hitherto unsuspected magnitudes and yet retain the old scale (thereby avoiding a wholesale revision of stellar catalogs) was to adopt a few negative magnitudes. As a result, Sirius, the brightest star in the sky aside from the sun, has an apparent magnitude of $-1.4$ and the star 61 Cygni A has an apparent magnitude of $+5.4$ and is just visible to the naked eye on a clear moonless night when viewed from a place far from the lights, smoke, and smog of our modern cities.

The upshot is that a magnitude scale set up some 2,000 years ago is still in use, although stretched somewhat on a mathematical rack. After the revisions, a zero-magnitude star is exactly 100 times brighter than a fifth-magnitude star, which is fairly close to Hipparchus' scale. Thus, for a magnitude difference of 5 there is a change in brightness by a factor of 100. Consequently, a difference in magnitude of 1.0 corresponds to a factor in brightness of $(100)^{1/5} = 2.512$. A star with an apparent magnitude of 2.0, for example, is about 2.5 times brighter than a star with an apparent magnitude of 3.0. That star with an apparent magnitude of 2.0 is about $(2.5)^2 = 6.3$ times brighter than a star with an apparent magnitude of 4.0, about $(2.5)^3 = 16$ times brighter than a star with an apparent magnitude of 5.0 etc. A magnitude difference of 10 amounts to a difference in brightness of $100^2$ or 10,000; a magnitude difference of 15 amounts to a difference in brightness of $100^3$ or 1,000,000. Figure 10-6 shows a star field with indications of magnitudes for some of the stars.

There are stars over the entire sky which have been chosen as standard stars for magnitude determination. The magnitude of any star can be determined by comparing it with one of the standard stars which appear in the same part of the sky as the star being studied. The magnitudes are determined by comparing the electric currents set up when the star's light strikes a photoelectric cell. The brighter the star the greater the electric current produced in the cell. If a star whose apparent magnitude

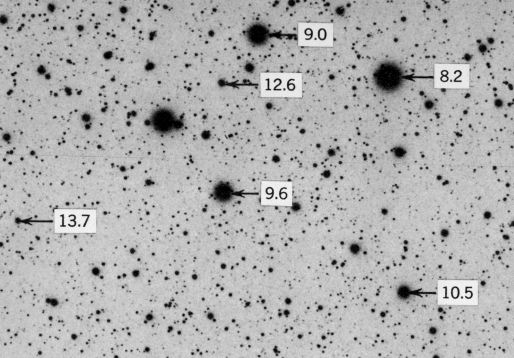

**Figure 10-6**

The same star field shown in Figure 2-13 with apparent magnitudes marked. The size and darkness (astronomers work with photographic negatives) of a star's image can be used to estimate the magnitude. (Courtesy of M. Walker, Lick Observatory)

is to be determined produces an electric current equal to one of the standard stars, it has the same apparent magnitude as that standard star.

**b. Luminosity and Absolute Magnitude.**   Before we can determine a star's *luminosity* (as distinguished from its brightness), its distance must be known. The scales used to measure luminosity are as arbitary as the scale of apparent magnitude, and indeed one of the scales of luminosity, called *absolute magnitude*, is based on the scale of apparent magnitudes. It is an indication of the total amount of light radiated by the star and not the amount we receive here on the Earth. Unlike apparent magnitudes, absolute magnitudes are independent of the star's actual distance. By definition *the absolute magnitude of a star is the apparent magnitude the* *star would have, were the star 10 parsecs from the sun.* On this scale, then, all stars are compared from the same standard distance. If a star is closer than 10 parsecs, its apparent magnitude is numerically less (that is, it is brighter) than its absolute magnitude. If a star is more than 10 parsecs distant, its apparent magnitude is numerically greater (fainter) than its absolute magnitude.

Sirius, for example, is 2.7 parsecs from the sun. Its apparent magnitude is −1.4 and its absolute magnitude is +1.5. This means that if Sirius were moved away to 10 parsecs from the sun its apparent magnitude would be +1.5. Pollux, with an apparent magnitude of +1.1, is just barely more than 10 parsecs from the sun and thus its absolute magnitude is +1.0. On the other hand, Rigel, with an apparent magnitude of +0.14 is 200 parsecs from the sun so its absolute magnitude is −6.4. Since Rigel has an absolute magnitude of about 8 less than Sirius, it is about 1,600 times more luminous. The sun is so close that it has an apparent magnitude of −26.7, even though its absolute magnitude is only +4.9. Rigel's absolute magnitude is therefore 11.3 magnitudes less than that of the sun, that is, it is 33,000 times as luminous.

Paralleling the scale of absolute magnitude is another scale whose reference point is the sun, which is given an arbitrary luminosity of 1.00. A star with a luminosity of 10 would be 10 times more luminous than the sun; one with a luminosity of $^1/_{100}$ is $^1/_{100}$ as luminous as the sun.

The range in the luminosities of the stars is staggering. The most luminous star is about $10^{11}$ times more luminous than the least luminous star. The most luminous stars are about 100,000 times more luminous than the sun, whereas the least luminous stars known have a luminosity about $^1/_{1,000,000}$ that of the sun. Bernard's star (see Figure 10-3) has a luminosity of only 0.00044. With an absolute magnitude of +13.2 and at a distance of only 5.9 light years, its apparent magnitude is +9.5.

For the study of *intrinsic* characteristics of the stars, luminosity (absolute magnitude) has a great deal of meaning, whereas brightness (apparent magnitude) has essentially no meaning because of the different distances involved.

## 10.7 The H–R Diagram

It is of interest to graph the stars by plotting their luminosities against their temperatures (Figure 10-7). Since a star's luminosity depends on both its temperature and size (see p. 272), a graph plotting temperature against luminosity separates the stars according to size. If two stars have the same temperature but different diameters the smaller will be less luminous, since it has a smaller surface area and thus is located lower on the graph. Conversely, if two stars have the same luminosity but different temperatures the cooler must be the larger. Since it is cooler it is located farther to the right on the graph.

Such a graph is called an *H–R diagram* in honor of Hertzsprung and Russell, the two astronomers instrumental in compiling the first such diagram. Because of its tremendous significance in astronomy we shall refer to the H–R diagram frequently.

We notice that there is a great number of stars forming a sequence from the upper left to the lower right of the diagram. This band is called the *main sequence*, because it includes the majority of the stars observed. Above the main sequence we see a rather large group of stars classified as *giants* and *supergiants*. The supergiants, as the name indicates, are more luminous and thus larger than the giants, which are in turn larger than the main-sequence stars of the same temperature. Below and to the

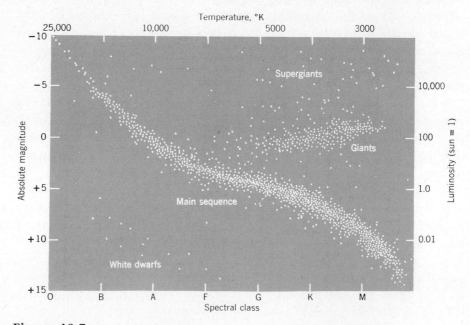

**Figure 10-7**

The H–R diagram.

left of the main sequence, we see a group of stars that must be smaller than the main-sequence stars of the same temperature and thus are called *white dwarfs*. The late main-sequence stars (the sun included) are often called dwarfs because of their smallness when compared with giants of the same temperature. Nevertheless they are larger than the white dwarfs and are not white in color.

That such a distinctive grouping of stars should exist when they are placed on a graph according to their temperatures and luminosities cannot be mere coincidence; there must be some meaning to this arrangement. It is well established that these groupings represent the stages that a star goes through in its evolution from birth to death. We shall examine this process in Chapter 12.

## 10.8 Spectroscopic Parallax

**a. The Inverse Square Law of Brightness.** One of the more beneficial results of the H–R diagram is that it can be employed to find the distances of stars too far away for heliocentric parallax determination. This is possible because it gives us the luminosity of a star, and once the luminosity is known we can find the star's distance by measuring its brightness; for brightness, luminosity, and distance are related very nicely by the inverse square law. By using luminosity units (as against the system of magnitudes) the brightness $B$ of a star in the sky is proportional to the star's luminosity $L$ and inversely proportional to the square of its distance $D$ from the sun:

$$B \propto \frac{L}{D^2}$$

This proportionality can be converted into an equality by inserting a constant of proportionality $K$:

$$B = K \frac{L}{D^2}$$

The numerical value of $K$ depends upon the choice of units for the brightness $B$ and $L$. If $L$ is based on the sun's luminosity, $B$ can be based on a group of stars just as the apparent magnitude is.* This equation, expressed in units of luminosity and brightness, permits us to see clearly how the

---

*An equivalent expression used by astronomers which relates the apparent magnitude $m$ of a star to its absolute magnitude $M$ and its distance $d$ (in parsecs), is logarithmic in nature: $M = m + 5 - 5 \log D$.

brightness of a star depends upon its luminosity and distance. If both the brightness and luminosity of a star are known, its distance can be found by algebraic manipulation of the equation

$$D = \sqrt{\frac{K \cdot L}{B}}$$

But the problem of determining the luminosity of a star without first knowing its distance remains. To do this we must find some way of determining, independently of its luminosity, whether the star is a dwarf (late main-sequence), a giant, or a supergiant. If this can be determined, a star's luminosity can be found by correlating its spectral type with its size. For example, if a K0 star is estimated to be a dwarf, its luminosity must be about 0.4; if a K0 star is a giant, its luminosity will be a little less than 100 (see Figure 10-8). But how is it possible to determine the size of a star from a study of its spectrum?

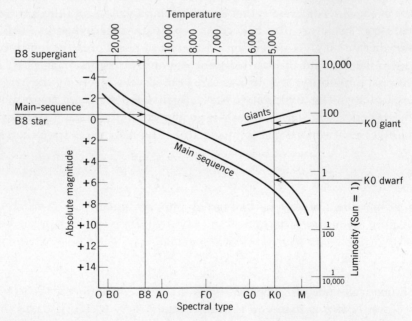

**Figure 10-8**

Spectroscopic parallax depends on the ability of the astronomer to determine from a star's spectrum alone whether a given star is a main-sequence star, a giant, or a supergiant. Once this is determined the H–R diagram yields the star's luminosity. The temperatures refer to the main sequence.

**b. Pressure Broadening and Luminosity.** It was pointed out that the diameters of stars vary by a factor of more than 100,000, but, as we shall see later in this chapter, their masses vary by a factor of only about 700 or so. Thus, the very large stars are large in size only and are not as massive as their sizes might suggest. They must, therefore, be expanded versions of the more numerous main-sequence stars. Because they are enlarged they must have smaller densities and thus lower pressures. It seems likely that this difference in pressure would affect the spectrum by producing different amounts of pressure broadening of the spectral lines (discussed on p. 61).

By placing the spectra of a main-sequence star and a supergiant of the same spectral type side by side, we do indeed find differences in the spectra (Figure 10-9). Not only are some of the lines of the main-sequence star's spectrum a little broader, but also some of the lines of ionized metals in the spectrum of the supergiant are more intense.

Both these results are to be expected. The main-sequence star has a higher pressure and some of its lines suffer some consequent pressure broadening. The supergiant has a lower pressure and more of its ions are able to remain ions; they and the electrons are farther apart, and thus it is more difficult for them to recombine and form neutral atoms. Since the intensity of ionized lines is one of the criteria for spectral classification and since ionization depends not only on the temperature of the star but also on the pressure, the temperature of a giant star is lower than that of a main-sequence star of the same spectral type. The giant star has more intense lines of the ionized atoms and is classified accordingly. A K0 dwarf, for example, has a surface temperature of 5,000°K; the temperature of a K0 giant, however, is 4,100°K, and of a K0 supergiant only 3,800°K.

It is therefore possible to distinguish between a dwarf, a giant, and a supergiant by an examination of their spectra. Once a spectrum has been classified into its spectral type and examined to determine whether it

B8 Supergiant

B8 Main–sequence star

**Figure 10-9**

The spectral lines in a B8 main-sequence star are noticeably broader than those in a B8 supergiant. The most obvious lines in each of these spectra are those of hydrogen. (Yerkes Observatory)

originates in a dwarf, a giant, or a supergiant, the star can then be placed on the H–R diagram. Once placed on the H–R diagram according to its ,spectral type and size, its luminosity can be read off directly.

The spectra shown in Figure 10-9 are both of B8 stars, one a main-sequence star and the other a supergiant. By referring to the H–R diagram (Figure 10-8) we see that the luminosity of a main-sequence B8 star is about 100. The supergiant, on the other hand, has a luminosity of about 30,000. Actually there is no sharp dividing line between the giants and the supergiants; the extent of pressure broadening and the extent of ionization can be used to classify the star with greater accuracy than merely placing it in one of only two classifications—giant or supergiant.

Once the luminosity has been calculated, we must obtain the brightness, either from a catalog of stars that lists the brightness or apparent magnitude, or by making direct observations of the star with either a photocell or photographic plate. Then, knowing both the luminosity and the brightness of the star, we can calculate its distance by means of the inverse square law given on p. 278.

Since this method of distance determination is based on a study of the spectrum, it is called the method of *spectroscopic parallax*. It has proved a very powerful tool in the astronomer's hands, although it involves certain difficulties. One difficulty is the determination of luminosity from the spectral lines. A slight error in judgment of line width and intensity means a large error in luminosity. Another difficulty lies in the fact that the inverse square law assumes that no light has been lost in transit from the star. But this is not always true, as will be seen in Chapter 15.

The determination of a star's spectroscopic parallax is limited mainly by the quality of the spectrum we can obtain. If a good spectrum can be obtained, the star is certainly bright enough for an accurate measurement to be made of its brightness, since more light is required to produce a good spectrum than to activate a photocell or to take a direct photograph.

## 10.9 Luminosity, Temperature, and Size

The luminosity of a star depends on both the temperature and the diameter of the star; therefore the luminosity and the temperature can be used to obtain some idea of stellar sizes for stars that are too small or too far away to be measured with the beam interferometer. If we assume that all stars radiate as nearly like a perfect radiator as does the sun, then we can determine their size by using the Stefan–Boltzmann law. We first need the luminosity of the star, which can be obtained from its brightness and distance or from its position on the H–R diagram. From this we know the total amount of energy it radiates into space. From the spectrum of the star we can learn its temperature which, with the help of the Stefan–Boltzmann law, tells us how much energy it radiates *per unit area* of its surface. If we know both how much energy the star radiates per unit area and the *total* amount of energy it radiates, then by simple division we can find the total surface area of the star. Assuming the star to be spherical (and the vast majority are spherical), we can then determine its diameter from the geometry of a sphere (Figure 10-10).

This method of determining stellar diameters is not without its difficulties. Not only is there likely to be an error in the calculation of luminosity, but also some stars may not radiate as closely to a perfect radiator as

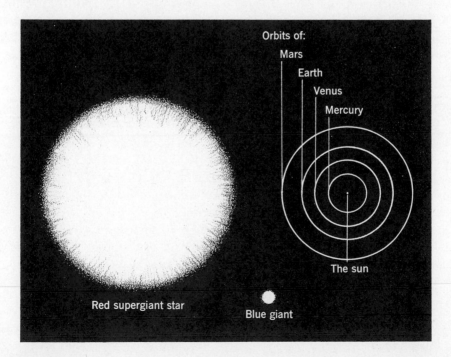

**Figure 10-10**

The relative sizes of some giant stars, the sun, and the orbits of the four terrestrial planets.

does the sun. Nevertheless, the method is an important factor in astrophysics, for it at least yields some idea of stellar sizes; and if we are to describe the stars we certainly need to know how large they are.

## 10.10 Mass, Luminosity, and Density

One of the most important and yet one of the most difficult characteristics to determine is the mass of a star. The only opportunity for making direct measurements of stellar mass is afforded by a star that is a member of a double-star system. Since the gravitational field about a star depends

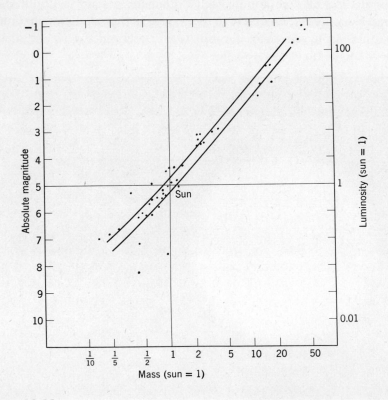

**Figure 10-11**

The mass–luminosity relation indicates that stars of increasing mass will be more luminous.

solely on its mass and since the motions of the two stars in a double-star system depend on their mutual gravitational attraction, their masses can be determined if we can describe their motions. This process is explained in Chapter 13; we may, however, discuss a closely related point at this time.

After the masses of a number of stars had been determined a close correlation between mass and luminosity was recognized. The more massive a star, in general, the more luminous it is. This *mass–luminosity relation* gives us a rough method for determining the mass of an isolated star. The mass thus determined is only an approximation, for we can see in Figure 10-11 that the stars plotted scatter about a mean curve that must be used to determine the mass. Since there is considerable deviation from this mean curve, the mass of any one star is always in doubt. When a large number of stars is used, however, the method can be employed very effectively to determine the average mass of a particular *group* of stars. For example, the masses of the main-sequence stars of each spectral type have been determined by this method and are given in Table 10-1, from which it appears that the masses vary less than the other characteristics.

Table 10-1 ● The Masses of Main-Sequence Stars

| Spectral type | Mass | Spectral type | Mass |
|---------------|------|---------------|------|
| B0 | 16 | G0 | 1.0 |
| B5 | 6 | G8 | 0.9 |
| A0 | 4 | K0 | 0.8 |
| A5 | 2 | K5 | 0.6 |
| F0 | 1.5 | M0 | 0.5 |
| F5 | 1.3 | M5 | 0.2 |

Since direct determination of mass can be made of only those stars that are members of binary systems, the mass-luminosity relation is based on such stars. If single stars differ from stars that are members of binary systems, this empirical mass-luminosity relation has limited value. It is known that some types of stars (for example, white dwarfs) do not follow the mass-luminosity relation and there are even special classes of binary stars that violate it. It is generally held, however, that most stars do follow this relation, which has become an important part of stellar astronomy.

The most massive star known is an O8 star which is a member of a double-star system. The name of that star is HD 47129 and each star in the system has a mass close to 50 times that of the sun. The least massive star known is less than 0.1 solar mass.

Once we know both the mass and the diameter of a star we can calculate its average density, for the density is equal to the mass divided by the volume. The densities of stars vary over a considerable range. The highest average densities are found in the white dwarfs (the subluminous stars) which are as much as 100,000 times more dense than water. A quart of material taken from such a white dwarf would weigh about 200,000 lb if placed here on the Earth. Such fantastic densities scarcely seem believable yet they have been verified both by many observations and from a theoretical point of view.

Densities as high as this can be found only in *degenerate* matter (see p. 343 ff), a state of matter that can best be understood through a consideration of atomic structure. A nucleus has a diameter of the order of $10^{-13}$ cm, whereas an atom has a diameter of the order of $10^{-8}$ cm. Thus an atom's diameter is about $10^5$ times that of the nucleus and its volume is $10^{15}$ times that of the nucleus; that is, an atom has sufficient volume to enclose 1,000,000,000,000,000 nuclei! In other words, atoms are composed mostly of empty space between very tiny atomic particles. If this space were eliminated atomic particles could be packed much more closely together, and densities could reach prodigious values.

Under the tremendous pressures encountered in white dwarfs the atomic structure does break down; the electrons and nuclei are packed together much more closely than is possible when the electrons occupy their usual energy levels. The space that occupied most of the atom has been largely eliminated and the same amount of matter that goes to make up a star like the sun, for example, can be packed into a volume less than that of the Earth.

On the other end of the scale of densities are the supergiants, which are very "inflated" stars. They have mean densities that would be considered a fairly good vacuum here on the Earth.

## 10.11 The Luminosity Function (or Stellar Census)

The question of stellar mass is very important, for on it hinges not only the process of stellar formation but also the life of the star. If stars form (as was intimated in the discussion on the origin of the solar system) from gaseous nebulae, this process of formation can be better understood if the astronomer knows the minimum amount of material necessary to form a star as well as the maximum. Furthermore, he wants to know if it is the less massive or the more massive type of star whose formation is more probable. But since the mass of an individual star can be determined directly only if it is a member of a double-star system, we must resort to luminosities to study how the stars in general vary in mass.

By making a graph of the number of stars of a given luminosity, there was formed what is called the *luminosity function*, which tells us what proportion of the stars have a given luminosity (Figure 10-12). It can be seen from the figure that the faint stars are much more numerous than the bright. Thus, the less massive stars are much more numerous.

Does this mean that they form more readily and more frequently from the nebulous material? Or does it mean that they are more numerous because they have a longer life? Houseflies are very abundant because they are born in quantities too numerous for comfort, despite the fact

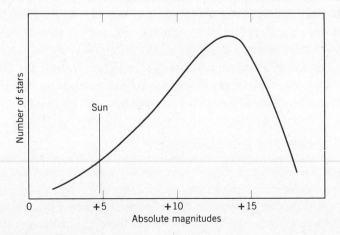

**Figure 10-12**

The luminosity function indicates that the less luminous stars are more numerous than the very luminous ones.

that they live for only a few weeks. Human beings, on the other hand, are numerous because they have a life expectancy of 65 to 70 years, even though their birth rate is so much lower than that of the fly. The answer to the question "Why are less massive stars so numerous?" may partly depend on their longer life (see Chapter 12). Whether they are more readily formed is a question that has yet to be answered, although it does appear that they are.

The formulation of the luminosity function was not an easy task. Essentially it involved a stellar census; that is, the number of stars in a given volume of space had to be counted and the luminosity of each determined. We are handicapped, however, for although we may know how many of each kind of star exist in the neighborhood of the sun (within a sphere whose radius is, say, 16 light years) it is difficult to go much further. Not only do heliocentric parallaxes fail farther out but also the less luminous stars soon become too faint to be seen.

Consequently, the portion of the curve dealing with the low-luminosity stars was formed from the region immediately about the sun—the "local swimming hole" as the late Walter Baade, of the Mount Wilson and Palomar Observatories, so quaintly described it. Unfortunately, however, the local swimming hole contains no very luminous stars. Hence, the portion of the curve dealing with very luminous stars had to be formed from studies of a region of space beyond the reach of the method of heliocentric parallax. Their distances were obtained by the method of spectroscopic parallaxes as well as by methods to be discussed in Chapter 15. Because of these inherent difficulties, the luminosity function represents only two regions of space: within about 16 light years of the sun for the least luminous stars and beyond this up to within about 1,500 light years for the most luminous. It does not represent the entire galaxy.

## 10.12 Chemical Composition

The last characteristic, chemical composition, is not startling; it simply answers the question, "What are stars made of?" No matter what substances may compose it, we are fairly certain that every star is gaseous throughout, even the white dwarfs with their fantastic densities. It is possible, although difficult, to determine the chemical compositions of some of the stars' atmospheres from an analysis of their spectra. But if we want to go inside a star, we must, for obvious reasons, forgo direct observation and use instead the fruits of theoretical studies.

These studies are in the forefront of astronomical thought. The conclusions reached indicate that stars begin life as balls of gas composed mostly of hydrogen. As they generate energy by converting hydrogen into helium, the amount of hydrogen decreases and the amount of helium increases. As they age, the helium is converted into the heavier elements. Thus a star's chemical composition depends on its age.

There are some stars, however, that have in their atmosphere an unusual abundance, as yet unexplained, of one element or another. Examples are the carbon stars of the spectral classes R and N, and the S-type stars with their relatively high percentage of zirconium and technetium. These are only some of the many deviant stars whose behaviors do not fall within the confines of this chapter's description of "normal" stars.

## BASIC VOCABULARY FOR SUBSEQUENT READING

Absolute magnitude
Angle of parallax
Apparent magnitude
Degenerate matter
Giant star
Light year
Luminosity
Main sequence

Parsec
Peculiar velocity
Proper motion
Spectroscopic parallax
Supergiant star
Tangential velocity
White dwarf

## QUESTIONS AND PROBLEMS

1. Given the angle of parallax, find the distance of the following stars in parsecs:
   (a) 61 Cygni $\quad p = 0.29$ second of arc
   (b) Capella $\quad p = 0.07$ second of arc
   (c) Barnard's Star $\quad p = 0.54$ second of arc

2. Find the distance in light years of each star in question 1.

3. A star has a proper motion of 0.5 second of arc per year. How many years will it be before the position of that star on the celestial sphere changes by $1°$?

4. From the absolute magnitudes given, determine how much more luminous one star is than the other.
   (a) Beta Canis Majoris, $M = -3.6$
   Mira (Omicron Ceti), $M = -2.6$

(b) Alpha Centauri, M = +4.1
Delta Velorum, M = +0.1
(c) Wolf 294, M = +10.9
Castor (Alpha Geminorum), M = +0.9
(d) Theta Aurigae, M = −0.2
Beta Cassiopeiae, M = +1.8

5. Find the absolute magnitude of the following stars whose apparent magnitudes $m$, and distances $d$, are:
(a) $m = 6.0$; $d = 5.0$ parsecs (log 5.0 = 0.70)
(b) $m = 6.0$; $d = 50.0$ parsecs (log 50.0 = 1.70)
(c) $m = 12.0$; $d = 5.0$ parsecs (log 5.0 = 0.70)
(d) $m = 12.0$; $d = 50.0$ parsecs (log 50.0 = 1.70)

6. Two stars have the same luminosity but one has a brightness in the sky equal to nine times that of the other. If the brighter star is 5 parsecs from the sun, how far away is the other star?

7. Describe how the distance of a star can be estimated from a study of its spectrum.

8. Describe how the following stellar characteristics are related:
(a) temperature, size, luminosity
(b) luminosity, brightness in our sky, and distance from the sun
(c) mass, luminosity, and temperature (for main-sequence stars)
(d) mass and diameter (for main-sequence stars)
(e) distance from the sun and proper motion

9. From the following list of stars plot an H–R diagram.

1. Sirius, A1, M = +1.5
2. Alpha Crucis, B1, M = −2.7
3. Ross 775, M4, M = +11.3
4. Beta Pegasi, M2, M = −0.9
5. 82 Eridani, G5, M = +5.3
6. Sirius B, A5, M = +11.4
7. Polaris, F8, M = −3.4
8. Rigel, B8, M = −6.5
9. Procyon, F5, M = +2.6
10. Canopus, F0, M = −5
11. Epsilon Indi, K5, M = +7.0
12. Antares, M1, M = −4

## FOR FURTHER READING

Kruse, W., and W. Dieckvoss, *The Stars*, The University of Michigan Press, Ann Arbor, Michigan, 1957.

Page, T., ed., *Stars and Galaxies*, Prentice-Hall paperback, Englewood Cliffs, N.J., 1962.

Struve, O., and V. Zebergs, *Astronomy of the 20th Century*, Crowell Collier and Macmillan, New York, 1962, Chapters X–XIII.

Anderson, J. H., "The Stars of Very Large Proper Motion," *Sky and Telescope*, p. 76 (Aug. 1969).

Barton, R., "Some Effects of Stellar Proper Motions," *Sky and Telescope*, p. 4 (Jan. 1963).

Boyce, P. B., and W. M. Sinton, "Infrared Spectroscopy With an Interferometer," *Sky and Telescope*, p. 78 (Feb. 1965).

Brown, H. R., "The Stellar Interferometer at Narrabri Observatory," *Sky and Telescope*, p. 64 (Aug. 1964.)

Hack, M., "The Hertzsprung–Russell Diagram Today," *Sky and Telescope*, part I, p. 260 (May 1966); part II, p. 332 (June 1966).

Mumford, G. S., "Distance Modulus," *Sky and Telescope*, p. 274 (May 1965).

Nielsen, A. V., "Ejnar Hertzsprung—Measurer of Stars," *Sky and Telescope*, p. 4 (Jan. 1968).

Sitterly, B. W., "Symposium on Abundances of Elements in Stars," *Sky and Telescope*, p. 11 (Jan. 1965).

Whitney, C. A., "New Trends in Spectral Classification," *Sky and Telescope*, p. 356 (June 1968).

"Extremely Cool Stars," *Sky and Telescope*, p. 195 (Oct. 1965).

"Star Sizes Measured," *Sky and Telescope*, p. 139 (March 1968).

CHAPTER 11 CHAPTER

# NONSTABLE STARS

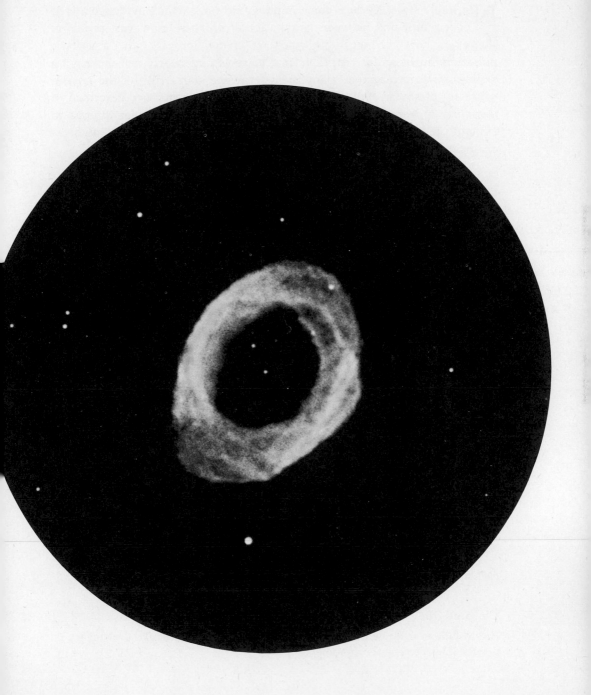

Nonstable stars, as their name implies, are stars which evidence a lack of equilibrium between gravitational pressure and gas pressure. This evidence is a change in brightness which may indicate either a periodic pulsation, an irregular eruption, or a massive and nearly self-destroying explosion. It may be that many main-sequence stars will, by the normal processes of stellar evolution, become unstable in a manner described in this chapter. For this reason, nonstable stars are of particular interest to us; and, since their study is fraught with difficulties, their problems are a challenge.

## 11.1 Cepheids

The most important class of these nonstable stars is called the *Cepheids*. The group derives its name from its most prominent member, Delta Cephei, a naked-eye star that misbehaves in that its luminosity varies in a periodic fashion. This variation of luminosity is detected by a variation in brightness or apparent magnitude. The apparent magnitude of Delta Cephei varies from about +4.3 at minimum brightness to about +3.6 at maximum. A variation of 0.7 in apparent magnitude amounts to a factor of almost 2 in the brightness. Thus, Delta Cephei at its maximum brightness is nearly twice as bright as at its minimum.

**a. The Light Curve.** The light variation of all the Cepheids is periodic in nature, repeating itself over a certain interval of time with only small deviations from the observed pattern. When the pattern of variation for a given Cepheid has been determined from repeated observations, it is possible to predict with considerable accuracy when it will be at its maximum brightness. Indeed, it is possible to predict when it will have any given magnitude within its range of variation. Such a prediction may be made from a graph on which the variation in brightness is plotted against time. The resulting graph is called the *light curve*.

The light curve for Delta Cephei is shown in Figure 11-1. The interval of time elapsing between two successive maxima or minima is called the *period*. Delta Cephei's light curve shows that its period is close to $5\frac{1}{2}$ days.

### Chapter Opening Photo

An unstable star ejected gases which now form the Ring nebula, a planetary nebula. (Lick Observatory photograph)

A light curve made from observations over a longer time interval enables us to measure its period at 5 days, 8 hours, 46 minutes, 38 seconds. But not all Cepheids exhibit this period of light variation. The shortest period for a Cepheid is just over a day and the longest is more than 50 days.

Since the periods are longer than 1 day, it is not possible to follow the variation of one entire period on any one night. The light curve, therefore, must be determined from many observations of brightness over many nights; all these observations must then be fitted onto a curve that can account for the past variations and be used to predict future ones. Thus it may take many months to obtain the light curve of a Cepheid with a period of 45 days.

**b. The Velocity Curve and Pulsations.**  The cause of light variations cannot be determined from the light curves alone; for this reason the spectra of the Cepheids have been investigated quite thoroughly and with startling results. There is not only a periodic change in the spectral lines but there is also a shifting of the lines; they oscillate back and forth across a mean position. Such a shifting of lines has been interpreted as a velocity or Doppler shift. If this interpretation is correct, and all indications are that it is, then there must be some periodic motion of the star; the star must either revolve in an orbit or pulsate. Both possibilities have been investigated, and astronomers have come to the conclusion that apparently the Cepheids are nonstable stars and pulsate like balloons that are partially deflated, then inflated, then partially deflated again, and so forth. Thus, the surface facing us periodically approaches and recedes from us, thereby causing the spectral lines to shift periodically to the blue and to the red.

When we plot a Cepheid's variations in velocity against time we obtain

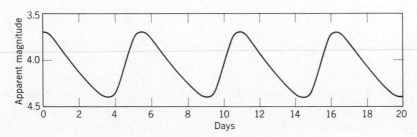

**Figure 11-1**

A schematic drawing of the light curve of Delta Cephei.

*a velocity curve* for the star. Figure 11-2 shows not only the velocity curve but also the light curve for Delta Cephei; when both curves are shown together it becomes apparent that one is the mirror image of the other. The light curve is at a maximum when the spectral lines are shifted farthest to the blue, and at a minimum when the spectral lines are shifted farthest to the red. But the spectral lines are shifted farthest to the blue (position 1) when the star is *expanding* at its maximum rate—not when the star is largest. Correspondingly, the lines are shifted farthest to the red (position 3) when the star is *contracting* at its maximum rate—not when it is smallest. Thus the star is brightest not when it has the largest diameter, but when it is in the stage of most rapid expansion. The star is faintest not when it is smallest but when it is contracting the fastest. We conclude that its changing size is not the principal cause of the light variations.

In view of this let us investigate temperature, the other factor influencing luminosity, to see whether its behavior might not be a more important cause of light variations. From its light we can see that a Cepheid is usually hottest when it is at maximum brightness and coolest at minimum brightness. This variation in temperature is accompanied by changes in the star's color that can be determined by means of a photocell and the proper use of color filters. The star is bluest when it is hottest. As we might expect, these changes in temperature also result in changes of the star's spectral type. Maximum brightness is accompanied by an earlier spectral type. For example, a Cepheid whose spectral type is F5 at maximum is likely to have a spectral type of F9 at minimum.

The change in radius of the star amounts to about 10% of its average radius. The change in temperature is about the same, roughly 600°K for a star whose surface temperature is 6,000°K. Since the Cepheids are brightest at maximum temperature, however, and at that same time are only intermediate in size, we can conclude that the pulsations are complex.

The cause of pulsation of a Cepheid is being investigated by theoretical studies of stellar interiors. It is becoming clear from these studies that as the hydrogen in the core is converted into helium, other changes occur as well. Eventually the helium starts to burn by converting into heavier elements. These changes cause the core of the star to change in size, and this in turn causes the outer layers to become unstable. The pulsation starts as a result of this instability, but the core itself does not take part in the pulsation. According to these theoretical studies, the pulsations

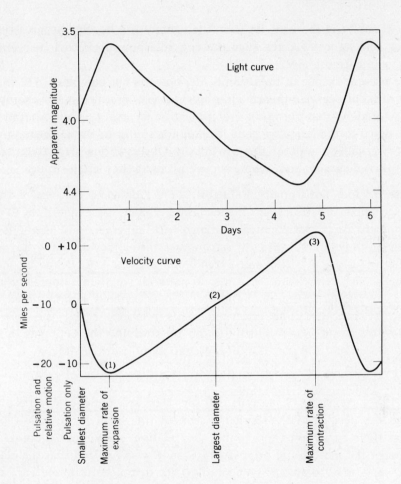

## Figure 11-2

The light curve (top) and the velocity curve (bottom) of Delta Cephei. The velocity curve shows the radial velocity changes due to pulsation alone, and since the star itself is moving toward the sun with a velocity of − 10 miles per second, its radial velocity relative to the sun is also given. The star is expanding whenever its radial velocity is between − 10 and − 20 miles per second; it is contracting whenever its radial velocity is between − 10 miles per second and 0. It is expanding at its maximum rate at point (1) and contracting at its maximum rate at point (3). The star ceases expanding and begins to contract at point (2), where the Doppler shift would change from a violet to a red shift if the star were at rest relative to the sun.

begin about $\frac{2}{5}$ of the way out from the center, and become progressively greater at distances greater from the center. They reach their maximum in the star's atmosphere.

The present theory of pulsation of Cepheids is not complete, but so far the agreement between theory and observation is gratifying. For example, theory predicts that the period of pulsation in any Cepheid should be about equal to the length of time it takes for sound to travel through the star. Observations support this prediction. Observations place these stars in the H–R diagram about where theory predicts they ought to be.

c. **The Period–Luminosity Relation.** The Cepheids, because of their light variation, constitute one of the classes of stars most important to the astronomer. Since astronomical distances are so great, and methods of measuring them are so difficult, astronomers are always looking for new and better ways to verify and extend our present knowledge of these distances. The Cepheids offer a unique method, for their luminosity and period are related so that once the period of a Cepheid is determined, its luminosity can easily be estimated. This relation, called the *period–luminosity relation*, was originally found by studying the Cepheids in the *Small Magellanic Cloud*, a large aggregation of stars visible as a faint patch of light in the southern hemisphere not too far from the South Celestial Pole.

In 1912, Henrietta S. Leavitt, of the Harvard College Observatory, plotted the periods of the Cepheids seen in the Small Magellanic Cloud against their brightness and found the relation shown in Figure 11-3: the longer the period, the brighter the star. It was realized that the cloud has a very small diameter compared to its distance from the sun and consequently all the stars in the cloud could be considered as being the same distance from the sun without introducing much error into measurements that compared the luminosities of these stars. Thus the period–*brightness* curve discovered by Miss Leavitt should mean that a period–*luminosity* curve exists, and if the luminosity of even one Cepheid could be determined along with its period, the luminosity of each Cepheid in the Small Magellanic Cloud could be determined by reference to this standard Cepheid. That is, the scale on the left of Figure 11-3 could be converted from apparent magnitude (brightness) to absolute magnitude (luminosity). If the luminosity of each Cepheid were known, the distance of the cloud from the sun could be determined by the inverse square law, as could the distance of *any* Cepheid in the sky whose period was known.

Unfortunately, there is no Cepheid close enough to the sun for a helio-centric parallax determination. The problem of finding the luminosity of a Cepheid or group of Cepheids took on major interest for astronomers. The explanation of the method used to find their absolute magnitudes will have to wait, however, until Chapter 15. But needless to say, once the absolute magnitudes of a number of Cepheids were finally determined, the distance of every Cepheid seen in the sky could be calculated. It was mainly through this discovery that the astronomer began to realize that the universe is really much larger than he had previously thought, and he began to speak in terms of millions of light years. Before that time (in the mid-1920's) it was not known with any certainty whether the so-called "spiral nebulae" were a part of our galaxy or were outside and enormously farther away and larger.

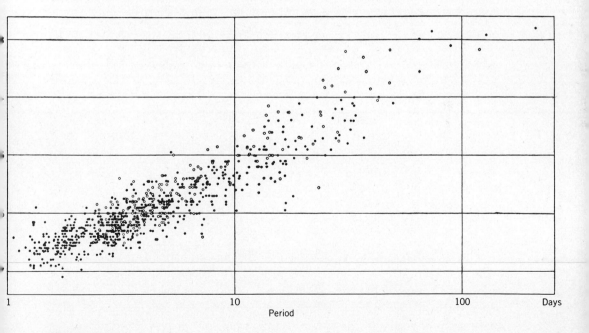

**Figure 11-3**

The period–luminosity relation for the Cepheids of the Small and the Large Magellanic Clouds. (Modified by permission from *Astrophysics*, edited by J. Hynek, Copyright 1951, McGraw-Hill Book Co.)

**d. Two Populations of Cepheids.** As fruitful as this discovery has been, all the distances based on the Cepheids had to be revised. In the early 1950's, Baade realized that stars can be divided into two basic types which he called Population I and Population II. Some of the Cepheids are Population I and others are Population II*, and each has a different period–luminosity relation.

It developed that Population I Cepheids are about 1.5 magnitudes more luminous than Population II Cepheids and the Cepheids studied by Miss Leavitt are actually more luminous than was earlier assumed. If they are more luminous, they must be farther away than the original calculations had led astronomers to believe: a magnitude difference of 1.5 amounts to a factor of almost 2 in distance. In fact this discovery indicated that the universe outside our own galaxy is twice as big as we had previously thought.

A neat clean period–luminosity relationship seems almost too good to be true. And so perhaps it is. Recent studies indicate that although there is a relationship between the period and the luminosity of Cepheids, it is complicated by the surface temperature of the stars. Thus, two stars with the same period but different surface temperatures will have slightly different luminosities. The result is that the period–luminosity relationship is spread out into a band about 1 magnitude thick, for both Population I and Population II Cepheids.

Now that the luminosities of the Cepheids are known, it is instructive to plot them on the H–R diagram. (Since both their luminosities and their spectral types vary, the average of the extremes is used for such purposes.) Figure 11-4 shows the position of the Cepheids (along with some other variables to be discussed in this and later chapters) on the H–R diagram. The Cepheids' luminosity places them in the giant and supergiant class, which is fortunate because they are visible even at great distances from the sun.

The H–R diagram shows also that among the Cepheids there is a relationship between spectral type and luminosity; the later the spectral type, the more luminous the star. An F-type Cepheid has a luminosity of about 100, and a G-type has a luminosity of about 10,000. At first glance this may seem contrary to the Stefan–Boltzmann law until we recall that this law only refers to luminosity *per unit area*. Thus, the later-type Cepheids must be very much larger than those of early type.

---

*The nature of stellar populations will be discussed in more detail later.

There is, then, a very definite relationship between the period of the Cepheids and a number of their other characteristics. An increase in period is accompanied by: (1) an increase in luminosity; (2) an increase in diameter; (3) a decrease in temperature; (4) a larger range in temperature variation; and (5) a larger range in luminosity variation. That all of these are so related is significant and must depend on the masses and sizes of the various Cepheids.

## 11.2 The RR Lyrae Stars

There are other stars related to the Cepheids in that they, too, appear to pulsate, and yet they differ enough to be classified separately. The first of these to be discussed are called the *RR Lyrae* stars after one of their

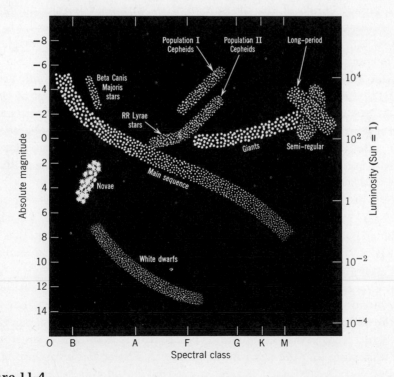

**Figure 11-4**

A schematic H–R diagram showing the location of many of the nonstable stars in relation to the main sequence and the giants.

members. Because they are often found in globular clusters (see p. 382 ff) they are sometimes called cluster-type variables. These stars belong to Population II and have periods ranging from about $1\frac{1}{2}$ hours to a little over 24 hours. The longer-period RR Lyrae stars blend into the type II Cepheids, since there are several stars that could fall into either group. The most luminous RR Lyrae stars have an absolute magnitude of 0.0 (close to 100 times the luminosity of the sun), and the least luminous have an absolute magnitude of $+1.2$ (Figure 11-4); the average is about $+0.6$. As with the Cepheids, the range in luminosity for a given period is attributed to differences in surface temperature.

The greater luminosity of the Cepheids results in their being seen at much greater distances than the RR Lyrae stars. The Cepheids can be used as distance indicators outside of our Milky Way Galaxy; RR Lyrae stars can be so used only inside our galaxy and its immediate environs. But, despite problems yet to be resolved, both groups of stars have been indispensable links in the measurement of distances within our galaxy and throughout the visible universe.

Most RR Lyrae stars are early A-type stars at maximum brightness, and become early F-types at minimum brightness.

Spectral classification of these stars, however, is rather difficult because of certain anomalies in their spectra. One such anomaly is that at certain phases during the period of an RR Lyrae star, double hydrogen spectral lines are observed. That is, each line that normally appears as one becomes two. A doubling of spectral lines can be explained by assuming that there are two layers of gas expanding and contracting out of phase with each other, each causing a different Doppler shift.

This speculation has been further strengthened by the observation of hydrogen emission lines at a time when the two layers of gas should collide. That hydrogen gas should emit bright lines at this time has been verified by aerodynamical experts who have recorded similar emissions from gases in powerful shock tubes.

RR Lyrae stars were one of the stepping stones toward the identification of Population I and II stars and, consequently, to the new distance scale. On the old distance scale the Andromeda galaxy, a neighboring galaxy (see Figure 16-3), was considered to be about 750,000 light years from the sun. Assuming that the Andromeda galaxy is similar to our own, if it were 750,000 light years away, Baade knew that he should be able to detect RR Lyrae stars in its midst; he became suspicious, however, when not a single RR Lyrae star could be detected. Detailed photographs and studies revealed that the nucleus of Andromeda contained no blue giant stars, only red giants; in its spiral arms Baade found blue giants and relatively few red giants. Their preferred location indicated to him that these stars were different in some fundamental respect, a difference he indicated by naming the blue giants Population I stars and the red giants Population II. It is this work that first pointed the way to a division of stars into various populations. The knowledge that the Cepheids are among the stars thus divided enabled astronomers to redetermine the distance of the Andromeda galaxy at about 2.2 million light years, too far for its relatively faint RR Lyrae stars to reveal themselves on photographs.

## 11.3 Long-Period Variables

*Long-period variables* form another group of nonstable stars. The stars in this group are red giants and have periods ranging from about 90 to over 700 days, with the majority around 300 days. Their light curves do not repeat themselves as accurately as those of the Cepheids or the RR Lyrae stars, and there are irregularities not only in the shape of such a light curve but also in the period of the star. The reasons for these irregularities are not known, but it has been suggested that they might be connected with the large size and low density of these red giants, whose diameters are several hundred times that of the sun and whose average densities are only about $\frac{1}{1,000,000}$ the density of water.

The range of luminosity for the long-period variables is about 5 magnitudes, considerably larger than for the Cepheids. Thus a star at maximum will be about 100 times brighter, visually, than at minimum. We know of at least one star of this type whose luminosity varies by a factor of 4,000.

Such a large change in the brightness of some stars accounts for the early discovery of this class of stars. In 1596 David Fabricius, a clergyman of East Friesland (now in The Netherlands), noticed a red star in the constellation of Cetus (the whale) that had not been seen before. In the next few months the star faded and finally disappeared from view. It was seen again a few years later, but no connection was made between these two isolated observations. It was not until 1667 that the regularity of its changes in brightness was observed. It has been observed with increasing interest since then. It was called *Mira* "The Wonderful" by the early observers, and even though it is known more technically as Omicron Ceti, the name Mira is still used.

The long-period variables have periods of more than 200 days, and temperatures that range from 3,400°K at maximum brightness to 1,700°K at minimum brightness, although the temperature of any one star does not have such a large temperature variation. Because the temperatures are so low at minimum, these stars nearly drop out of sight. Figure 11-5 indicates that stars with temperatures lower than 2,000°K do not emit much visible light. Most of the energy emitted by these stars is in the infrared region of the spectrum, and they can be effectively studied by using equipment sensitive to infrared.

Mira, itself, has a period of light variation of 331 days, or about 11 months, and is visible to the naked eye for about half of this period. At maximum it is a second magnitude star or brighter and consequently easily observable. It is a favorite of many amateur astronomers, and the changes in its spectrum make it a favorite of many professional astronomers as well.

Since the long-period variables are so cool, molecular bands appear in their spectra; in fact, a number of these stars are classified as R-, N-, or S-type stars. During minimum temperature the intensity of their molecular bands increases to such an extent that they effectively block

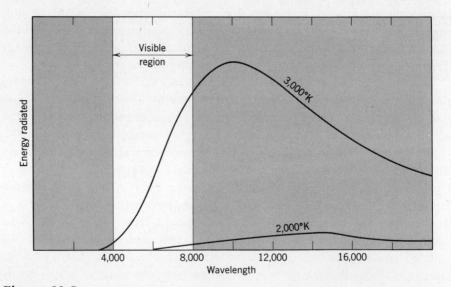

**Figure 11-5**

The radiation curves of stars at 2,000° and 3,000°K showing the large difference in the amount of visible light each radiates.

out a considerable part of the visible spectrum. Long-period variables have been likened to a star surrounded by a molecular cloud or cloak which obscures the star beneath. Their spectra will often show bright lines near the time of maximum luminosity, but these lines fade as the luminosity decreases. It has been suggested that these bright lines may be caused by a series of shock wave phenomena that move out from the center of a star. The energy thus carried causes the bright hydrogen lines.

The irregularities in the variations of long-period variables are to a certain extent characteristic of all the red giants. In fact, red giants with large irregularities are often called *semiregular* variables. Their periods (or cycles, since "period" implies regularity) seem to average around 100 days, with another group averaging around 350 days. Thus, there appear to be two types of stars composing the semiregular variables. Their spectra classify them all as M-, N-, R-, or S-type stars, and most of them show bright lines in their spectra. They cannot be comfortably classified in either Population I or II; it has been suggested that they may form a bridge between the two populations.

The distribution curve of stars that pulsate in a more-or-less regular fashion is shown in Figure 11-6. The long-period variables are the most common, with the RR Lyrae stars next. The actual ranges of periods for the groups overlap, and, as we have seen, an arrangement according to

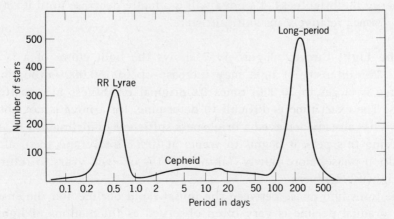

**Figure 11-6**

The distribution of pulsating variables.

period alone does not give us an arrangement according to other properties of these stars. It does yield the relative number of stars of each period, however, and this will be significant for our discussion of a star's life history (Chapter 12).

## 11.4 Novae

Some obviously nonstable stars erupt or burst with a display of energy so large it is difficult to imagine. Such outbursts have been noted in history when a star so increases in brightness that it becomes easily visible where no star was seen before; it is called a *nova* (new), even though we now know that novae are not new at all.

Many novae have been recorded in the history of Western civilization. One of the more notable, which occurred in 134 B.C., stimulated Hipparchus to compile his catalog of the stars by position and apparent magnitude.

Bright novae which have occurred since the development of modern equipment have been observed as thoroughly as possible. Nova Aquilae (1918) was nearly the brightest star in the sky and is the first nova which had been recorded before it became a nova. Direct photographs show that Nova Aquilae was perhaps an A-type star of variable brightness before it became a "new" star. Thus, novae are really existing stars that increase in brightness many thousands of times. After such an outburst of energy the brightness of a nova will gradually decrease until it returns to its former relatively insignificant state.

**a. The Light Curve.** Figure 11-7 shows the light curve of a typical nova. The intensity of light may increase up to 160,000 (although the average is closer to 50,000) times its original brightness in a matter of hours. The exact time is difficult to determine, for a nova is not noticed until it has already achieved a brightness sufficient to distinguish it. After achieving this peak it begins to wane, at first rapidly and then, after a transition phase, more slowly, taking months or even years to return to its original brightness.

The transition phase between the initial rapid decline and the ensuing more gradual decline is very often observed as fluctuations of light, as if the star were pulsating. But a single and greater decrease in brightness has been observed during this phase in other novae, which then brighten again before declining gradually to their original magnitude.

**b. Spectral Changes.** That such an outburst of energy would be accompanied by changes in the spectrum of a nova seems a foregone conclusion, and indeed, all novae seem to go through nearly the same spectral changes. During the rapid increase in brightness the spectrum is usually that of a late B- or early A-type star, although it contains emission lines, something that normal stars of these spectral types do not manifest. The absorption lines are considerably broadened and displaced toward the violet, indicating that the region of the star facing us is approaching and turbulent. At maximum brightness, the star apparently cools, for the spectrum becomes that of a slightly later-type star, and the emission lines begin to fade.

As the light begins to fade, the spectrum becomes similar to that of a late F-type star; it develops very broad emission bands, each with an absorption line on its violet side. Eventually the absorption lines fade and the spectrum becomes a continuous spectrum with broad emission bands superimposed on it. Then, long after the star has returned to its normal brightness, the emission lines begin to fade and the spectrum may consist only of a continuous spectrum with a very few faint absorption lines.

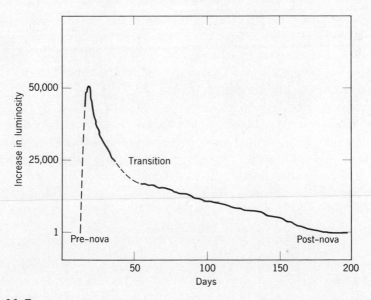

**Figure 11-7**

A schematic light curve for a typical nova.

**c. Spectral Evidence for Expanding Shell of Gas.** The idea of a stellar explosion best explains these changes in the spectrum, as well as the initial phenomenal increase in luminosity. Since a star continues to shine many years after the explosion with nearly the same luminosity as before, the explosion is apparently a surface phenomenon. The star literally "blows its top." Let us now interpret the spectral changes in the light of such an explanation.

As the star's surface begins to expand, the spectral lines are shifted far to the violet. For Nova Aquilae this shift amounted to a velocity of 1,100 miles per second. The widening of the lines may result from the turbulence to be expected in such an explosion. As the shell of expanding gas grows larger, it begins to produce emission lines since it is a gas under very low pressure hot enough to emit its own light. Only that portion of the gaseous shell immediately between the star and the Earth (region *a* in Figure 11-8*a*) continues to produce absorption lines in the star's spectrum observed from the Earth. The rest of the shell (regions *bcd* and *b'c'd'*) produces bright lines in spectra recorded on the Earth.

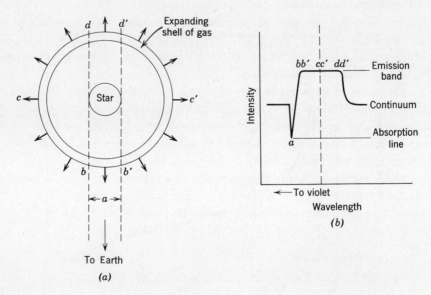

**Figure 11-8**

The expanding shell of gas about a nova (*a*) gives rise to a bright band in the spectra (*b*) with an absorption line on its violet edge.

That portion of the expanding shell immediately between the star and the Earth not only causes absorption lines but is also traveling toward the Earth faster than any other portion. This results in having those absorption lines shifted farther to the violet than the lines originating in any other region. The regions marked *b* and *b'* emit bright lines; they do not cause any absorption lines because they do not lie between the star and the Earth. These regions are also traveling toward the Earth, but more slowly than region *a*; consequently the spectral lines they emit are not shifted as far to the violet as the absorption lines from region *a*. Regions *c* and *c'* are traveling at right angles to the line of sight; as a result the bright lines they emit are shifted neither to the violet nor to the red unless the whole star is moving toward or away from the Earth. Since the shell is very transparent to its own light, we receive light emitted by the regions marked *d* and *d'*. But these regions are traveling away from the Earth and their bright lines are therefore shifted to the red.

The net result of these factors is an emission band (Figure 11-8*b*) given off by the shell from *bcd* and *b'c'd'*, and all the intervening regions. If these regions are considered in succession, the spectral lines they emit are shifted ever less toward the violet as we approach *c* and *c'*, from which point on they become shifted ever more to the red. The blending of all these immediately adjacent bright lines produces the solid band. Again, it is region *a* alone that causes the absorption line that appears on the violet side of each emission band.

As the shell progressively expands, the spectrum changes noticeably, going through a stage where it includes some of the same lines as the aurora spectrum of the Earth's upper atmosphere. Later the nova spectrum contains bright lines that are seen in the spectra of gaseous nebulae.

During this explosion the photosphere is not quiescent. Indeed, the suggestion has been that it also expands until the star reaches maximum luminosity, at which time the gases of the photosphere start to fall back into the star. The gas above the photosphere, however, having achieved escape velocity, would continue to expand and form the shell.

**d. Observations of Ejected Gas.** That novae do eject gases is confirmed by photographs of Nova Persei (1901); one of these, recently taken at the Palomar Observatory, is shown in Figure 11-9.

When the shell of expanding gas is visible it gives us a method for finding the distance of the nova. From the Doppler shift of the absorption lines we can determine the velocity with which the shell is expanding.

**Figure 11-9**
The expanding nebulosity about Nova Persei (1901). (Photograph of the Hale Observatories)

Then, when after a number of years the shell becomes visible, we can compute its actual radius in miles because the radius must equal the velocity of the ejected material multiplied by the interval of time between the explosion and the appearance of the observed shell. Since we now know the linear diameter of the shell we can calculate its distance by simply measuring its angular diameter.

By this method astronomers have been able to determine the absolute luminosity of some of the novae. Nova Aquilae (1918) had a luminosity of about 40,000 times that of the sun. Before it became a nova it had been only slightly more luminous than the sun. Such an increase in luminosity obviously required the sudden release of a tremendous amount of energy.

## 11.5 Supernovae

Novae, on the average, achieve an absolute magnitude of about −6 or −7 at maximum. This is, momentarily at least, very bright indeed, but there are other stars which in exploding achieve an absolute magnitude as great as −13 or even −16. In other words, an ordinary nova may achieve a luminosity of about 50,000 suns but the star that reaches an absolute magnitude of −16 achieves a luminosity of about 200 million. Stars that reach this latter, far greater magnitude are called *supernovae*. A supernova indeed, when a star gives off as much light at maximum as 200 million suns! This amount of light may be similar to the amount given off by the entire galaxy in which the star resides, and represents a positively staggering expenditure of energy. Certainly an explosion of this sort must leave a star somewhat the worse for wear (Star Plate 2).

**a. The Crab Nebula.**   The Crab nebula (Figure 11-10) is the result of such an explosion observed in 1054. The connection between the Crab nebula of today and the supernova of more than 900 years ago is quite definite. We can observe and measure the rate of expansion of such a nebula by two methods. The first is to determine the radial velocity from the Doppler shift of a portion of the nebula that is approaching us. This gives us a linear velocity of expansion of 800 miles per second. We can also measure the angular rate of expansion by taking two pictures separated by a considerable interval of time. The combination of these two observations tells us that the nebular gas is about 4,000 light years away, is about 3 light years in diameter, and is expanding at a rate such that it should

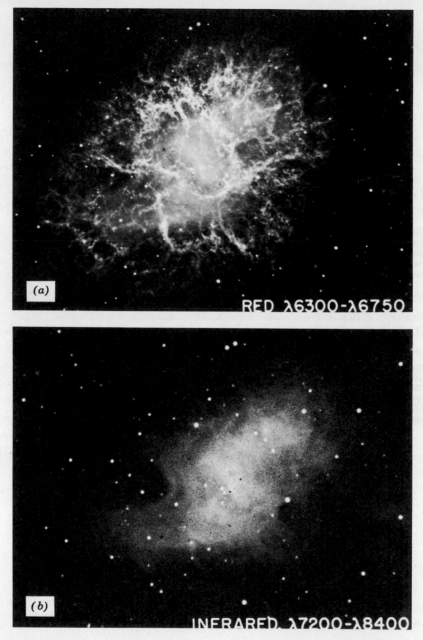

RED λ6300–λ6750

(a)

INFRARED λ7200–λ8400

(b)

**Figure 11-10**

The Crab nebula as seen (a) in the red light of hydrogen and (b) in green light. (Photograph of the Hale Observatories)

have begun expanding about 900 years ago. The exact date was derived from the chronicles of ancient Chinese astronomers who recorded a "guest" star of the brightness of Venus in the celestial region now occupied by the Crab nebula. This guest star, which was also recorded by the Japanese, was visible during the daytime for 23 days and at night for more than a year.

Pictures of the Crab nebula, taken in different colors with color filters and photographic plates sensitive to the colors transmitted by each filter, reveal that the gaseous filaments are rather complex. The photograph taken in red light is centered on the $H\alpha$ line of hydrogen and it appears that the nebula's complex structure is caused principally by gases emitting this spectral line. The photograph taken in green light not only reveals that there is an amorphous structure in the center but also that the light emitted by this amorphous structure is polarized (see p. 17 f).

The nebula also emits strongly in the radio region of the electromagnetic spectrum. But its radio emission is not *thermal*; that is, the intensity of radiation at each wavelength does not correspond to Planck's law of radiation (see p. 226). The amount of energy emitted decreases with decreasing wavelength, so that more radio energy is emitted than microwave energy, and more microwave energy is emitted than infrared, etc. (Figure 11-11.) This observation, coupled with the fact that both the light radiation from the amorphous gas and the radio-energy radiation are polarized, leads to the conclusion that this is what we call *synchrotron radiation*.

In a synchrotron (a high-energy particle accelerator) charged particles are propelled at speeds very close to that of light, and these particles are forced by magnetic fields to follow a curved path. Moving at such high speeds in a magnetic field results in the emission of energy. The greater the speed and the stronger the magnetic field, the more energy these particles will emit; but as the speed and magnetic field are increased, the wavelength of the radiation decreases. Synchrotron radiation is polarized and corresponds to the radiation we receive from the Crab nebula. We conclude, therefore, that in the Crab nebula electrons (electrons emit synchrotron radiation more easily than protons) are traveling at speeds close to that of light in a magnetic field. These electrons must move in corkscrew paths about a given magnetic line of force.

**b. Pulsars.** The star that remains in the center of all this gas appears to be a most fantastic object. It is one member of a class of objects called *pulsars.* These objects were discovered because their radio brightness varies, which in itself is not surprising, but their period of variation is surprising. Most pulsars have a period of variation of less than one second. The pulsar in the Crab nebula has a period of 0.033 seconds (see Figure 11-12). It flashes in radio, visible light, and X-rays about 30 times each second. The period is amazingly consistent, so it can be determined very accurately as 0.033099522 second. The main difficulty is that the period for most pulsars is changing slightly: their periods are becoming longer. The pulsar in the Crab nebula is increasing by $3.652256 \times 10^{-8}$ second each day! Even the rate of change of the period is itself changing.

The best ideas concerning the identity of these objects seem to center around a compact mass of neutrons. Apparently the core of the star is compressed during a supernova explosion to such an extent that the electrons and nuclei are pushed into one another leaving only neutrons.

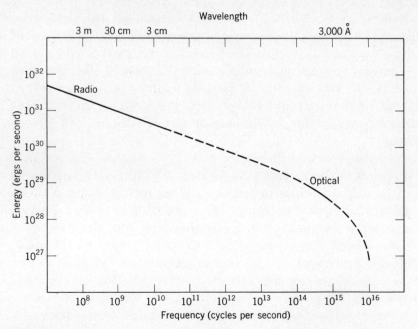

**Figure 11-11**

The continuous spectrum of the Crab nebula extending from the radio to the optical region. Solid lines represent observed regions. (After Steinberg, J. L., and J. Lequeux, *Radio Astronomy*, McGraw-Hill Book Co., New York)

This new form of matter has been called a *neutron star*. The mass of a neutron star must be close to that of the sun, but its diameter cannot be more than about 15 miles! Such an extremely compact blob of matter would rotate with the very short periods observed in pulsars.

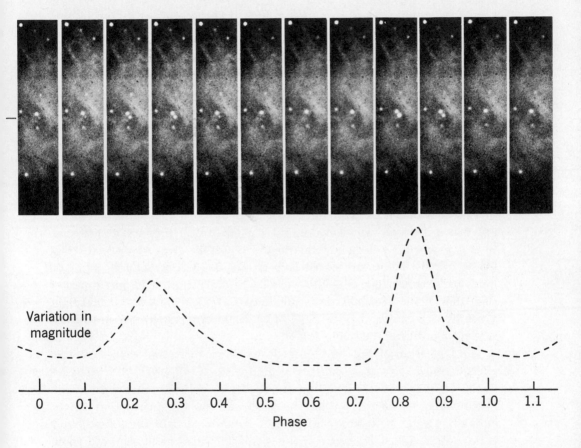

**Figure 11-12**

The variations in the brightness of the pulsar in the Crab nebula (NP 0532) have been photographed with the help of a stroboscopic device and an electronic image intensifier attached to the telescope. These attachments enabled the camera to make a 6 second exposure while viewing only a single phase of the pulsar's variation. One full period (phase 0 to 1.0) of the pulsar is 0.033 second. (Kitt Peak National Observatory, photographed by H. Y. Chiu, R. Lynds, and C. P. Maran)

Whether one portion of the star radiates more energy than another (which may be caused by a magnetic field), making it blink as it alternately exposes its brighter and then its darker side to us, or whether the star pulsates is not yet known with certainty. However, because of their recent discovery, in 1967, there is a great deal of work yet to be done on these fascinating objects. We may, in fact, find that pulsars do not pulsate at all. This is one problem astronomers meet: they must often apply names to objects about which they know very little, and these names often prove inappropriate at a later date.

c. **Supernova Remnants.**   Bolstered by the successes of theories and observations of the Crab nebula, astronomers have searched the records and the skies for other supernova remnants. Their efforts have been rewarded because they have found the remnants of Tycho's nova, Kepler's nova, and many others. The most intense discrete radio source in the sky, Cassiopeia A, is now known to be a remnant of a supernova explosion which occurred about the year 1700. There is no reference in historical records to a nova at that time, and the records of those years are fairly adequate, but it is a distant object and behind a good deal of interstellar material, so it may not have been obvious in the nighttime sky. At present, fragments of this nebula are expanding at the staggering velocity of 4,600 miles per second. Just why this velocity of expansion is so much greater than that of the Crab nebula is not known. At a distance of 1,000 light years it is estimated that an amount of material equal to the sun's mass was ejected during the explosion.

The Loop (Veil) nebula in Cygnus is another nonthermal source of radio energy, and it, too, is a remnant of a supernova (Figure 11-13). Material in the Loop nebula is expanding at the rate of only 56 miles per second. Presumably, its expansion has been slowed by the interaction of the expanding gases with interstellar gases. The diameter of the Loop nebula is now about 120 light years, so this explosion must have occurred thousands of years ago. The center of the nebula is about 2,300 light years from the sun.

It is not yet clear just why the electrons in these supernova remnants have such high velocities. It may result from the original explosion, or it is possible that electrons are being accelerated by mechanisms not yet understood. In any event, every nonthermal source of radio energy in our galaxy is immediately suspected of being a supernova remnant.

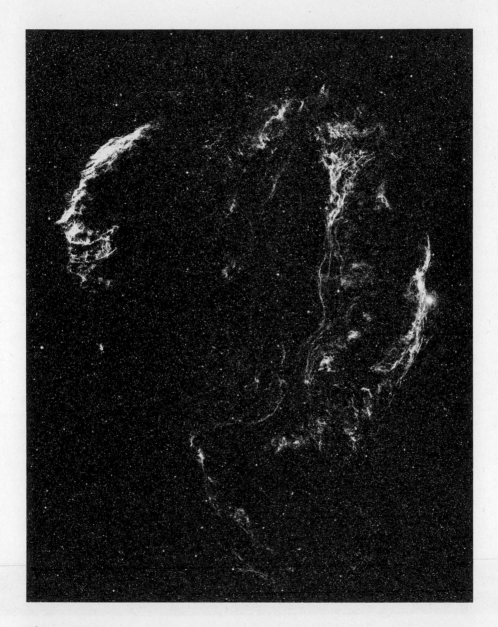

**Figure 11-13**

The Loop nebula in the constellation Cygnus is the result of a nova that exploded many thousands of years ago. (Photograph from the Hale Observatories)

## 11.6 Other Stars with Expanding Shells

There are stars that have spectra similar to the spectra of novae but that do not vary much in brightness. One such group is called the *P Cygni* stars (Figure 11-14). These stars have broad emission bands with absorption lines on their violet edges, indicating that each of them, too, is surrounded by an expanding shell of gas. The absorption lines of P Cygni itself give a velocity of expansion of about 70 miles per second, considerably less than that of a nova. The P Cygni stars seem to be continually losing material at a rate that would amount to the mass of the sun in 100,000 years.

The *Wolf–Rayet* stars are similar to P Cygni stars in that they have broad emission bands with absorption lines on their violet sides, evidence of an expanding shell of gas about each star. The displacement of the absorption lines in a Wolf–Rayet spectrum indicates a velocity of expansion for the shell of close to 2,000 miles per second. The lines of ionized helium in their spectra show that these stars are hotter than the P Cygni stars; with temperatures in excess of 50,000°K they are among the hottest objects in the sky.

## 11.7 Planetary Nebulae

There are stars in the sky that have very obvious shells of gas about them (Figure 11-15). When these were first observed visually with smaller telescopes they gave the appearance of planetary disks and consequently were named *planetary nebulae*. (To be sure, this is a misnomer, for they have scarcely anything in common with the planets that revolve about the sun.) The gas in the shell is greenish in color because of several strong emission lines in the green and the lack of a continuous spectrum. The shells are observed to expand at the rate of 6 to 30 miles per second, considerably less than the expansion of the shells about even the P Cygni

**Figure 11-14**

The spectrum of P Cygni showing the emission bands with an absorption line on the violet side of each. (Yerkes Observatory)

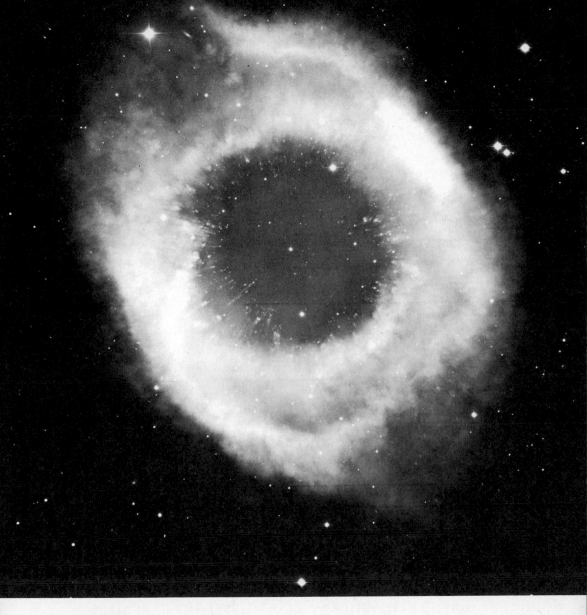

**Figure 11-15**
The planetary nebula NGC 7293. (Photograph from the Hale Observatories)

stars. The diameters of planetary nebulae range from $\frac{1}{3}$ to 3 light years.

The gas in the shell is so tenuous that it would be considered a good vacuum on the Earth, yet we see the shell. Why? The overall dimensions of the planetary nebula in Aquarius (NGC 7293) are about 2 by $2\frac{1}{2}$ light years, and the shell is approximately $\frac{1}{2}$ light year thick. Since the nebula looks like a smoke ring but is actually a shell, it is apparent that we cannot see the gases very well where they are only $\frac{1}{2}$ light year thick. This is the "hole" in the middle. When we observe the edge of the nebula, however, we look through a layer of gas thick enough to be easily visible.

Photographing planetary nebulae with plates sensitive to blue and violet light discloses a central star that may be very faint, if not invisible, when viewed through a telescope directly with the eye. Since these stars are very hot (their temperatures range perhaps from 50,000° to 100,000°K), they emit most of their light in the blue and violet region where the eye is not very sensitive. There is much evidence to support the suggestion that the central stars are actually white dwarfs. This leads to the tentative conclusion that a planetary nebula is a white dwarf in the making. Its shell eventually expands to invisibility and all that remains is the central white dwarf.

But what is a planetary nebula before the shell of gas forms? The 700 known planetaries are concentrated strongly toward the center of our galaxy and are not confined to the central plane of the Milky Way. Such a distribution is typical of Population II objects. Although this seems to eliminate all Population I stars as possible predecessors of planetary nebulae, it still leaves a choice between the many different kinds of Population II stars. Possible candidates for the job of supplying the universe with planetary nebulae are novae, Wolf–Rayet stars, RR Lyrae stars, the red-giant irregular variables, and perhaps a few more. But none of these seems to fill the bill completely; astronomers are left with another puzzle to solve. Because of the shell's slow rate of expansion, it is generally believed that planetary nebulae do not result from novae or Wolf–Rayet stars. Nevertheless, since they have undoubtedly ejected some material from their surfaces, they are related to the nova-shell star class.

## BASIC VOCABULARY FOR SUBSEQUENT READING

Cepheid

Light curve

Neutron star

Nova

Planetary nebula

Population types

Pulsar

RR Lyrae star

Shell-type star

Synchrotron radiation

Velocity curve

Thermal radiation

## QUESTIONS AND PROBLEMS

1. Draw a typical light curve and velocity curve for a Cepheid, the former directly above the latter. Now below the velocity curve draw a third curve to show generally how the diameter of the star changes.

2. What observations must be made before the distance of a Cepheid can be estimated?

3. What observations led astronomers to believe that Cepheids pulsate?

4. Why do Cepheids make good "yardsticks" for the astronomer?

5. Why can't the distance of the Andromeda galaxy be determined by using RR Lyrae stars?

6. Explain how the astronomer detects an expanding shell of hot gas about a star. How can he determine its velocity of expansion?

7. What evidence indicates that the Crab nebula was a supernova in the year 1054?

## FOR FURTHER READING

Struve, O., and V. Zebergs, *Astronomy of the 20th Century*, Crowell, Collier and Macmillan, New York, 1962, Chapters XV and XVI.

Burbidge, G., "Dissecting the Crab," *Natural History,* p. 66 (Oct. 1970).

Cox, A. N., and J. P. Cox, "Cepheid Pulsations," *Sky and Telescope*, p. 278 (May 1967).

Gorgenstein, P., and W. Tucker, "Supernova Remnants," *Scientific American*, p. 74 (July 1971).

Green, L. C., "Pulsars Today," *Sky and Telescope,* p. 260 (Nov. 1970); p. 357 (Dec. 1970).

Hack, M., "RU Camelopardalis—A Unique Cepheid Variable," *Sky and Telescope,* p. 350 (June 1967).

Hewish, A., "Pulsars," *Scientific American*, p. 25 (Oct. 1968).

Kraft, R. P., "Pulsating Stars and Cosmic Distances," *Scientific American*, p. 48 (July 1959).

Liller, W. and M. Liller, "Planetary Nebulae," *Scientific American,* p. 60 (April 1963).

Mumford, G. S., "The Dwarf Novae," *Sky and Telescope*, part I, p. 71, (Feb. 1962); part II, p. 135 (March 1962); part III, p. 190 (Oct. 1963).

Struve, O., "The Pulsating Star RR Lyrae," *Sky and Telescope*, p. 311 (June 1962).

"A Rapidly Pulsating Radio Source," *Sky and Telescope*, p. 207 (April 1968).

"Alcock's Nova in Delphinus," *Sky and Telescope*, p. 150 (Sept. 1967); p. 300 (Nov. 1967).

"Barnard's Loop Nebula," *Sky and Telescope*, p. 145 (Sept. 1967).

"Distances of Pulsars," *Sky and Telescope*, p. 378 (Dec. 1968).

"Expansion of the Crab Nebula," *Sky and Telescope*, p. 13 (Jan. 1969).

"Further Observations of Pulsars," *Sky and Telescope*, p. 339 (June 1969).

"The Pulsar Industry," *Scientific American*, p. 43 (June 1968).

"Visible Pulsar," *Scientific American*, p. 46, p. 49 (March 1969).

# THE EVOLUTION OF STARS

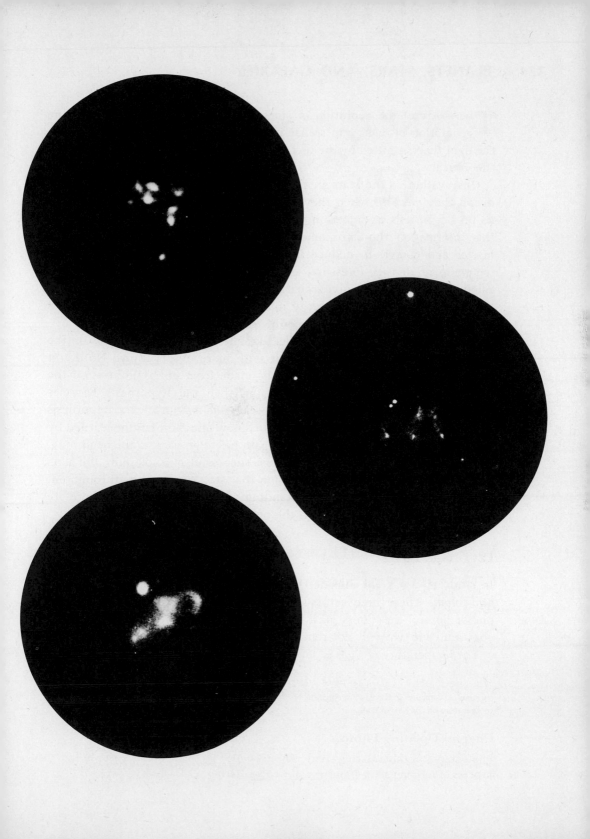

The study of the evolution of stars is based largely on studies of the generation of energy in stellar interiors. Thus, since we can see only the light that is emitted by the surface, evolutionary studies must be largely theoretical.

Observation of the stars in their present condition (actually the condition they were in when they emitted the light that only now reaches us on Earth) tells us something of their structure and characteristics. To this information the astronomer must apply his general knowledge of matter and energy to deduce information about stellar interiors: their temperature, density, pressure, and chemical composition. Then he must calculate how a star of given characteristics would theoretically react to the conversion of nuclear energy into heat and radiant energy. He must also try to find some way to account for stellar formation.

When he has built up what looks like an acceptable hypothesis on the evolution of stars, he must check it with more observations. No doubt there are some predictions that can be made from his hypothesis; if the predictions are verified by observation, then the hypothesis becomes more acceptable. On the other hand, observation may prove the hypothesis incorrect and the astronomer must build a new one. But, having learned by each step in the procedure, he never again has to go back to the beginning.

The hypotheses that have been most successful in explaining the stars have all been based on the assumption that stars are gaseous throughout. In order to obtain a better understanding of the stars we should therefore investigate the laws that explain the behavior of gases.

## 12.1 Gas Laws

In studying a gas we consider three main factors: *temperature*, *pressure*, and *volume*. These three factors are related in such a way that if one changes, one or both of the others must change. Any one of the three may remain constant, but two will not remain constant while the third changes.*

The temperature of a gas is dependent upon the kinetic energy (energy of

*The general gas law is given as: $PV = NkT$; where $P$ is the pressure of the gas, $V$ its volume, $N$ the number of molecules in the gas, $k$ a constant of proportionality called the Boltzmann constant, and $T$ the temperature on the Kelvin scale.

### Chapter Opening Photo

Photographs of three Herbig–Haro objects, which are presumably gases in the process of forming stars. (Courtesy of George Herbig, Lick Observatory)

random motion) of the atoms composing the gas. For a given gas this means that an increase in the velocity of the atoms manifests itself as increased temperature.

The pressure of a gas is dependent on the momentum (mass multiplied by velocity) of the atoms. This means that if the velocity of the atoms in a constant volume of gas is increased, their momentum increases, and consequently the pressure of the gas increases.

The volume of a gas depends on the internal pressure of the gas as well as on the pressure confining it. For a gas in a balloon or in a cylinder the confining pressure is supplied by the rubber in the balloon or by the walls of the cylinder. In the stars it is supplied by the gravitational field of the star itself. An atom cannot escape from a star unless it reaches the escape velocity.

If the temperature of a gas is increased by heating, the velocity of the atoms increases. Consequently their momentum increases and thus also the pressure. If the confining pressure is less than this increased internal pressure the gas will expand like an inflated balloon placed in a warm room. Conversely, if a gas is cooled the pressure will decrease and the gas will contract, if allowed to do so, like a balloon placed in a refrigerator.

On the other hand, if a gas is compressed to a smaller volume by the confining pressure (for example, by a piston in a cylinder) its temperature increases. If it is allowed to expand (blowing out breath rapidly as against breathing it out gradually) its temperature decreases.

## 12.2 Energy and Pressure Balance in a Stable Star

A stable star is one that is neither expanding nor contracting. Such a star is said to be in equilibrium, for all the forces acting on it are in balance. A system of checks and balances is necessary to maintain such a stable equilibrium; the thermal energy (i.e., heat) generated in the core of the star must be allowed to escape in some way or other as rapidly as it is generated. So we must concern ourselves with the sources of energy and the ways by which this energy, once generated, may escape from the star.

The main source of energy in stars is the thermonuclear reactions (the proton–proton reaction and the carbon cycle discussed on p. 229 ff.) that convert hydrogen into helium, releasing heat and radiant energy. The rates at which both these reactions proceed is dependent on the temperature: the higher the temperature, the faster each reaction takes place. The reaction rates also depend on the density of the gas, and again, the higher the density the faster the reaction takes place. If the gas is not hot enough, neither of the reactions will take place and no energy will be generated from this source.

There is a second source of energy in the stars, but it is used only when required. This is the *gravitational potential energy* of the gas itself. If a star were allowed to contract and thus compress the gases in its interior, the temperature of those gases would increase as gravitational energy is converted into thermal energy.

To maintain equilibrium, however, the energy generated in the interior must be allowed to escape. This may happen in one of two ways. First, the star could radiate electromagnetic energy (light, etc.) into space in the manner of all stars. But a star may also rid itself of some of its thermal energy by expanding and converting heat into gravitational potential energy. In expanding, the gases cool and the star becomes larger. We might recall that the star is not in equilibrium while it is expanding or contracting.

It is the combination of these sources of heat energy and of outlets for the disposal of the energy that constitutes the system of checks and balances. The gravitational force acting on the gas acts to compress the star. If the star were allowed to contract, the temperature and density of the interior would increase, and an increase in the reaction rate of the thermonuclear reactions would result. The increased reaction rate would generate more heat and the temperature would increase even more, causing the star to expand.

The internal pressure acts to force the star to expand. If the star were allowed to expand, the temperature and density of the interior would decrease, and a decrease in the reaction rate of the thermonuclear reactions would result. Consequently, less heat would be generated, the temperature and pressure would drop, and the star would contract back to equilibrium size.

The star's gravitational potential energy, its rate of energy generation by the thermonuclear reactions, and its rate of energy loss by radiating energy out into space are all in balance. Practically all of the energy generated in the interior is radiated out into space; the gravitational potential energy acts only as a safety valve.

## 12.3 Energy Transport and Stellar Structure

A star's energy is generated deep in its interior where the temperature is highest; however, the energy must reach the surface before it can radiate into space. If a star were completely opaque to all forms of energy transport, the core would heat up to such an extent that the star would eventually have to explode to rid itself of the energy generated. On the other hand, if a star were very transparent, its energy would leak out faster than it is being generated by the thermonuclear reactions. The star would then have to contract to supply the necessary heat to maintain the temperature and pressure required to support the overlying layers. A stable star adjusts itself so that the amount of energy generated equals the amount of energy radiated.

The energy generated in the interior is transported to the surface in three ways: *conduction, convection,* and *radiation.* Conduction is the familiar form of heat transport along a metal rod. If one end is placed in a fire the other end will heat up, because the atoms in the hot end transfer their momentum to those in the cooler end simply by knocking into and jostling them.

Heat transport by convection is the actual movement of the hot object or gases. A central heating system in a house transports heat by convection if the hot material (air, water, or steam) itself is moved from the furnace into the rooms.

The transport of heat by radiation is accomplished when a hot object emits electromagnetic waves. An example is the familiar electric heater without a fan. If a fan is attached to the heater, heat will be transported by convection as well as by radiation. Heat energy is transported from the sun to the Earth by radiation.

A star's center contains a core through which heat may or may not be transported by convection currents (Figure 9-3). If there are convection currents in the core, it is evident that the material in the core will be thoroughly mixed. Surrounding the *convective core* is the *radiative zone* where energy can be transported only by radiation, and surrounding all this is the *convective envelope.* The convective envelope is the outer layer of the star; from the outside surface of this convective layer the energy is radiated into space. The convective envelope must be rather homogeneous in composition, since convection currents act to mix the gases.

According to current theories, not all stars have convective cores and envelopes. Of the stars on the main sequence, the sun, a G2 star, seems to be of the latest spectral type to have a convective core. Main-sequence stars later than F0 have convective envelopes, and thus the sun has a convective envelope as well.

Once the thermonuclear reactions begin in the hot center of the star they convert hydrogen into helium, increasing the amount of helium in the convective core and depleting the amount of hydrogen. Since the reactions demand a supply of hydrogen, the energy is eventually generated on the surface of the growing helium core where the high core temperature meets the hydrogen.

The radiative zone does not permit convection currents. Consequently the gases in this region are not well mixed, nor do they mix with the helium core. Since the temperature is so very high on the surface of the helium core, the energy is emitted in the form of X-rays that are successively absorbed and re-emitted by the atoms of the radiative zone as they proceed outward. The X-rays become longer in wavelength (and are then often called *softer* X-rays) as they are re-emitted, until they reach the convective envelope. Here they are absorbed by the gases, which consequently heat up, expand, and thus rise to the surface of the star, radiating their energy into space. In radiating their energy they cool and therefore sink back into the envelope to become heated again. The granulations on the surface of the sun (the only star whose surface we can see) are direct evidence of these convective currents in the envelope (Figure 9-4).

All the energy must be transported by radiation at least part of the way to the star's surface, even though convection may transport it over other parts of the distance. Radiation exerts a pressure and the intense X-ray radiation in the stellar interiors can exert a significant amount of pressure to help support the overlying layers of gas. The radiation pressure increases as the temperature of the core increases, and consequently becomes important in the very hot stars.

Energy transport by conduction is not effective in stellar interiors until the gases become highly compressed and completely ionized, and thus degenerate. Under this condition, heat conduction becomes the dominant form of energy transport, for the very fast-moving electrons conduct the heat from the hotter to the cooler regions of the degenerate gas. This is called *electron conduction* and is important only in the cores of red giants and throughout the interiors of white dwarfs.

## 12.4 Theory of Stellar Evolution

a. The Protostar. The process by which stars form from a gaseous nebula is not well known, but it has been speculated that a parcel of gas must by some method, perhaps by a whirling motion, detach itself from the rest of the nebula and commence the process of contraction. Such a contracting parcel of gas is called a *protostar*, for it is not hot enough to support thermonuclear reactions and thus does not emit visible radiation. As the gas continues to contract, gravitational potential energy is converted into thermal energy and the temperature increases. When the central temperature approaches 500,000°K, a reaction involving deuterium (hydrogen atom with mass 2) may begin if the primeval gas contains deuterium. The deuterium would combine with a proton to produce helium 3, and this reaction would help increase the temperature of the protostar until the proton–proton reaction is initiated.

Soon after the proton–proton reaction begins, the protostar presumably becomes a full-fledged star and can be placed on the H–R diagram according to its luminosity and surface temperature. Since the central temperature and thus the luminosity depend on the mass of the star, the H–R diagram will separate new stars according to their mass. If the mass of the protostar is, let us say, 20 times that of the sun, it contains more gravitational potential energy to convert into heat than the sun. In fact, it *must* produce a hotter interior to create an internal pressure high enough to support the greater amount of gas.

For the more massive stars the contracting process is not stopped by the relatively slow proton–proton reaction. Contraction continues until the central temperature becomes high enough to start the carbon cycle, which proceeds more rapidly and is thus able to produce enough energy to support the gas above the core. The proton–proton reaction does not stop when the carbon cycle begins. Both reactions proceed simultaneously, but if the temperature becomes greater than about 25,000,000°K the carbon cycle begins to supply more energy. Presumably, both reactions take place in the sun, with the proton–proton reaction supplying more energy than the carbon cycle.

**b. The Zero-Age Main Sequence Star.** The proto-star enters the H–R diagram on the right-hand edge (Figure 12-1). As soon as the nuclear reactions in the interior produce enough energy to stop the contraction by balancing the pressure there is a slight gravitational adjustment; the star drops slightly in luminosity and settles down on the main sequence. The line which represents the positions of stars newly arrived on the main sequence is called the *zero-age main sequence*.

Since a very massive star will have a high temperature and great lumi-

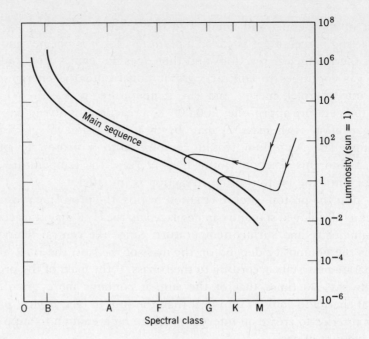

**Figure 12-1**

The evolutionary tracks of stars of moderate mass in the initial contracting stage.

nosity, it will begin its life as a full-fledged star somewhere in the upper left-hand corner of the H–R diagram. A star of 16 solar masses will become a B0 star. A star of about $\frac{1}{3}$ solar mass will become a late M-type star. The stars of intervening mass will fall at positions between these two and form the bulk of the main sequence. Thus, the main sequence represents stars that are in the first stage of their development as full-fledged stars. Stars more massive than 16 or 20 solar masses are rare and become the O-type stars.

Thus far, we have neglected possible variations in the chemical composition of the stars. We have considered them to be composed very largely of hydrogen. However, differences in chemical composition do occur, even in young stars. These consist of a difference in percentage of helium, or of some of the heavier elements. These differences raise or lower the star's path on the H–R diagram. The larger the percentage of hydrogen, the lower the curve will appear on the diagram. The possibility of differences in chemical composition makes us uncertain about the placement of the curve for any one star even though any differences that may exist would be slight, since hydrogen is by far the most abundant element.

A star spends a relatively short time in the protostar stage. A star the mass of the sun will contract from a portion of a nebula into a star on the main sequence in perhaps 50,000,000 years. As will be seen later, even such a span, long compared to a human life, is short in comparison with the entire life of a star. It is estimated that a star of 9 solar masses takes about 21 million years to contract and find a place on the zero-age main sequence. A very massive star, because of its stronger gravitational field, must contract even faster, perhaps in 500,000 years. A star with a 0.2 solar mass is estimated to take as long as 600 million years to contract from its initial stages to a full-fledged star. It is only because a star spends such a short time on this portion of its *evolutionary track* (its path on the H–R diagram as it evolves through various stages of its life) that we do not find many stars in this region of the diagram.

Once on the main sequence the star remains there for a considerable length of time, but not indefinitely. A star whose mass is equal to that of the sun may remain there for 10 billion years. A star with a mass less than that of the sun will remain on the main sequence longer. A very massive star will convert its hydrogen into helium faster and consequently not last as long.

For example, a star whose mass is 20 times that of the sun is an early B-type or late O-type star. Such a star has a luminosity in the order of 10,000 times that of our sun. Consequently, it must burn its hydrogen fuel 10,000 times faster than does our sun. But it has only 20 times the amount of fuel. A simple division indicates that the sun will live 500 times longer than this massive star, whose life would be roughly 10 million years.

**c. After the Main Sequence.** When about 10% of the hydrogen fuel has been burned, the core begins to contract, releasing energy that forces the outer regions of the star to expand and hence cool. However, the increase in luminosity resulting from an increase in the size of the star exceeds the decrease in luminosity caused by the drop in surface temperature, so the star's net luminosity increases, moving it upward on the H–R diagram. The decrease in surface temperature simultaneously carries it to the right; its high luminosity and low surface temperature have made it a red giant (Figure 12-2). At a certain point, depending on its mass, it ceases to expand. In establishing this temporary equilibrium it may pulsate slightly, presumably becoming an irregular variable.

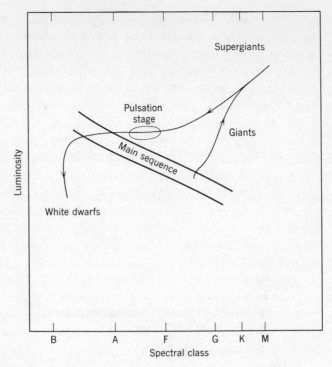

**Figure 12-2**

The evolutionary tracks of stars of moderate mass after leaving the main sequence.

Oddly enough, the outer regions of the star in this stage are expanded and thus have a low temperature; at the same time the contraction of the core results in a higher central temperature and density. Because of this higher temperature and density in the core, some of the helium is converted into carbon and other heavier elements. These new nuclear reactions, of course, supply even more energy for the star, preventing it from contracting immediately.

How long the star remains a red giant is not well known, but it cannot retain this position as long as it does its main-sequence position. This follows from the fact that there are many more stars on the main sequence than anywhere else on the H–R diagram, which indicates that the average star spends a longer part of its life there than it spends in any other stage of its early evolutionary track.

The reasons why a star leaves the main sequence and becomes a red giant are known to the extent that astronomers feel they are at least (and at last) on the right track. Calculations by R. Härm and M. Schwarzschild at Princeton University for a star with a mass 1.3 times that of the sun indicate that, by the time the star reaches the red-giant stage, its helium core has grown to nearly 40% the mass of the star. By this time the core material has been subjected to very high pressures and temperatures. Under these extreme conditions of temperature and pressure, the core material is compressed into an exceedingly compact form of matter— *degenerate gas*. Degenerate gases (see p. 284) have exceedingly high densities and are not described by the same laws that describe the gases we have been discussing. The hydrogen gas overlying the core continues to "burn," but the radiative zone is not able to transport all of the energy generated by the core. As a result the core temperature increases. When it reaches 80,000,000°K (!), the helium starts to burn by converting into more massive elements. Because the core material is degenerate, it cannot yet expand; therefore, the temperature increases still more, and the helium-burning proceeds more rapidly. This rapid helium-burning is limited, however, for when the core reaches the fantastic temperature of 350,000,000°K it is forced to expand and the gases become nondegenerate.

**d. The Helium Flash.** The very rapid burning of helium before the core expands is called the *helium flash*, since it lasts for only a few thousand years. The expansion of the core causes the temperature to drop and the helium flash ends.

These same calculations indicate that the star's position on the H–R diagram does not remain constant during the helium flash. The star apparently moves from the red-giant stage slightly downward and to the left along the nearly horizontal giant branch of the H–R diagram. During part of this transition, instabilities arise and the star becomes a pulsating variable; which pulsating variable it becomes depends in part upon its mass. Presumably a more massive star becomes a long-period Cepheid, a less massive star either a short-period Cepheid or an RR Lyrae star. But the star's chemical composition may also be a factor in determining which type variable it becomes. There is theoretical evidence to suggest that the star may have a pass through the helium-flash stage more than once. Consequently, the star may move back and forth along the giant branch, becoming a variable more than once.

**e. The White Dwarf.** About the time the star crosses the main sequence, its surface temperature reaches a maximum and its luminosity decreases. The star sinks into the last stage of stellar evolution and becomes a white dwarf!

The general shape of the evolutionary track described is the same for all stars of moderate mass, even though the location of the track differs for stars of different masses. As the position of a star along the main sequence depends on its mass (the least massive star appears at the bottom), so does the location of the star's evolutionary track. With increasing mass, the evolutionary tracks appear ever higher on the H–R diagram. A more massive star will be more luminous at any point along its track than a less massive one at a corresponding point. Furthermore, the more massive a star is, the faster it will evolve along its evolutionary track. The stages through which a star passes during its evolution are represented in part by the different types of stars already discussed. If one stage of a particular evolutionary track is populated by more stars than any other, we can assume that stars of that particular mass must remain in that stage of evolution longer than in any other stage.

## 12.5 Observational Evidence of Stellar Evolution

The observational evidence for such a hypothesis of stellar evolution is quite strong. If stars form from gaseous nebulae, then we ought to examine nebulae in search for evidence of recent star formation. If in that search we observe the Rosette nebula we will find a large aggregate of gas with many very luminous stars in the center and with many dark globules of gas projected against the bright nebula (Star Plate 3 and Figure 12-3). Continued search would reveal other nebulae that also exhibit these globules, which, it might be speculated, are protostars in the process of contracting. Their central temperatures are not yet high enough to support thermonuclear reactions; in fact the gas is still quite distended. They are simply blobs of gas and dust which may be contracting.

**a. Before the Main Sequence.** Evidence that those dark blobs of gas and dust might indeed be protostars is given by the *Herbig–Haro objects*. These are also found in nebular gases and each has a stellar or semistellar nucleus surrounded by a small emission nebula. Figure 12-4 shows a

**Figure 12-3**

The Rosette nebula showing dark globules that may be protostars in an early stage of contraction. (Photograph from the Hale Observatories)

group of Herbig–Haro objects as they appeared in 1947, then again in 1954, and still again in 1959. The photographs clearly reveal new centers from which light is emitted. Presumably these are newly forming stars! Observations of their spectra indicate that these objects resemble in many ways another class of stars called the T Tauri stars.

In the constellations of Taurus and Orion there are other nebulae that contain *T Tauri stars*—a class of stars that are not very luminous, that have bright lines in their spectra, that are irregularly variable in light, and that are invariably associated with nebular material. All of the more than 500 T Tauri stars known are of a later spectral type than F8, and each is more luminous than a main-sequence star of the same spectral type. These two characteristics indicate that they may be to the right of, as well as above, the main sequence. Such a position on the H–R diagram is predicted for a newly contracting star following an evolutionary track that leads to a point on the main sequence to the left of, and slightly below, its present position as a T Tauri star.

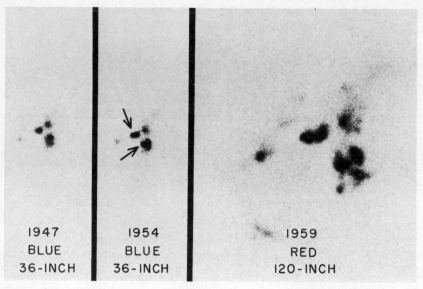

1947
BLUE
36-INCH

1954
BLUE
36-INCH

1959
RED
120-INCH

**Figure 12-4**

Herbig–Haro No. 2 in Orion taken in three different years: two in blue light (with a 36-in. telescope) and one in red light (by the 120-in. telescope). There has been further brightening of one of the new nuclei since 1959. (Courtesy George Herbig, Lick Observatory)

T Tauri stars have bright lines in their spectra which indicates that they have either extensive atmospheres or nebulous material immediately surrounding them. The fact that they are invariably associated with nebular gases places them in what present theory considers the necessary environment for stars in the early stages of formation. However, they do move with respect to the clouds in which they are now immersed, which indicates that they will eventually leave these clouds. Since T Tauri stars are generally not found outside gaseous clouds, we must conclude that they lose their identifying characteristics before or when they leave the clouds. Thus, the stage of development represented by the T Tauri stars may be a short one, embracing perhaps only 10% of the time required for a star to contract onto the main sequence.

**b. On the Main Sequence.** Recent studies indicate that it may be possible to learn something about the length of time stars have spent on the main sequence. The time indicator is lithium. It is well known that lithium is present in the atmosphere of stars, but it cannot survive the high temperatures encountered in stellar interiors. It is also known that young stars such as T Tauri stars have a great deal more lithium in their atmospheres than has the sun. But George Herbig of the Lick Observatory has observed a range in the lithium content of 60 G-type stars in the solar neighborhood.

If the gases in the atmosphere mix to some extent with the gases deeper in the star, the lithium content in the atmosphere should decrease with time, for as the gases are mixed some of the lithium is carried into the deeper and hotter regions of the star where it is destroyed. At an age of 4.6 billion years the sun has very little lithium left. Younger G-type stars presumably have more. Before ages can be assigned, however, something must be known about the percentage of lithium in the atmosphere of the new star, the rate at which the atmospheric gases mix with the gases beneath the photosphere, the temperature under the photosphere, and the rate of destruction of lithium at those temperatures. Continued study of this problem promises to help our understanding of stellar evolution considerably.

**c. After the Main Sequence.** During their life on the main sequence, the stars burn about 10% of their hydrogen content; they deplete the hydrogen in their core. Evidence of what happens to the very massive stars after they have depleted this hydrogen is given by the *Beta Canis*

*Majoris stars.* This is a group of stars composed of early B-type stars that are more luminous than their main-sequence counterparts (Figure 11-4), and that are variable in every sense of the word.

The brightness of a Beta Canis Majoris star is evidently related to its period; the stars of earlier types are more luminous and have longer periods than those of later types. The range in periods for the whole class is from $3\frac{1}{2}$ to 6 hours. They also exhibit periodic fluctuations in radial velocity, indicating that they pulsate. One of the more remarkable characteristics of a Beta Canis Majoris star is that its period increases in a regular fashion. Even though this increase is small, less than 3 seconds a century, it is nonetheless observable. Because the rate of pulsation is so rapid the star goes through as many as 2,000 oscillations each year, which enables us to determine its period to an accuracy of 0.03 second. Since at least one Beta Canis Majoris star has been observed for 30 years, a change of 3 seconds in a century or 1 second in 33 years is relatively easy to measure.

The late Otto Struve suggested that the change of period of a Beta Canis Majoris star may actually be interpreted as an evolutionary change. Struve proceeded from the assumption that a Beta Canis Majoris star once existed on the main sequence to the left of where it now resides (Figure 12-5). As its hydrogen was converted into helium, it expanded;

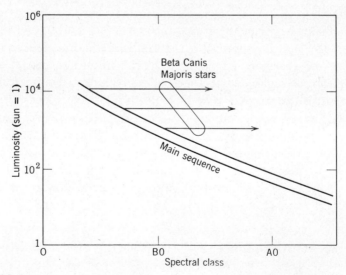

**Figure 12-5**

The Beta Canis Majoris stars may represent a stage in the evolution of massive stars.

thus, its surface temperature decreased, taking it to the right of its original position on the main sequence. Pulsation then set in when it entered the region defined by the Beta Canis Majoris stars. As the expansion continued, the density of the star decreased and the length of the pulsation period increased correspondingly.

If this interpretation is correct, this is, among the stars, the only evolutionary change that can now be followed and studied in some detail. It is not surprising to find that our first observation of what we suspect to be an evolutionary change occurs in the very massive stars, for these are the stars that evolve the most rapidly.

If the very massive stars on the main sequence are less than 1,000,000 years old, they must have been born only recently on the astronomical time scale. (The Earth is about 4.6 billion years old.)

**d. Evidence from Star Clusters.** As we see in Chapter 14, stars are frequently found in large groups called *clusters*, each held together by the gravitational field set up by the many stars. Since individual stars, like the sun, are so very far apart, the gravitational effect of any one on another is almost negligible. Consequently, clusters could hardly have been formed after the individual stars, separated by great distances, were formed. It is therefore generally assumed that all the stars in a cluster were formed at the same or nearly the same time and in a region of space roughly the same size as the cluster now occupies.

If every star does go through an evolutionary process the rate of which depends on its mass, in any one cluster we would expect to find the more massive stars in a more advanced stage of evolution than the less massive ones. That this is actually true is demonstrated by the H–R diagram (Figure 12-6) of a globular cluster (see compact cluster in Figure 14-5). The stars that have already strayed from the main sequence in their evolutionary track are the more massive stars; those still left on the main sequence are less massive. That point on the main sequence where the stars branch off toward the red-giant stage is called the *turn-off point*. It should be noted that very few stars remain on the main sequence above the turn-off point; however, the fact that some stars are on the main sequence above the turn-off point leads us to suspect that these stars may have been formed later than the bulk of stars in the cluster. Therefore, not all the stars in the cluster are of precisely the same age. This diagram, then, does not represent the evolutionary track of any one star of given mass, but identifies points on evolutionary tracks of stars of differing masses whose ages are nearly the

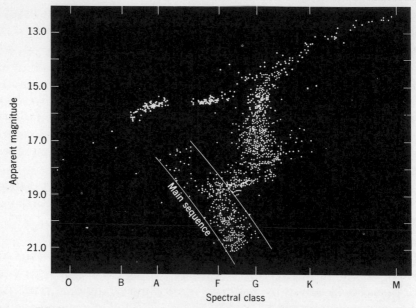

**Figure 12-6**

An H–R diagram of the globular cluster M3 according to H. C. Arp, W. Baum, and A. Sandage.

same. The stars that are still on the main sequence will eventually evolve into red giants, but these less massive stars will presumably be less luminous than the red giants appearing in the cluster at the present time.

The gap in the upper branch of the diagram is of particular interest; it is here that the RR Lyrae stars in the cluster would be located, were they included on this diagram. According to current theory, the stars pass through the RR Lyrae gap (or Cepheid gap) during their helium flash when the stars move horizontally along the giant branch. It is of great interest to realize that a well known Cepheid, RU Camelopardalis, has become an ex-variable star! The variability of this star was first recognized in 1907 and since that time its period has been established as just over 22 days, with a range in apparent magnitude of 8.2 to 9.1. In 1961 and 1962 its variations were normal, but in 1963 the extent of its variations decreased. Its light variations continued to decrease in 1964 and finally in 1965 they essentially stopped. After that time only slight variations (less than 0.04 magnitude) have taken place and these are irregular. This is the first observation of a cessation of a regular variable, and it has been confirmed at a number of observatories. Certainly this star will be watched closely for a number of years.

The evolutionary tracks of the very massive stars are not at all well known. There is some indication that they do evolve to the right of the main sequence and are similar in at least this respect to the less massive stars. One suspicion is that they do not evolve very far to the right of and above the main sequence. Evidence for this is offered by the fact that there are no super red giants which would correspond to the evolution of an O-type star that is evolving along a track *similar in shape* to the evolutionary tracks of the moderately massive stars. When a moderately massive star evolves off the main sequence it becomes more luminous. The most luminous stars in the universe (except novae), however, are the O-type stars and not the red giants. Accordingly, the O-type stars do not become more luminous when they leave the main sequence.

If the stars leave the main sequence by expanding into red giants, it is not yet entirely clear just what happens to them immediately after the red-giant stage. Does the helium-flash stage apply to stars of all masses? In which direction do stars move along the horizontal giant branch? How many times do stars move back and forth along the horizontal giant branch? These and other questions have yet to be answered.

It is clear, however, that most stars eject vast quantities of gas back out into space. The sun has a solar wind which presumably is typical of all stars on the main sequence. Observational evidence makes it clear that both the blue giants and the red giants eject gas in the manner of a "stellar wind," but eject much more gas than the sun. Furthermore, it is clear that at some time near the completion of the helium flash, many stars eject matter more violently—recall the P Cygni stars, the Wolf–Rayet stars, the planetary nebulae, and the supernovae. There is evidence to support the idea that as the degenerate helium core builds up and its temperature increases, the increased radiation pressure literally blows gases away from the star. A recent study of the central stars of 65 old planetary nebulae by G. O. Abell of the University of California at Los Angeles indicates that these stars are just entering the white-dwarf stage. They have an average surface temperature of 50,000°K and an absolute magnitude of +5, placing them in the white-dwarf region of the H–R diagram. Their average radius is estimated to be about 0.1 the radius of the sun. It seems that eventually all of the gases overlying the degenerate core are blown away, and the core becomes the white dwarf.

It now appears that a very massive star will evolve into a supernova, blow off most of its gas and leave a neutron star, a pulsar, behind. There is also good evidence to conclude that each regular nova is a member of a double-star system. In fact, some novae erupt more than once giving rise to a class of stars called dwarf novae (or recurrent novae).

**e. The White Dwarfs Again.**   Once a star has become largely degenerate we recognize it as a white dwarf. Because white dwarfs are composed mostly of degenerate matter, they do not obey the laws that explain the actions and structure of other stars. The laws explaining the white dwarfs were derived to a large extent by S. Chandrasekhar of the Yerkes Observatory, University of Chicago.

As we noted on p. 284, atoms no longer exist in degenerate matter— the nuclei and electrons become closely packed together. This limits the amount of space available to each particle, and its motion is consequently reduced. Some of the electrons manage to reach velocities that approach the velocity of light, but many are bound to a very small volume of space within the star and are not free to move at random. Since the nuclei are more massive, they are not rigidly bound and, although still restrained, are freer to move. If any protons (hydrogen nuclei) remain, they are squeezed out of the central regions of the degenerate gas by the more massive nuclei and float on the surface. The electrons remain inside the star because the electric field set up by the nuclei attracts them.

One of the seemingly strange characteristics of degenerate gas is often called the *Chandrasekhar's limit*. Chandrasekhar pointed out that the size of a degenerate core or white dwarf depends almost entirely upon the amount of matter it contains, but that if the *mass is increased the diameter will decrease*. An increase in the amount of matter increases the gravitational field and simply squeezes everything closer together. Consequently as the star ages and the mass of the degenerate core increases, the volume of the core actually decreases! But there is a limit. Chandrasekhar showed that a degenerate helium core cannot exceed a mass equal to 1.2 times the solar mass. With this mass it would shrink down to a diameter of zero! It is supposed that as the core approaches this limit, its temperature reaches such a high value that the core sheds its overlying layers of nondegenerate gas. If the star begins its life on the main sequence as a B-star, its mass is about 16 times that of the sun. During its succeeding evolution it ejects gases, first as a main-sequence star, then as a giant, then (perhaps) more dramatically, to leave exposed only the naked core of degenerate material. That core must have a mass less than 1.2 solar masses, yet be more nearly the size of our moon than the sun.

Such a high concentration of mass in a small volume gives rise to gravitational forces at the surface that exceed the gravitational force on the Earth by as much as 3,000,000 times. Thus, a man weighing 180 lb on the Earth would weigh nearly 270,000 tons on the surface of a white dwarf and, needless to say, would not survive even if the temperature were bearable.

This tremendous force of gravity affects the surface of the white dwarf. The theory of white dwarfs indicates that the degenerate gas comprising most of the star ends rather abruptly and is surrounded by an envelope of nondegenerate gas about 65 miles thick. Above this there is an atmosphere of sorts, but it is not more than a few hundred *feet* thick! This atmosphere probably contains any residual hydrogen left over from the star's earlier stages, which not only converted most of the hydrogen into helium but also, as a last resort, ejected the remaining hydrogen, perhaps as a shell-type star.

The pressure inside a white dwarf no longer follows the same laws that prevail in other stars and is called *degenerate pressure* rather than *gas pressure*. Unlike gas pressure, degenerate pressure is unaffected by temperature and therefore does not permit the star to convert thermal energy to gravitational potential energy by expanding. Any significant increase in thermal energy in the star would have to be reckoned with by an explosion. However, both theory and observation indicate that white dwarfs do not explode; it is presumed, therefore, that they have used up all their available nuclear energy in transforming one element into another. When a star becomes a white dwarf it no longer generates thermal energy; it is left to cool by radiating energy from its hot surface into cold space.

Observational evidence of the white dwarfs supports the theory of degenerate gases. The observed masses of white dwarfs are all under the theoretical limit. The larger white dwarfs have diameters of 10,000 miles; the smallest ones known can have a diameter not bigger than 1,000 miles. The observed surface temperatures range from 50,000°K to 4,000°K.

The spectra of white dwarfs differ markedly from those of other stars. In general, if any hydrogen lines are present they are weak, and indicate a small amount of residual hydrogen only. The white dwarfs with observed temperatures of less than 8,000°K have no hydrogen lines in their spectra, only a few metallic lines. At least six white dwarfs have no lines whatsoever in their spectra. The spectral lines that do appear must originate in the atmosphere; consequently when no spectral lines appear at all we assume that the star has no atmosphere.

One star, WZ Sagittae, seems to have entered the white-dwarf stage since observational astronomy began recording data. When this dwarf nova exploded in 1913 and again in 1946, its luminosity increased some 1,000 times. Its spectrum, except for some emission lines, now classifies it as a white dwarf. The emission lines are presumed to be caused by the ejected material that is still expanding, yet remains close enough to the star, and thus is hot enough to emit its own light.

According to theory and observation, WZ Sagittae has entered the last act of the drama of stellar evolution. Since it no longer generates thermal energy, the white dwarf is left to cool by radiating its energy into space. As its surface temperature decreases, it radiates less energy. Consequently its temperature decreases more slowly, and it may require 3 billion years to cool to 7,000°K from its initial temperature of more than 12,000°K.

A star with a surface temperature of 7,000°K, however, no longer radiates white light—its light is a yellowish-white. Five billion years after this, its surface temperature will have dropped to 4,000°K and it will be red in color. The term "white dwarf" is retained, however, for this name has come to imply a star whose mass, size, and composition classify it as a ball of degenerate gas. The term "red dwarf" would imply that the star resides on the lower part of the main sequence (Figure 12-7).

Thus it takes about 8 billion years for a white dwarf to cool to a temperature of 4,000°K, and even then it is not dead. This time span amounts to nearly twice the age of the solar system and even so does not include the time the star has spent in the evolutionary stages which it has had to undergo before it became a white dwarf. Since white dwarfs with this surface temperature have been observed, we can conclude that our galaxy (in which they reside) is considerably older than the sun.

If stars spend such a long time in the white dwarf stage, then, following our supposition that the longer they remain in any given stage the more heavily populated that stage will be, there should be many more white dwarfs than any other type of star. But it appears that the white dwarfs comprise only 3% of the stars in our galaxy. This seeming inconsistency is eliminated, however, if we assume that our galaxy is no older than just the age required for the most massive stars formed early in its life to have evolved into such cool white dwarfs.

Nevertheless, a star whose temperature is 4,000°K still radiates energy, and heat will continue to flow from it until it reaches the temperature of space, since, according to the second law of thermodynamics, heat flows from a higher to a lower temperature. As the white dwarf cools it will do so ever more slowly. After billions of years its surface will have cooled to such an extent that it will radiate no energy in the visible region of the spectrum; it will become first an infrared white dwarf and, when it has cooled completely, a black white dwarf. At this stage the nuclei and electrons reach their lowest possible energy level and the white dwarf (now emitting no light) becomes essentially a single huge molecule. Since no more energy is available from either contraction or nuclear trans-formations, it must remain in this stage indefinitely, a celestial tombstone.

Our present state of knowledge about stellar evolution is far from complete and will require a good deal more theoretical as well as obser-vational work before we can determine the stage of evolution for any one

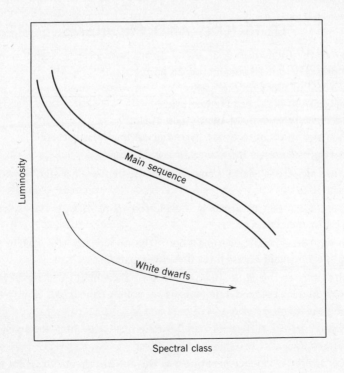

**Figure 12-7**

The evolutionary track of white dwarfs.

particular star. In other words, for any single star we cannot yet tell accurately its age, its past, and its future existence. It is comforting, however, to find that some observations do fit in very well with present theoretical work.

In summary, stars form from very tenuous gases composed chiefly of hydrogen. They contract to the main sequence, begin converting hydrogen into helium, expand and eject matter back into space, and then pass through various stages before they finally contract into extremely compact balls of degenerate gas completely devoid of hydrogen.

## BASIC VOCABULARY FOR SUBSEQUENT READING

Beta Canis Majoris star

Herbig–Haro object

Evolutionary track

T Tauri star

## QUESTIONS AND PROBLEMS

1. For each of the following situations, indicate how a star's central temperature, central pressure, and diameter would change:
   (a) a decrease in the central pressure
   (b) an increase in the central temperature
   (c) an increase in the rate of energy generation
   (d) an increase in the rate of energy radiation from the surface

2. What two pressures are in balance in a stable star?

3. Why must all of the energy generated in the center of a star be radiated out into space?

4. What would happen to a star if it radiated more energy into space than it generated in the center?

5. Explain why stars with larger mass form the upper part of the main sequence, and those with smaller mass form the lower part.

6. With the help of the H–R diagram explain the changes in temperature and luminosity of a star of moderate mass as it evolves from a parcel of gas and dust in a nebula.

7. What would be the differences in the brightest stars of a very young and a very old galaxy?

8. Cite observational evidence in support of the current theory of stellar evolution.

## FOR FURTHER READING

Baade, W., *Evolution of Stars and Galaxies,* Harvard University Press, Cambridge, Mass., 1963.

Page, T., and L. W. Page, ed., *The Evolution of Stars,* The Macmillan Co., New York, 1968.

Ezer, D., and A. G. W. Cameron, "Early Solar Evolution," *Sky and Telescope,* p. 328 (Dec. 1962).

Greenstein, J. L., "Dying Stars," *Scientific American,* p. 46 (Jan. 1959).

Herbig, G. H., "The Youngest Stars," *Scientific American,* p. 30 (Aug. 1967).

Sitterly, B. W., "Symposium on Abundances of Elements in Stars," *Sky and Telescope,* p. 11 (Jan. 1965).

Struve, O., "Variable Stars of Low Intrinsic Luminosity," *Sky and Telescope,* p. 67 (Aug. 1962).

"An Ex-variable Star?" *Sky and Telescope,* p. 323 (June 1966).

"Dark Interstellar Clouds," *Sky and Telescope,* p. 227 (April 1968).

"Infrared Glows Around Stars," *Sky and Telescope,* p. 158 (March 1965).

"Lithium and Star Aging," *Sky and Telescope,* p. 82 (Aug. 1963).

"Lithium in Red Giant Stars," *Sky and Telescope,* p. 146 (March 1968)

"Smallest White Dwarf?" *Sky and Telescope,* p. 17 (Jan. 1964).

"Some Early Results from Celescope," *Sky and Telescope,* p. 280 (May 1969).

CHAPTER
13
CHAPTER

# DOUBLE STARS AND PLANETS

Our discussion of the evolution of stars has indicated that the stars are, over their lifetimes, very active and changeable, and that we have difficulty in observing their changes only because our life span is so short. This chapter and the next will show that stars are also gregarious; single stars are actually less common than multiple stars. Some stars travel in pairs, others break the rule "two's company, three's a crowd," still others travel in fours, fives, sixes, tens, hundreds, and even thousands. All the stars in any one group are held together by their mutual gravitation. Some of these groups form very stable configurations and others are breaking up because the gravitational forces are not strong enough to hold the stars together. In a loosely held group each star goes its own way, parting company with its companions of travel through space.

## BINARY STARS

Consider, for the present, only the *binary stars*, those that travel in pairs. We can learn a great deal more about them than we can about single stars. Some of the stars that form a binary system are so close together they seem to be in contact with one another and cannot be resolved into separate stars by even the largest telescopes. Others are so far apart, they seem almost like single stars until their motions through space demonstrate that they are traveling together.

Most binaries in our galaxy, however, do not present either of these extremes. The most common separation of binary stars is somewhat less than 20 A.U.—about the distance of the Jovian planets from the sun! Perhaps if the nebular disk around the sun had been different in some as yet unknown respect, Jupiter or Saturn might have become stars, and the sun would have had quite different companions. It has been suggested that all the stars on the main sequence may be accompanied by planets.

## 13.1 Visual Binaries

The binary stars that can be seen as two stars when viewed with a telescope are called *visual binaries*. Most of these need the resolving power of

### Chapter Opening Photo

The eye end of the 36-inch telescope of the Lick Observatory. This telescope has been used to make thousands of measurements of double stars. (Lick Observatory photograph)

a telescope to separate them, but there are a few in which each member can be seen with the naked eye. Mizar and Alcor, at the bend in the handle of the Big Dipper, appear to the naked eye to be a binary system, but when viewed with even a small telescope Mizar is seen as two stars and Alcor is a third in the same field of view. With only a modest telescope many other stars that seem to be single when viewed with the naked eye are revealed to be double; Polaris is another example.

Some stars may appear to be visual binaries when in reality they are separated in space by many light years. Their near coincidence on the celestial sphere results from their being nearly lined up when viewed from the Earth. Such apparent binaries, called *optical doubles*, are fairly uncommon. The difference between an optical double and a visual binary can be noted by following the motions of the two stars. A visual binary forms a physical system, in that the stars are under one another's gravitational attraction. Therefore they travel through space together as they revolve around each other, or more precisely, as each revolves around the common center of mass of the system. Stars forming an optical double, on the other hand, usually have different space motions and do not comprise a physical system.

**a. Stellar Masses.** Since the force of gravity holding two stars into a physical system depends on their masses, we have an approach to the problem of measuring stellar masses. Fortunately this approach is not completely new to our discussion. The laws that explain the motions of two stars about their center of mass are the same laws that explain the motions of the Earth and moon about their common center of mass and the motions of the planets about the sun. We can use Kepler's three laws.

The first two laws can be restated as they were given on p. 112: the orbits of each star about the center of mass must be elliptical, and the line joining the star with that center of mass (the radius vector) must sweep out equal areas in equal intervals of time. The third law, however, must be given in a more complete form. As stated on p. 113 it does not take into account the masses of the sun and planets. Those masses could be neglected because the sun is so much more massive than any of the planets. But the stars in a binary system are likely to have more nearly the same mass. Consequently, it is Kepler's third law that gives us information about the masses of the two stars.

Kepler's third law was originally given as a direct proportion between the squares of the periods and the cubes of the orbital radii of any two planets. The more complete law is also given as a proportion, but rather than comparing any two planets it compares any binary system with the sun–Earth system.

$$\frac{M + m}{M_s + m_e} = \frac{s^3/P^2}{s_e^{\,3}/P_e^{\,2}}$$

Fortunately this equation can be greatly simplified by the proper choice of units of measurement. Since the separation of the stars in most binary systems is similar to the distance of the planets from the sun, we can measure these distances in astronomical units. Thus the radius of the Earth's orbit $s_e$ becomes equal to 1. We can also select the year for the unit of time and by so doing make the period of revolution of the Earth $P_e$ equal to 1. This enables us to eliminate two of the factors from the equation.

The sun's mass, 332,958 times that of the Earth, almost completely overwhelmes the Earth's mass; there is no appreciable difference between the sun's mass $M_s = 332,958$, and the combined sun–Earth mass $M_s + M_e = 332,959$. Therefore we may neglect the Earth's mass and not feel that by doing so we will have damaged the validity of our calculations.

Lastly, we can measure stellar masses in terms of the sun. Thus the mass of the sun $M_s$ equals 1.

The formula now takes on a simpler form:

$$M + m = \frac{s^3}{P^2}$$

where $M$ is the mass of one star (usually the more massive of the two) and $m$ is the mass of the other, both in solar masses; $P$ is the period of revolution of one star about the other in years; and $s$ is their mean separation (the average orbital radius of one star about the other) in astronomical units.

In using this formula it is convenient to assume that the *primary* (the brighter star) does not move, and to establish by observation the orbit of the *secondary* (the fainter star) with respect to the primary component. In doing this we neglect the center of mass of the system, but Kepler's laws still hold for this *relative orbit*.

Assuming for the time being that the mean separation $s$ and the period $P$ can be measured, we are left with the sum of the masses and not with each individual mass. If the sum of the masses of a given binary system is found to be 10 solar masses, for example, we do not know whether the star's individual masses are 1 and 9, 5 and 5, or any other combination that adds up to 10.

We therefore need to establish some other relationship which when combined with the sum of the masses will yield each individual mass. This can be done only if we consider the motion of each star around the center of mass. The average distance of each star from the center of mass is equal to its mean orbital radius and depends on the individual mass of each star. The larger the mass, the smaller the mean radius. This can be expressed as an inverse proportion:

$$\frac{M}{m} = \frac{r}{R}$$

where $M$ and $m$ are the masses of the two stars, $M$ being the more massive star, and $R$ and $r$ are the mean orbital radii, $R$ referring to the more massive.

From this expression we see that the mass $M$ is equal to the mass $m$ multiplied by the ratio of their orbital radii.

$$M = \frac{r}{R}\, m$$

Since the mean orbital radius of the less massive star $r$ must be larger than that of the more massive star $R$, the quotient $r/R$ must be greater than 1.

To demonstrate how the equation for the ratio of the masses can be combined with the equation for the sum of the masses, let us consider the double star Sirius. The sum of the masses of Sirius and its companion is 3.26 solar masses:

$$M + m = 3.26$$

The mass of Sirius alone, however, is 2.33 times that of its companion:

$$M = 2.33m$$

Therefore,

$$2.33m + m = 3.26$$

and,

$$3.33m = 3.26$$

So the mass of the white dwarf companion is

$$m = 0.98 \text{ solar mass}$$

and the mass of Sirius is

$$M = 2.28 \text{ solar masses}$$

A mathematician will say that if we have two unknowns we can only find the value of each if we have two equations. This we did. Our two unknowns were the individual masses of the stars, and the two equations were the expressions involving the sum of the masses and the ratio of the masses. Thus the mathematics involved are not too impenetrable, but the same cannot be said for the observations that must be performed to supply the data needed for the calculations.

We have three quantities that must be observed: the period $P$ of the secondary component about the primary, the mean separation $s$ of the two stars in astronomical units, and the ratio of the mean orbital radii $r/R$ of the two stars about the center of mass of the system.

**b. The Necessary Observations**   The observational part of the problem is complicated because the orbits of the visual binaries are oriented more or less at random in the galaxy. Thus we see them at different angles. Some we see "face-on," that is, the plane of the orbit makes a right angle with the line of sight, and we can get a true picture of the shape of the orbit of the secondary about the primary. Most of the visual binaries, however, are inclined from this most desirable situation, and the shape of their orbits becomes distorted by foreshortening. If we look at the top of a round table situated at a short distance from where we are, its outline will be elliptical rather than circular (unless we are directly above the table). Elliptical orbits seen at an angle will also appear to change shape, although they will still appear as ellipses. At the other extreme some of the orbits may be seen "edge-on," and the stars seem to move in straight lines back and forth in front of and behind each other.

The plane of the orbits of a binary system that makes a right angle with the line of sight must be parallel to the *plane of the sky*. The plane of the sky is that portion of the celestial sphere in the region of the orbit. To illustrate, we may place a small ring flat on the inside surface of a large globe with the observer at the center. The ring touches the globe at all points and thus the two are parallel.

The angle between the plane of the orbit and the plane of the sky is called the *inclination*. When the two planes are parallel the inclination is zero. If the plane of the orbit is tipped from the plane of the sky, the inclination is greater than zero. Such an orbit resembles the ring on the inner surface of the globe when one side of the ring is raised from the surface. The true shapes of the orbits are not seen when they are so oriented in the sky, and the astronomer must deduce the *true orbit* from its *apparent orbit* (the projection of the true orbit onto the celestial sphere). Fortunately, Kepler's laws hold true even for the apparent orbit. When the plane of the orbit is seen "edge-on" it makes a right angle with the plane of the sky and its inclination is 90°. A ring placed "upright" against the inner surface of the globe is in a comparable position.

To relate these observations to the mathematics already discussed let us use a simple example—when the inclination of the orbit is zero. In practice, the motion of the secondary component is observed and referred to the primary star which is assumed fixed. This yields the *relative orbit*. Kepler's laws again apply. The period of the secondary component about the primary component is the same as the period of each about their common center of mass, for as the stars revolve, a line joining the stars always passes through that center of mass (Figure 13-1). Thus, the true period can be determined by observations of the motion of the secondary relative to the primary.

The necessary observations are of the angular separation of the two stars (measured in seconds of arc) as well as of the orientation of the stars in the sky. At one time the two stars may be 5 seconds of arc apart with the secondary due south of the primary. At another time their separation may be 7 seconds of arc, and the secondary may be 10° east of south with respect to the primary.

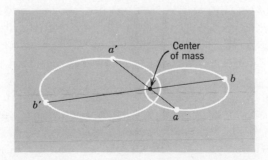

**Figure 13-1**

The motion of two stars in a binary system about their common center of mass.

The period can be determined quite easily from either the true orbit or the relative orbit. The observed positions must be joined by the ellipse that best fits Kepler's second law, the law of equal areas. From this ellipse the period can be determined.

To determine the mean separation of the two stars in astronomical units, however, we must know how far the stars are from the sun. If they are close enough we can measure their heliocentric parallax and from it obtain the distance. More often than not, however, the stars are too distant for parallactic measurements. When they are, we must resort to a study of their spectra and determine their distance by spectroscopic parallax. When we know the distance and the apparent size of the true orbit in seconds of arc we can obtain the actual size of the orbit and thus the mean separation of the two stars in astronomical units. The period and the mean separation enable us to find the sum of the masses.

But we have only the sum of the masses. In order to obtain the ratio of the mean radii of each orbit we must follow the motion of *each* star as it moves relative to the background stars; that is, we must observe the proper motion of each star. The best example is the proper motion of Sirius, an A0 star, and its white-dwarf companion. In fact it was by following the proper motion of Sirius that its faint companion was discovered, for the motion of Sirius is not a straight but a wavy line. Figure 13-2 shows not only the proper motion of each star but also gives the relative orbit of the secondary about the primary as well as the orbit of each about their common center of mass. The inclination of Sirius is not 0° as in our earlier ideal example; it is 43°.

The size and shape of each apparent orbit is enough, however, to determine the ratio of the mean radii, for this ratio is not altered by foreshortening. After we find the combined mass from Kepler's third law, the ratio of the radii shows the individual masses of Sirius to be 2.28 solar masses for the primary and 0.98 for the white-dwarf secondary. The true shape of an apparent orbit can be determined by the geometry of the situation since the inclination of the orbit foreshortens the distances involved in the same ratio. The true orbit is needed to get the true separation of the stars in astronomical units.

Thus for a visual binary, if the true relative orbit of the secondary about the primary can be found along with the binary's distance from the sun, the sum of the masses can be determined. The individual masses can be found only if the ratio of the mean orbital radii is known, and this ratio can be determined only if the motion of each star is referred to the background stars.

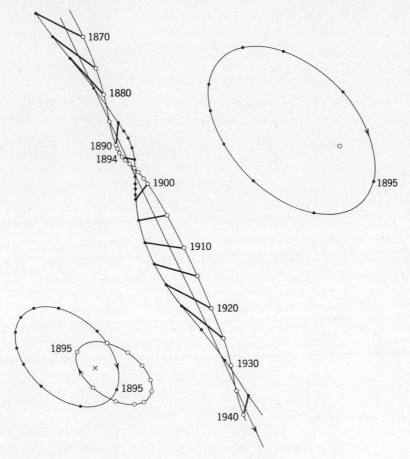

## Figure 13-2

The motions of Sirius and its white-dwarf companion. Sirius is shown by a circle and its companion by a dot. (Upper right): The apparent relative orbit of the companion about Sirius. Dots are at five-year intervals. (Lower): Apparent orbits of Sirius and its companion about their common center of mass. Note that the center of mass (the focus of the elliptical orbits) is not at the focus of the apparent orbit. (Center): The motions of Sirius and its companion relative to the background stars. The center of mass moves on a straight line. The scale is not the same for the three drawings. (Cecilia Payne-Gaposchkin, *Introduction to Astronomy*, Copyright © 1954, Prentice-Hall, by permission.)

## 13.2 Spectroscopic Binaries

If the inclination of an orbit is large, the motion of the stars will be revealed by a periodic Doppler shift of their spectral lines. Since a line joining the stars always passes through the center of mass of the system, one star will be approaching while the other is receding from us, and their spectral lines will be shifted in opposite directions. Consequently, since the Doppler shift of the spectral lines is independent of the apparent separation of the stars in the sky, a binary system will always be disclosed by the periodic Doppler shift in the spectra of two stars, although these may be so close together that they cannot be resolved even with the largest telescope. A system that can be recognized by its spectrum is called a *spectroscopic binary*.

In Figure 13-3*a* we see two spectrograms of a binary system, each taken at a different time. In the lower stellar spectrogram the spectral lines are all shifted toward the violet, indicating that the brighter component of the pair is approaching the Earth; the upper stellar spectrogram reveals that it is receding. The brighter component is so much brighter than the fainter component that the spectral lines of the fainter do not even appear. When the radial velocities are plotted on a graph against time, the velocity curve is obtained (Figure 13-3*b*).

The corresponding motions of the two stars are shown in Figure 13-4: (*a*) shows the brighter component receding from the Earth while the fainter is approaching; (*b*) shows the two stars traveling at right angles to the line of sight (at this time neither star shows a Doppler shift unless the entire system is moving relative to the sun); (*c*) shows the brighter component approaching while the fainter is receding.

If the two stars are of nearly equal magnitude, the spectral lines of the stars will be of nearly equal intensity, and both spectra will be visible. When the lines of one star are shifted to the violet the lines of the other star will be shifted to the red, but they will both be at their mean position at the same time. Therefore, as the stars continue to revolve about their common center of mass, the spectral lines of each star will oscillate back and forth across their mean position so that we sometimes see double lines and sometimes only single lines.

Since the velocities of the two stars are proportional to their distances from the common center of mass and since these distances are inversely proportional to the individual stellar masses, the ratio of their maximum velocities is inversely proportional to the ratio of their masses. We can

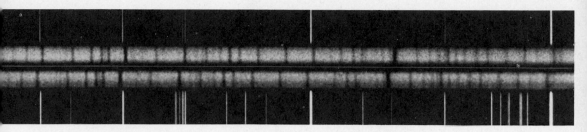

### Figure 13-3 (a)

Two spectra of the brighter component of the double star Castor, alpha-Geminorum, are included between two bright-line comparison spectra (*above*). This star is a spectroscopic binary. The shifting of the spectral lines from the longer wavelength (the upper of the two stellar spectra) to the shorter is evident.

### Figure 13-3 (b)

A schematic drawing of the velocity curve for this spectroscopic binary (*below*).

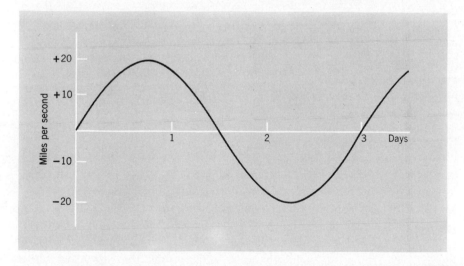

therefore obtain the ratio of the masses when the spectral lines of both components are visible. But unless the binary system is actually a visual binary we cannot obtain their individual masses, for we cannot determine the inclination of the orbit. Hence we cannot determine the mean separation of the two stars, without which we are unable to determine the sum of the masses.

If one of the components of a spectroscopic binary system is much

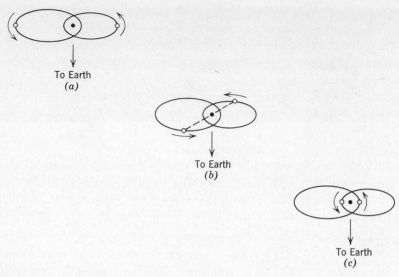

**Figure 13-4**

The motion of the two components in a spectroscopic binary. In (a) the primary star is receding and therefore its spectral lines are shifted to the red and those of the secondary (if visible) are shifted to the violet. The two stars in (b) are traveling in a direction perpendicular to the line of sight and thus their spectral lines are not shifted from their mean position. (c) shows the primary approaching and the secondary receding. The orbits are coplanar.

more luminous than the other, we see the spectral lines of only the more luminous star shifting back and forth. We can obtain its velocity curve but we can learn very little about the fainter component. We are nevertheless aware of the fainter star's presence, because the brighter component reveals a periodic shifting of its spectral lines without the same variation in brightness that occurs with the Cepheid. If the velocity curve of a Cepheid were mathematically treated as though it were a spectroscopic binary with only one velocity curve observable, the supposed fainter component would have to be inside the Cepheid. There are spectroscopic binaries, however, that have a Cepheid as one of their components.

## 13.3 Eclipsing Binaries

There are spectroscopic binaries with inclinations so large that one star passes in front of the other, causing an eclipse. These are called *eclipsing binaries.* As with the moon eclipsing the sun it is possible for the star nearest us to eclipse all or only a part of the star behind, and we can witness a total, a partial, or an annular eclipse. The type of eclipse can be determined from the light curve—the record of brightness of the system plotted against time.

Figure 13-5 shows a binary system with an inclination of 90°, in which the larger but fainter star can completely hide the smaller brighter star causing a total eclipse. An annular eclipse occurs when the smaller star passes between us and the larger star.

In order to describe the shape of the light curve, let us assume the larger star to be stationary with the smaller one revolving about it. When the stars are not eclipsing we receive the maximum amount of light and the system is at its brightest (Figure 13-5a). The partial phase of the total eclipse begins when the smaller star just begins to go behind the larger star (b). The amount of light reaching the Earth starts to decrease until the smaller star is completely hidden from our view (c) and the total eclipse begins. The bottom of the light curve for a total eclipse is flat, for during

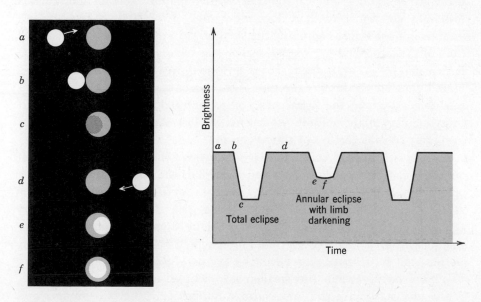

**Figure 13-5**

Eclipsing binary with total and annular eclipses showing the relative positions of the stars at various phases of the light curve.

this time we receive light only from the larger star. The total eclipse ends when the smaller star first comes into view from behind the larger star, and the amount of light reaching the Earth then begins to increase until both stars are completely in view (d). Actually all eclipsing binaries are so close together that we cannot resolve them into separate stars; thus the words "in view" refer to the stars as they are drawn in Figure 13-5.

After the total eclipse the smaller star revolves in its orbit to a point at which it goes in front of the larger star. When it begins to hide some of the larger star, the partial phase of the annular eclipse begins. The amount of light reaching the Earth decreases until the smaller star conceals an area of the larger star equal to itself (e). Since the area of the larger star that is concealed during an annular eclipse is always the same, namely, the area of the smaller star, the amount of light received during that phase is constant provided that the eclipsed star does not exhibit limb darkening (see p. 243).

In an annular eclipse the amount of light that the smaller star blocks depends not only on its area but also on the temperature of the larger star. If the larger star does exhibit limb darkening, its limb is cooler and radiates less light per unit area than does the center of the disk. When the full annular eclipse is just beginning the smaller star conceals a region near the limb of the larger star (e). As the smaller star moves directly in front of the larger star (f) it continues to cover an equal area, but this area radiates more light than the limb. Therefore, if the larger star exhibits limb darkening, the bottom of the light curve of an annular eclipse is not flat but dips slightly in the middle when the smaller star covers up the more luminous central portion of the larger one.

In most eclipsing binaries the larger star has the lower surface temperature and radiates less light per unit area than the smaller star. Since the area eclipsed is the same (namely, the cross-sectional area of the smaller star) no matter which star is in front, more light is cut from our view when the smaller, hotter star goes behind, and the light curve may have two unequal minima. The deepest minimum, which occurs when the smaller, hotter star is eclipsed, is called the *primary minimum*. The *secondary minimum* occurs when the smaller star eclipses the larger, cooler one. If the two stars have the same surface temperature the two minima will have the same depth.

A partial eclipse results when the smaller star is never completely hidden by the larger (Figure 13-6). Thus the area of the smaller star concealed by the larger one always changes and the bottom of the minimum of the light curve is not flat. The light curve reaches a minimum when the maximum area of the eclipsed star is concealed, but at this instant the eclipsed star begins to move out from behind the eclipsing star, and the amount of light we receive increases once more. A partially eclipsing binary system is of less value to the astronomer than a totally eclipsing one, for the amount of area concealed in a partial eclipse is uncertain.

Total eclipsing binaries in which both stars are of about equal luminosity are extremely valuable to the astronomer, for they yield a wealth of information. If the stars have nearly the same luminosity, the system is also a spectroscopic binary with the spectra of both stars visible. Thus we can obtain the velocity curve for each star moving in its orbit. Since the inclination can be determined (for totally eclipsing binaries it is very nearly 90°), the maximum velocity of each star yields the actual velocity (in miles per second) of the star in its orbit. Most of the orbits of eclipsing binaries are nearly circular and hence the velocity of the star in its orbit is nearly constant. The actual velocity multiplied by the period (the time for one complete revolution) thus gives the circumference of the orbit. Once the circumference of each orbit is known, it is easy to find the orbital radii. With this information it is possible to determine the masses according

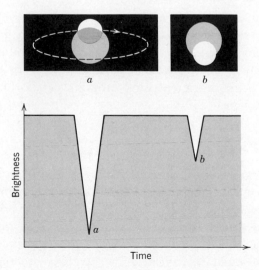

**Figure 13-6**

The light curve for a partially eclipsing binary.

to the method used with visual binaries. The period is known from the light or the velocity curve, the mean separation is the sum of the mean orbital radii, and since the radius of each orbit is known the ratio of the radii is easily determined.

For close binaries the duration of the eclipse depends on two factors: the size of the larger star and the velocity of each star in its orbit. The velocities can be determined if both spectra are visible, and by means of these velocities we can find the diameter of each star. The diameter of the smaller star is equal to the time required for it to cross the limb of the larger star (the time from the beginning to the end of the partial phase of a total or annular eclipse, *b* to *c* in Figure 13-5) multiplied by the velocity of the smaller star relative to the larger. The diameter of the larger star is equal to the time it takes the smaller star to traverse one diameter of the larger star (from the beginning of the partial eclipse to the end of the total eclipse) multiplied by the velocity of the smaller relative to the larger.

It is not uncommon to find one of the components of an eclipsing binary to be a red giant. Red giants, it will be recalled, are inflated stars with extended atmospheres. As the smaller hotter star goes behind the red giant, its light must travel through the atmosphere of the cooler companion, yielding of course a changing spectrum that gives us information about the atmosphere of the giant.

More than half of the 3,000 or so known eclipsing binaries have periods of less than 10 days, and more than three-fourths of them have periods of less than 100 days. Those stars with the shortest period are usually early type stars, mostly O- and B-types. A short period means that the stars are close together. Some of the eclipsing binaries are so close together they seem almost to touch.

## 13.4 Stars, Planets, and Life

It seems significant that the average distance of spectroscopic binaries is about 20 A.U. Furthermore, it seems reasonable to presume that whatever forms in the nebular disk about a star depends upon the conditions of that disk. It is, consequently, reasonable to assume that either stars or planets may form in such a disk. Since double stars have formed, it is again reasonable to assume that planets have also formed about other stars.

There is, in fact, convincing evidence that planets are revolving about other stars. Work toward discovering such planets has been carried out at

Sproul Observatory, Swarthmore College, under the directorship of Peter van de Kamp. Faint companions of stars have been discovered by the wavy nature of the brighter companion's proper motion. Recall that the white dwarf companion of Sirius was discovered by this technique (see Figure 13-2). So, reasoned the astronomers, planets might also be discovered if those stars with large proper motion (and consequently close to the sun) and with small mass are studied in detail. The proper motion of stars with large mass would not be appreciably affected by planets.

**a. Barnard's Star.** Barnard's star is not only an M5 dwarf but it has the largest known proper motion of any star in the sky: 10.27 seconds of arc per year (see Figure 10-3). At this rate it will move one apparent diameter of the moon (30 minutes of arc) in 170 years, and it has already been observed for 50 years.

The proper motion of Barnard's star has been studied in detail to determine the nature of its invisible companion. Its proper motion is more subtle and complex than can be shown in Figure 10-3. So far the best solution to that motion is to assume that it has not one but two planets accompanying it and causing very slight wavy motions in its proper motion.

If the mass of Barnard's star is assumed to be 0.15 solar mass, then one of its planets, called B1, would have a mass of 1.1 times the mass of Jupiter and be revolving about Barnard's star in an orbit with a radius of 2.8 A.U. The second planet, B2, would have a mass 0.8 times that of Jupiter and be 4.7 A.U. from the central star.

There are other stars with large proper motions that also appear to have planets about them. One, known as BD+68°946, appears to have a companion with a mass of only 0.026 solar mass. (Jupiter's mass is about 0.001 solar mass.) Another star, 61 Cygni, may also have planets about it.

**b. More Planets than Stars?** The sun and apparently three of the stars nearest the sun have planets; how about the rest of the stars in the sky? We know from our stellar census (the luminosity function, see Figure 10-12) that the number of stars in the solar neighborhood increases as we consider stars of decreasing luminosity. This trend continues to about absolute magnitude +13, then the number begins to fall off for still less luminous stars. A star with a luminosity of +13 has a mass of about 0.07 solar mass.

A gaseous sphere with a mass greater than about 0.07 solar mass will, during the initial contraction stage, generate a high enough central temperature to ignite and sustain thermonuclear reactions. The energy so generated permits the object to radiate light into space and we call it a star. A gaseous sphere with a mass less than about 0.07 solar mass will not generate a high central temperature during the initial contraction; thermonuclear reactions will not be ignited and sustained. After contraction ceases the object cools down. Such a cold object we call a planet. Mass is the critical factor that differentiates a planet, such as Jupiter, from a star.

It would appear, therefore, that the number of faint stars in the solar neighborhood begins to decrease at about absolute magnitude +13 only because objects with mass less than about 0.07 solar mass become planets and are not visible from the Earth.

We are left to ask, then, what would the curve for the luminosity function look like if it were redrawn to represent the number of objects—stars and planets—against their mass. Would the number of objects continue to increase as shown in Figure 13-7? Are there more planets than stars in our galaxy? There certainly are more planets than stars in our own solar system! and Barnard's star? and the rest?

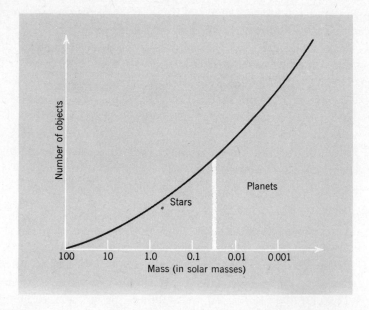

**Figure 13-7**

If the luminosity–function curve were redrawn as the number of objects against their mass, perhaps it would look like this. Are there more planets than stars in our galaxy?

**c. Stellar Rotation and Planets.**   Still another line of attack to this problem is brought forth from a study of stellar rotation. It will be recalled that the sun contains only 2% of the angular momentum of the solar system (see p. 210), even if it has 99.9% of the mass. To explain this, it has been proposed that the magnetic field of the sun reacted with newly ionized gases on the rotating nebular disk to slow the sun's rotation. That is, angular momentum was transferred from the sun to the nebular disk. This leads us to the conclusion that if the proto-sun had not formed a nebular disk, it would now be without planets and it would be rotating much faster. Therefore, the rate of stellar rotation may be a clue to help us decide whether other stars have planetary systems.

Stars reveal rotation by their spectral lines. A rapidly rotating star will have spectral lines that are broadened. Portions of the star are approaching the Earth while other portions are receding, and the resulting Doppler shifts yield broad spectral lines. This broadening can be distinguished from other forms of line broadening, such as pressure broadening. The average observed rotational speed at the equator of main sequence stars of different spectral types is shown in Table 13-1.

**Table 13-1 ● Stellar Equatorial Rotational Velocities**

| Spectral type | Main sequence stars |
|:---:|:---:|
| B0 | 125 mi/sec |
| B5 | 130 |
| A0 | 120 |
| A5 | 100 |
| F0 | 60 |
| F5 | 16 |
| G0 | $< 10$* |
| K, M | $< 10$ |

*The symbol $<$ means less than.

It is clear from the table that late-type main sequence stars, F0 and later, rotate slowly. The implication is that main sequence stars later than about F0 have formed a nebular disk about each. And presumably in every such nebular disk planets formed. (Barnard's star is an M5 star.) In addition, the luminosity function makes clear that there are many more stars of spectral type F, G, K, and M than of type B and A. The least massive stars far outnumber the more massive ones. So from this line of reasoning, we can only conclude that most of the stars in our galaxy have planets about them.

**d. Radio Signals from Clouds with Molecules.** Recent observations with radio telescopes operated as long base-line interferometers indicate that radio signals resulting from molecular vibrations originate in clouds of gas no larger than 2 A.U. in diameter. These radio signals result from molecules such as water, $H_2O$; carbon monoxide, CO; the hydroxyl radical, OH; ammonia, $NH_3$; hydrogen cyanide, HCN; formic acid, HCOOH; and cyanoactylene, $HC_3N$. It is tempting to assume that these rather small gaseous clouds are protostars in the initial contraction stage.

From the observations of the separation of spectroscopic binary stars and the planets about Barnard's star, it is logical to assume that in most planetary systems, there should be one planet with a surface temperature such that water can exist as a liquid. The chemicals observed in gaseous clouds in space when mixed with that water, with ultraviolet radiation from the central star, or with lightning, will form amino acids and protein molecules just like those formed on Earth. Our logical argument, then, leads us to the conclusion that given a planet with conditions similar to those on Earth, life will develop on that planet as it has developed here on Earth.

The fantastic variety of life forms on this Earth, in the ocean, in caves, in the air, living and in fossil form, indicates that innumerable varieties have evolved, but only those which can succeed have survived. The same should apply on any planet with similar conditions and similar chemistry. Most stars in our galaxy ought to have at least one planet with these conditions and this chemistry.

## BASIC VOCABULARY FOR SUBSEQUENT READING

| | |
|---|---|
| Apparent orbit | True orbit |
| Eclipsing binary | Visual binary |
| Spectroscopic binary | |

## QUESTIONS AND PROBLEMS

1. Why must the distance of a visual binary be known before the sum of the masses of the system can be determined?

2. What must be observed before the masses of the individual stars of a visual binary can be determined? Why must this observation be made?

3. (a) Capella is a visual binary system with a period of 0.285 year. The separation of the two stars is 0.85 A.U. What is the sum of the masses of the two stars?

   (b) The mean orbital radius of the brighter member of the Capella system is 1.27 times that of the fainter. What is the individual mass of each member?

4. The terms relative orbit, apparent orbit, and true orbit are used in the study of binary systems. What is the meaning of each term and why must three such orbits be considered?

5. When would the terms relative true orbit and relative apparent orbit be applicable?

6. What would be the difference in the light curve of an eclipsing binary system if the major axis of its orbits were at right angles to the line of sight rather than parallel to it?

7. A spectroscopic binary is composed of a main sequence B-type star and a main sequence K-type star. Would this be a single-line or a double-line spectroscopic binary?

8. Under what conditions do we observe: (a) a single-line spectroscopic binary, and (b) a double-line spectroscopic binary?

9. (a) Must a spectroscopic binary also be an eclipsing binary?
   (b) Must an eclipsing binary also be a spectroscopic binary?

10. Under what conditions can a binary system cause:
    (a) a total eclipse
    (b) a partial eclipse
    (c) an annular eclipse

11. For what kind of eclipsing binary systems is:
    (a) the primary minimum of nearly the same depth as the secondary
    (b) the primary minimum much deeper than the secondary

12. Suppose that the larger component of an eclipsing binary system has "limb brightening." How would the shape of the secondary minimum appear if it results from an annular eclipse?

13. What observations have led to the belief that Barnard's star is accompanied by planets?

14. What observations indicate that stars of small mass are more common than stars of large mass?

15. What evidence indicates that late-type stars are more apt to have planets than early-type stars?

16. Discuss the likelihood that life forms on other planets would be very similar to life forms on earth.

# FOR FURTHER READING

Struve, O., and V. Zebergs, *Astronomy of the 20th Century*, Crowell, Collier and Macmillan, New York, 1962, Chapter XIV.

Brown, H., "Planetary Systems and Main Sequence Stars," *Science*, p. 1177 (Sept. 11, 1964).

Eggen, O. J., "Stars in Contact," *Scientific American*, p. 34 (June 1968).

Hack, M., "Stellar Rotation and Atmospheric Motions," *Sky and Telescope*, p. 84 (Aug. 1970); p. 143 (Sept. 1970); p. 208 (Oct. 1970).

Huang, S.–S., "The Origin of Binary Stars," *Sky and Telescope*, p. 368 (Dec. 1967).

Limber, D. N., "The Pleiades," *Scientific American*, p. 58 (Nov. 1962).

Lippincott, S. and M. D. Worth, "The Double Star Sirius," *Sky and Telescope*, p. 4 (Jan. 1966).

Lovell, B., "Radio-emitting Flare Stars," *Scientific American*, p. 13 (Aug. 1964).

van de Kamp, P., "Barnard's Star: The Search for Other Solar Systems," *Natural History*, p. 38 (April 1970).

Vandervoort, P. O., "The Age of the Orion Nebula," *Scientific American*, p. 91 (Feb. 1965).

Walker, M. F., "New Observations of AE Aquarii," *Sky and Telescope*, p. 23 (Jan. 1965).

Worley, C. E., "Measuring Double Stars with a Filar Micrometer," *Sky and Telescope*, p. 73 (Aug. 1961); p. 141 (Sept. 1961); p. 261 (Nov. 1961).

# STAR CLUSTERS

Binary stars may travel alone through space or they may be accompanied by other stars. Epsilon Lyrae is composed of two pairs of binaries. The naked eye can see two stars close together, but the telescope reveals that each of these is a visual binary, one star of which is a spectroscopic binary. Castor, the brightest star in the constellation of Gemini, is seen as one star to the unaided eye, in the telescope it appears as two, but with the spectrograph each is revealed to be a spectroscopic binary. There is another spectroscopic binary close to Castor that is moving with the same speed and in the same direction as Castor and therefore must be a physical member of the same system, bringing the total to six.

The gregarious nature of stars is revealed most startlingly, however, in the clusters, which contain as many as 100,000 stars all held together by the mutual gravitation of the stars in the cluster.

Clusters can be divided very nicely into two distinct kinds: *open clusters* (also called *galactic clusters*—a term easily confused with clusters of galaxies, so we prefer the more descriptive term) and *globular clusters*. The most obvious differences between these two types are the number of stars contained in each as well as their general shape and appearance, and location in the galaxy.

Open clusters have from around 20 to several thousand stars in a loose assemblage of rather indefinite shape. Globular clusters, on the other hand, may be composed of as many as 100,000 stars held together in a rather compact, spherical, or ellipsoidal shape.

## 14.1  Open Clusters

a. Cluster Membership. A number of open clusters are visible to the naked eye; for example, the Pleiades, the Hyades in the constellation Taurus, and most of the stars in the Big Dipper (Ursa Major). We are so close to the Ursa Major cluster that one of its members, Sirius, appears in quite a different part of the sky. The sun, however, is not a member of this cluster or of any other known cluster.

Without exception the open clusters are found very close to the central plane of our Milky Way galaxy. This striking conformity makes it difficult to identify open clusters at very great distances from the sun. They simply blend in with the profusion of background stars in the Milky Way.

**Chapter Opening Photo**

The double cluster, h and χ Persei. (Lick Observatory photograph)

Members of a cluster may be identified by their motion. If the cluster is close to the Earth, a common proper motion will identify the stars that physically belong to the group. If the proper motion of a particular star in the group is different, it may be either behind the cluster or between the cluster and the Earth. Although each star in a cluster moves with respect to the cluster, this velocity is small compared to the velocity at which the cluster is moving through space.

The Hyades cluster illustrates the principle of common proper motion better than any other because it is close enough to reveal good proper motions and yet compact enough to appear as a cluster on a photograph. From Figure 14-1 it can be seen that all the stars in the cluster are moving toward a common point called the *convergent point*, which indicates that the cluster is moving away from the sun. As the cluster recedes from us, its apparent size will decrease, and the paths of the individual stars converge like railroad tracks leading across flat country. The average radial velocity of the stars in the cluster is about +21 miles per second, which confirms our observation of the cluster's recession.

The brightest star in the constellation of Taurus is the binary, Aldebaran. Its proper motion, however, indicates that it is not a member of the Hyades cluster. Its lack of membership becomes even more convincing when we realize that its parallax gives a distance of only 58 light years, and the cluster is 130 light years away. All the stars that belong to the Hyades are included in a volume of space about 33 light years in diameter.

Proper motion studies of the more distant clusters are impractical as a means of identifying cluster members, however, for the greater distances mean that the proper motions are correspondingly reduced. It is nevertheless possible to identify members by radial velocity measurements. For example, the radial velocity of Aldebaran is +30 miles per second and not +21, which is the radial velocity of the Hyades cluster.

**b. Color–Luminosity Diagrams.** With the utilization of the photoelectric cell for obtaining very accurate measurements of the brightness and color of a star, it has become convenient and instructive to make an H–R diagram of stars in a cluster. The stars in the field that fit the expected pattern are considered members of the cluster in good standing; the stars that appear on the diagram in a position quite different from the bulk of the cluster stars are rejected from membership. Other criteria, such as radial velocity and spectroscopic parallax, may be employed to differentiate those stars whose membership is in doubt.

The use of the photoelectric cell to determine color replaces the spectral sequence in the H–R diagram. Since both the spectral type and the color of the star are dependent on the temperature, they are often used inter-

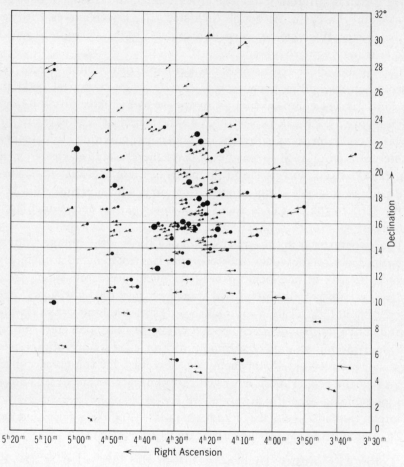

**Figure 14-1**

The proper motions of the stars in the Hyades cluster converge to a point, indicating that the cluster is receding from us. The size of the dots on this drawing by Van Bueren, of Leiden Observatory, indicates the magnitudes of the stars. The arrows show the proper-motion displacement expected in the course of the next 18,000 years. (Reprinted by permission of the publishers from Bart J. and Priscilla F. Bok, *The Milky Way*, Cambridge, Mass., Harvard University Press, Copyright © 1957 by the President and Fellows of Harvard College.)

changeably. However, since spectral classification depends also on the pressure of the gases in the atmosphere of a star, a giant and a dwarf of the same color belong to slightly different spectral classes.

The color of a star is rated by its *color index*, a number indicating the relative brightness of the blue and the yellow regions of the spectrum.* By use of appropriate filters in front of a photocell, measurements of apparent magnitude can be made which will approximate measurements made with the eye. These magnitudes are called visual magnitudes *V*. Other filters approximate magnitudes established by photographic means; these are called *photographic magnitudes B*, for blue. The visual magnitudes depend upon the color sensitivity of the eye which reaches a maximum in the green-yellow region of the spectrum. Photographic magnitudes are based upon the sensitivities of commonly used photographic emulsions which are most sensitive to the blue, violet, and the near-ultraviolet. A blue star will appear brighter in the photographic magnitude system than in the visual magnitude system, and the reverse is true for a red star. By agreement, the two systems have been adjusted so that the magnitude of an A0 star is the same in both.

The color index of a star is the difference between the photographic and the visual magnitudes. It is calculated by subtracting the visual magnitude *V* from the photographic magnitude $B$: $B - V$. Consequently, an O- or B-type star will have a negative color index; since it will appear brighter on the photographic system, its photographic magnitude will be numerically *less* than the visual magnitude. Any star later than A0 (for example, a G-type star) will have a positive color index, for it will appear fainter in the blue and thus have a larger number on the photographic magnitude scale than on the visual scale. Table 14-1 gives the color indices of main-sequence stars of different spectral types. This table is based on an overall temperature of 5,730°K for the surface of the sun, a G2 star. This temperature includes limb darkening to keep it in line with stellar temperatures. The color index of the sun is +0.65.

Since the color index can be determined so much more quickly and more objectively than the spectral type, it is common practice to plot a color–luminosity diagram for a star cluster. A color–luminosity diagram represents essentially the same thing as the spectral–luminosity (H–R) diagram.

*There are other color indices relating other regions of the spectrum; for example, the ultraviolet to the blue.

Table 14-1

| Spectral type | Temperature | Color index |
|:---:|:---:|:---:|
| B0 | 25,000°K | −0.32 |
| B5 | 15,600 | −0.16 |
| A0 | 11,000 | 0.00 |
| A5 | 8,700 | +0.15 |
| F0 | 7,600 | +0.30 |
| F5 | 6,600 | +0.44 |
| G0 | 6,000 | +0.60 |
| G5 | 5,520 | +0.68 |
| K0 | 5,120 | +0.82 |
| K5 | 4,400 | +1.18 |
| M0 | 3,600 | +1.45 |

Not only has the color–luminosity diagram enabled astronomers to identify the members of distant clusters, it has also yielded valuable information concerning the evolution of stars within the many clusters. Figure 14-2 shows a color–luminosity diagram for a number of open clusters. All the clusters have stars on the main sequence, the main branch in the diagram, but each cluster is represented by a different branch off the main sequence. The double cluster in Perseus (H and Chi Persei) has many very hot blue luminous stars. The hottest stars in the Pleiades, although very luminous, are not O-type stars. The hottest stars in NGC 188 are not much hotter than the sun. These differences are suspected to result from differences in the clusters' ages. The youngest, NGC 2362, is perhaps less than one million years old. In NGC 188, however, even the moderately massive stars are starting to evolve off the main sequence toward the red-giant stage; it must, therefore, be the oldest cluster on this diagram. Estimates place the age of Hyades cluster at 900 million years, i.e., 0.9 billion years old.

c. **The Pleiades.** The Pleiades cluster (Figure 14-3) is 400 light years from the sun and, with its early B-type stars, is young on the astronomical scale. The time required for a B-type star of 9 solar masses to contract to the main sequence, deplete its core hydrogen, and start its expansion into a red giant is estimated to be about 21,000,000 years. This, then, should be

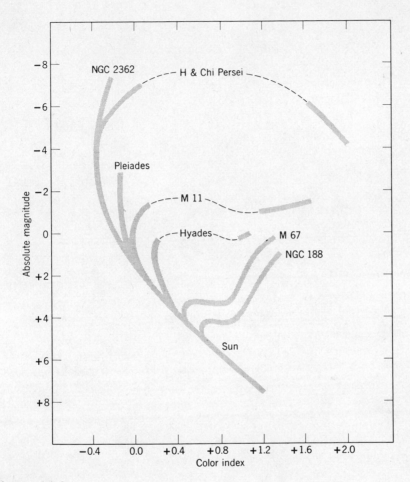

**Figure 14-2**

A composite color-luminosity diagram for several open clusters. (After Sandage)

the age of this open cluster. A color–luminosity diagram of the Pleiades alone, however, makes it clear that stars of only 0.2 solar mass have already passed through their initial contraction stage and are essentially on the zero-age main sequence. Current theories of stellar evolution estimate that it takes roughly 600,000,000 years for a star of 0.2 solar mass to contract onto the zero-age main sequence. So how old is the Pleiades cluster, 21,000,000 years or 600,000,000?

Actually the discrepancy is not so great as it may appear. It has been

**Figure 14-3**

The Pleiades cluster is embedded in nebulous dust. (Lick Observatory photograph)

suggested by Herbig that stars of small mass form from the nebular material before stars of large mass form. If the late-type main-sequence stars form first, they may be able to reach the zero-age main sequence in about the same time interval required for the early-type stars to contract onto the main sequence, burn their core hydrogen and start the expansion to the red-giant stage. Theory seems to be supported by observation. Additional evidence for the early formation of low-mass stars is seen in the Taurus–Auriga dark clouds which have a high population of faint red stars and no luminous blue stars.

All this indicates that the Pleiades, with roughly 300 stars, started to form 600,000,000 years ago and continued until the B-type stars finally formed. These very luminous stars, which radiate ultraviolet radiation copiously, have ionized the gases and literally blown all but the remaining vestiges of gas out into the galaxy. With this gas gone, stellar formation was brought to a halt. It is interesting to speculate that the Rosette nebula (Star Plate 3 and Figure 12-3) with its cluster of very luminous stars at the center may be caught in the act of blowing the gases away from the central region. This may account for the ring-like structure of this very interesting aggregate of gas and stars.

**d. Stellar Association.**   All of these theories of star formation are well supported by the Orion complex. The Orion complex, it may be recalled, has some Herbig–Haro objects and an overabundance of T Tauri stars. The Orion region is also liberally populated by O- and B-type stars that form a very loose group called an *association*. An association is not a cluster but a large number of stars of similar type in the same region of the galaxy. The association of stars in the Orion region is embedded in a huge tenuous cloud of hydrogen gas some 1,500 light years from the sun. Only a small part of this cloud is visible to the naked eye as the Orion nebula (Figure 2-9), because only in its central region is there a group of very hot luminous stars. The entire cloud may have a diameter as large as 300 light years while the visible portion is about 20 light years in diameter. Radio observations indicate that the gas is distributed in a symmetrical pattern around the group of hot blue stars. The uneven appearance on optical photographs results from interstellar dust in the foreground.

Studies of the entire Orion complex suggest that it is expanding. The hot blue stars in the central portion (Figure 14-4a) cannot be very old, and their motions indicate that they all formed in an even smaller volume of space some 10,000 years ago. Studies of the extent of ionization and the motions of the gas indicate the nebula itself has an age of some 10,000 to 20,000 years. Presumably it existed as a dark nebula long before that, perhaps a million years ago, during which time less massive stars formed, which now appear as the faint red stars in Figure 14-4b. However, once the very massive stars of the central part formed with surface temperatures ranging from 20,000° to 50,000°K, the gases became ionized and the expansion started. As the expansion continues, star formation in the central regions must decrease.

The color–luminosity diagram of the Orion association of stars shows that the most massive stars have not only had time to evolve onto the main sequence, but that they have burned enough of their hydrogen to start their evolutionary process which carries them to the right on the H–R diagram. The stars of moderate mass are on the zero-age main sequence, but the stars of very small mass have not yet reached the zero-age main sequence. Their slow contraction assures that they will reach the main sequence last of all—even if they started to form first. Their handicap does them no good!

The Orion complex, profusely populated with young stars all embedded in a matrix of hydrogen gas, convincingly supports the belief that stars are born of gaseous clouds.

It is not well understood why an association of stars forming in a nebula either breaks up shortly after formation or remains to become an open cluster. One suggestion deals with the percentage of the cloud that condenses into stars. The cloud, originally held together by the gravitational attraction of the gases that compose it, may condense largely into stars with only a little gas left over. In this situation most of the mass remains and the gravitational field remains strong enough to hold the stars in a more permanent cluster. If, on the other hand, a goodly amount of the cloud does not condense into stars, it is likely to get blown away by the particle pressure of those stars that have formed; consequently the total gravitational field will be reduced and the association is likely to break up sooner.

## 14.2 Globular Clusters

Unlike open clusters, globular clusters are characterized by extreme stability. Not only are they very compact, with many more stars than the

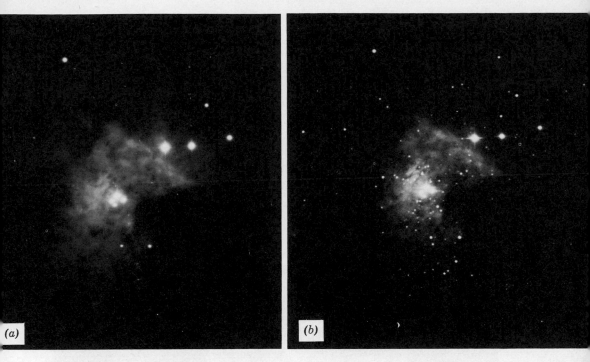

(a)          (b)

**Figure 14-4**

The Orion nebula (*a*) in ultraviolet light and (*b*) in infrared light. (Courtesy of G. Herbig of the Lick Observatory)

open clusters, but their ages are much greater—in the neighborhood of 5 billion years or more. This follows from the observation that globular clusters contain many red giants and RR Lyrae stars. One globular cluster even contains a planetary nebula. Each of these objects indicates a much greater age than do the blue giants in open clusters. The relative ages of the two types of clusters are made clear by the differences between the typical color–luminosity diagram of each (Figure 14-2 for the open cluster and Figure 12-6 for the globular cluster).

Globular clusters contain few if any early-type stars on the main sequence. These stars, if they once existed in the globular clusters, have long since passed through their evolutionary track and may exist as white dwarfs too faint to be seen in the more distant globular clusters. The color–luminosity diagrams of globular clusters also differ considerably from those of the stars in the neighborhood of the sun. Most of the stars near the sun are main-sequence stars of types later than A0; few are red giants, RR Lyrae stars, or planetary nebulae.

Not only do globular clusters contain stars of a type different from those found in open clusters and in the neighborhood of the sun, but globular clusters are distinguished even further by their location in our galaxy.

There are no globular clusters close to the sun; the brightest, Omega Centauri, is 22,000 light years from the sun. Despite its great distance it is visible to the naked eye in the southern hemisphere as a faint patch of light in the sky, because it has many thousands of stars. The globular cluster in the constellation Hercules (Figure 14-5) is 34,000 light years away and can be seen with the aid of only a small telescope.

There are about 121 globular clusters known. Of these 30 (25% of the total) reside in a very small portion (2%) of the sky in the constellation Sagittarius and the rest are spread throughout the entire sky. This con-

**Figure 14-5**
The globular cluster M13 in Hercules which must contain more than 50,000 stars. (Photograph from the Hale Observatories)

centration in Sagittarius is especially significant, for the center of our galaxy lies in that direction. Since both the 30 globular clusters and the galactic center lie at about the same distance from the sun and in the same direction, these clusters must be concentrated about the center of our galaxy.

The remaining globular clusters lie at the fringes of our galaxy and form a nearly spherical halo concentric with the center and contrasting with the flatness of our galaxy (Figure 14-6). Open clusters, on the other hand, lie close to the plane of our galaxy.

Because of their location in our galaxy the clusters can be easily classified into the two main stellar populations. An open cluster generally consists of young, blue-giant stars, lies close to the galactic plane, and is therefore of Population I. A globular cluster consists of stars such as red giants and RR Lyrae stars that are well advanced in their evolution. In addition, it is either far from the galactic plane or near the galactic center. Consequently it is of Population II.

Since there are no globular clusters of Population I we can only assume that they were all formed billions of years ago at a time when our galaxy was quite different from what it is now. Similarly, since there seem to be

**Figure 14-6**
Our galaxy is surrounded by a halo of globular clusters.

very few extremely old open clusters we can only assume that open clusters are too unstable to enjoy a life of billions of years. But speculations on the age and evolution of our galaxy must wait until Chapter 16.

In apparent contradiction to theories of stellar evolution, the globular clusters, with their red giants and RR Lyrae stars, also have a few fairly hot blue stars on the main sequence. The color-luminosity diagram of M3 (Figure 12-6) reveals some of these early-type stars—those stars on the main sequence above the *turn-off point*. All the other fairly massive stars that once existed on the upper part of the main sequence have expanded and evolved to the right. Why should these few stars remain? After all, a star's location on the main sequence and the length of time it remains there are both dependent on its mass. Therefore, if these early-type stars were formed at the same time all the other stars in the cluster were formed, we see no reason why they, too, should not have evolved off the main sequence.

One possible solution to this dilemma is to assume that these stars, somehow, formed after the bulk of the stars in the cluster had formed. If so, they are younger and behind the rest in their evolution. In support of this, there is some evidence that globular clusters contain intracluster gases. The dark patch in the upper left of M13 (Figure 14-5) may be some of this gas, that presumably could have given rise to stars younger than the cluster. It is speculated that this intracluster gas could come from the more massive cluster stars that have already ejected gases to become white dwarfs.

It does not seem likely that intracluster gases could have been picked up from our galaxy, for globular clusters are never found associated with interstellar gases. Each is located in a region of the galaxy that is quite devoid of nebulae, whereas the more recently formed open clusters are very commonly associated with nebular gases. These gaseous clouds, along with interstellar gases, form a goodly portion of the mass of our galaxy. Although they conceal a large percentage of the stars, they have given us valuable information on the structure of our galaxy.

## BASIC VOCABULARY FOR SUBSEQUENT READING

| | |
|---|---|
| Color index | Photographic magnitude |
| Globular cluster | Stellar association |
| Open cluster | Visual magnitude |

## QUESTIONS AND PROBLEMS

1. How do the following characteristics differ in the open and globular cluster?
   (a) Location in the galaxy
   (b) Number of stars
   (c) Color–luminosity diagram
   (d) Type of star
   (e) Age
   (f) Stability
2. How is the age of an open cluster estimated?
3. How is membership for stars in an open cluster determined?
4. What is the principal difference between an open cluster and a stellar association?

## FOR FURTHER READING

Struve, O., and V. Zebergs, *Astronomy of the 20th Century*, Crowell, Collier and Macmillan, New York, 1962, Chapter XIII.
Boyko, A., "Inside a Globular Star Cluster," *Sky and Telescope*, p. 269 (Nov. 1964).
Saunders, J., "The Globular Cluster Omega Centauri," *Sky and Telescope*, p. 133 (Sept. 1963).

# BETWEEN
# THE STARS

A good photograph of the Milky Way not far from the galactic center (see Figure 15-1) strikingly reveals the extent to which the stars near the galactic plane are immersed in gaseous clouds. But even such an excellent photograph does not reveal the gases and dust that are invisible though present in many regions of our galaxy. Observational methods more refined than those possible with direct photography have disclosed the existence of such interstellar matter. All interstellar matter can be divided into three main groups: the *bright nebulae*, the *dark nebulae*, and the gas and dust between the nebulae, generally called *interstellar gas* and *dust*.

## 15.1 Bright Nebulae

Bright nebulae (also called *diffuse nebulae*) are found only in close association with fairly luminous stars. It is these stars which cause such nebulae to become bright. They do so by one of two processes, which may be illustrated by a consideration of the Orion nebula (Figure 2-9) and the nebulosity in the Pleiades (Figure 14-3). A spectogram of the Orion nebula discloses a bright-line spectrum, indicating that the gas in this nebula is at a low pressure and emits its own light. Such a nebula is therefore called an *emission nebula*. The spectrograms of the nebulosity about the Pleiades cluster, however, indicate that this nebulosity is merely reflecting the light of the embedded stars. The light from the nebulosity is recorded as an absorption spectrum similar to that of the brightest stars in the cluster. Such a nebula is therefore called a *reflection nebula*.

**a. Reflection Nebulae.**  A pure gas will not reflect light, since the atoms composing such a gas are too small. Astronomers therefore assume that the light-reflecting agent in the reflection nebulae consists of small particles of "cosmic dust." The exact composition of this dust is not known with certainty, but we are able at least to estimate the size and nature of its particles by the light they scatter.

The light of the reflection nebulae is bluer than that of the stars whose light is scattered, although the difference in color is not as extreme as the difference between the red setting sun and the blue sky (see p. 80). The quality of this scattered light in the reflection nebulosity has led astrono-

**Chapter Opening Photo**
The nebula and cluster M16. (Lick Observatory photograph)

**Figure 15-1**

Close association of bright and dark nebulae near the star Rho Ophiuchi. Two globular clusters can be seen just below the center of the bright nebulosity. (Yerkes Observatory)

mers to believe that the dust particles causing the scattering are about $10^{-5}$ in. in diameter. (A simple molecule is about $10^{-8}$ in. in diameter.) Further studies have shown that the albedo of these dust particles is almost as high as that of white paper or snow. Such a high albedo indicates that they cannot be metallic but must be composed of hydrogen, carbon, nitrogen, oxygen, along with other elements, all perhaps combined chemically into molecules which join to form the dust particles.

**b. Emission Nebulae and Forbidden Lines.** The stars associated with reflection nebulae, although very luminous, are never of a spectral type earlier than B2. A nebula that has stars of spectral type B1 or earlier associated with it does not reflect very much light. Consequently the resulting nebular spectrum is not the same as the absorption spectrum of the brightest stars, but is instead a faint continuous spectrum upon which brighter emission lines are superimposed. These very hot luminous stars apparently reduce the number of light-scattering particles to a minimum either by blowing them away or by evaporating them with their intense radiation.

Gases in an emission nebular emit a bright-line spectrum by means of a process called *fluorescence* (the same process that gives rise to commercial fluorescent lighting), which is activated by the very hot stars of B1 or earlier. Since these stars are so hot, they emit most of their light in the ultraviolet region of the spectrum. When the atoms of a material absorb energy of one wavelength (such as ultraviolet) and emit energy of a longer wavelength (such as visible light), the material is said to fluoresce. Ultraviolet radiation has more energy than visible light and thus is more capable of exciting the atoms of the gas. When an electron in an orbit around an atom is struck by a photon of ultraviolet light having just the right amount of energy (that is, just the right wavelength), the electron will absorb the photon and make a transition to a higher energy level within the atom. It will then fall back to the ground state and in so doing, it emits light (see p. 56 f). This light need not be of the same wavelength as the ultraviolet radiation absorbed, as the electron may return to the ground state by cascading down from one energy level to another until the lowest energy level is reached. In cascading the electron will emit photons of various but discrete amounts of energy, yielding the characteristic emission spectrum.

Emission lines in the spectra of gaseous nebulae, with the exception of the hydrogen lines, are not the usual spectral lines found in stellar spectroscopy. Although lines of nitrogen, oxygen, helium, and other elements are also found in nebulae, they result from transitions peculiar to emission nebulae. When the density of a gas is so extremely low as it is in gaseous nebulae, transitions occur that do not occur in the atoms of stars or in terrestrial laboratories. These transitions are so improbable in gases under pressures greater than those found in emission nebulae that they have been termed "forbidden transitions." Spectral lines resulting from such transitions are called *forbidden lines*.

Each electron transition is governed by what we call selection rules. These selection rules (derived in large part from the spectral line intensities), give the probability that any particular transition will occur. The most probable transitions result in the most intense lines in the spectrum of any given element. Transitions which are extremely improbable are called *forbidden transitions*. An undisturbed electron will remain in an energy level for a long time if to reach the ground state it must make a forbidden transition. Such an energy level is called a *metastable state* (meta-stable). An undisturbed electron in a metastable state may remain there for seconds, minutes, or even hours before making the forbidden transition to the ground state. Many such transitions result in a *forbidden spectral line*. An electron will remain in the more common energy level for only about $10^{-8}$ second. If an atom finds itself in a region of relatively high pressure such as the lower atmosphere of a star or a gas-filled tube in a college physics laboratory, it is being bombarded billions of times per second by the other atoms in the gas. Consequently its electrons cannot remain in metastable states long enough to make a transition down. The bombardment will knock an electron from the metastable state into a higher energy level from which it may make a downward transition within $10^{-8}$ second. For this reason forbidden lines are not usually seen in the terrestrial laboratory or in stellar spectra.

The gases in the nebulae, however, are under such low pressure that an atom may travel for days, weeks, or even years before it collides with another atom. If during this time one of its electrons falls into a metastable state, it may remain there long enough to make the downward transition that causes a forbidden line. Forbidden lines can result not only from neutral atoms but also from atoms that have been ionized once or twice. In emission nebulae forbidden lines result principally from ionized oxygen, nitrogen, neon, and sulfur.

It is interesting to compare the density of the Orion nebula with terrestrial densities. The Orion nebula has about 5,000 atoms (mostly ionized) per in.$^3$ One cubic inch of atmosphere at the Earth's surface contains about $4 \times 10^{20}$ molecules. It is this lack of "congestion" in emission nebulae that creates the condition under which the atoms emit forbidden lines. However, the density of the Orion nebula is not the same throughout. In the brightest central part it is as "high" as $3.2 \times 10^5$ (320,000) atoms per in.$^3$

Despite this extremely low density, emission nebulae are visible. This is due to their tremendous size, which varies from nebula to nebula, although a diameter of 25 light years is fairly common. Thus, even though there may be only about 5,000 atoms per in.$^3$, an entire cloud contains many atoms, for there are about $5 \times 10^{52}$ in.$^3$ in a cubic light year. To take our example, there are almost 7,000 cubic light years in the Orion nebula, which thus must comprise about $10^{60}$ atoms—enough material to make about 1,500 suns.

As was pointed out on p. 381, the visible phenomenon that we call the Orion nebula is only a small portion of a much larger gaseous cloud, a large part of which is invisible to direct photography. It thus appears that the photographic boundaries of the Orion nebula are not the boundaries of the entire gasous cloud, only the extent to which the immersed stars excite the gases to emit their own light. The same may be true of any bright nebula.

## 15.2 Dark Nebulae

The gaseous nebulae that neither reflect the light of embedded stars nor emit their own light are called dark nebulae. Their existence can be determined by the fact that they partially conceal luminous objects behind them. The Horsehead nebula in the Orion region is an excellent example (Star Plate 4). Not only does the dark gas outlining a horse's head conceal the emission nebula behind it, but by comparing the number of stars on the bottom of the photograph with the number on the top it can be seen that the Horsehead nebula is only a projecting branch of a much larger dark nebula that conceals the more distant stars on the bottom.

Evidence of the coexistence of dark and bright nebulae in the same region is also seen in Figure 15-1, a photograph in the region of the star Rho Ophiuchi. Lanes of dark nebulosity can be seen extending from the central portion of the photograph where gases partially conceal stars behind them. In this nebula the bright portion is largely reflection nebulosity.

There are many other dark nebulae appearing as "voids" in space that are actually large masses of gas and dust with no bright stars nearby to illuminate them. The Coalsack in the southern hemisphere is an excellent example (Figure 15-2). A photograph of the star "clouds" in Sagitarrius (Figure 15-3) not only tells us that our sun is not alone (anyone for star counting?) but also tells us that between the Earth and this prodigious number of stars intervene concealing dark clouds. The true extent of the dark nebulae can best be seen, however, by a contemplative look at the Milky Way when one is free of the stifling effect cities have on astronomical observing. Running nearly the entire length of the Milky Way is the "Great Rift" which seems to divide the Milky Way into two parts. This is not a real rift in the groupings of stars, but an apparent rift caused by intervening gas and dust.

The sizes and distances of the dark nebulae present a new problem, for all the methods of distance determination discussed so far rely on the fact that objects emit light. The dark nebulae do not. We can, however, fall back on the stars to give us some idea of the distances and thus of the sizes of dark nebulae.

It is possible to estimate the distance of a dark nebula by comparing the number of stars of each apparent magnitude in the region of the dark nebula with a nearby region which appears to be unobscured. In general,

**Figure 15-2**

The Coalsack can be seen easily with the naked eye in the southern Milky Way. Just above and to the right of the Coalsack are the four stars of the Southern Cross. Bright nebulosity can be seen to the right. (Harvard College Observatory)

**Figure 15-3**

Dark clouds in the region of Sagittarius photographed in red light with the 48-in. Schmidt telescope. (Photograph from the Hale Observatories)

the number of stars increases with decreasing brightness. The rate of increase of the number of stars which lie between us and the dark nebula will be the same as the rate of increase for the unobscured region. This is illustrated in Figure 15-4 as the line which extends unbroken to about the eleventh magnitude. For stars fainter than the eleventh magnitude, however, the rate of increase for the region of the nebula (the dashed line) is less than the rate of increase in the unobscured region. The reason, evidently, is the dimming effect of the nebula; and, if we can estimate on a statistical basis the average absolute magnitude of the stars involved, we can estimate the distance of the nebula. By knowing the distance and the angular diameter of the dark nebula, we can calculate the linear diameter by the use of trigonometry.

From studies of this kind the Coalsack is estimated to be only 400 light years away and to have a diameter of roughly 40 light years. Since it can be seen easily with the naked eye, it is interesting to speculate how it would appear if there were a cluster of O and B stars immersed in it.

Dark nebulae seem to be composed of about the same material as bright nebulae. The stars whose light shines through the dark clouds are reddened to such an extent by having their light scattered that it appears likely that the dust particles are about the same size as those in the reflection nebulae.

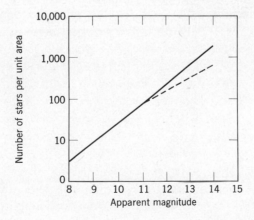

**Figure 15-4**

The distance of dark nebulae can be determined by comparing star counts in adjacent obscured and unobscured regions.

## 15.3 Interstellar Gas and Dust

Now that we recognize the existence of dark gases, how can we be sure that the unobscured regions are really unobscured? If there is material between the stars that does not reveal itself as obvious dark nebulae, we might be able to detect it by the effect it would have on the light passing through it. There are two possibilities. If the material were to contain dust particles, the light could become slightly reddened by scattering; or if the material were to contain gas, it could cause additional absorption lines in the spectra of the stars whose light shines through the material.

To find evidence for this interstellar material we should study stars that we know are at large distances from the Earth, for if such an effect is detectable it is more likely to be noticeable in distant stars whose light would have to travel through more interstellar material. Furthermore, the stars we study should lie near the galactic plane, where the vast majority of nebulae appear; for example, we could observe the faint B-type stars. B-type stars have a high luminosity; consequently, if they appear faint in a region of little obscuration they must be far from the sun.

Such studies do in fact reveal the existence of both interstellar gases and dust. The faint B stars are considerably redder than they should be. The amount of reddening can be measured by comparing their observed color index, as determined by photoelectric studies, with their expected color index, as determined from their spectral type. The difference between the two color indices is called the *color excess*. This term is derived from the fact that the color index increases in a positive sense for increasingly cool stars. Thus stars reddened by interstellar dust have greater color indices than they would normally have (see p. 377). Color excess is a misleading term, however, since the increase in the color index is really a measure of the amount of blue light that has been *subtracted* from the star's light by scattering. The term is thus descriptive of no more than the reddening of the light that reaches us.

The existence of interstellar gas (but not dust) was first noted in the spectra of spectroscopic binaries. The lines in the spectra of spectroscopic binaries oscillate back and forth as the stars revolve about their common center of mass. If we line up a series of spectra according to wavelength (Figure 15-5), we can see this oscillation of spectral lines. Of particular interest is the line (or lines) with a wavelength of 3,394 Å (Ca II, which is calcium ionized once). The darkest line is caused by the star Beta Lyrae, and its wavelength changes periodically. There is, however, an absorption line whose wavelength does not change. Not only does its wavelength not change, but this line is sharper than the other lines in the spectra. This line is caused by the interstellar calcium between Beta Lyrae and us.

The complex nature of the line (or lines, since it appears as both an absorption and as a bright line) with a wavelength of 3,889 Å (He I, neutral helium) is interpreted by astronomers as evidence that Beta Lyrae is ejecting

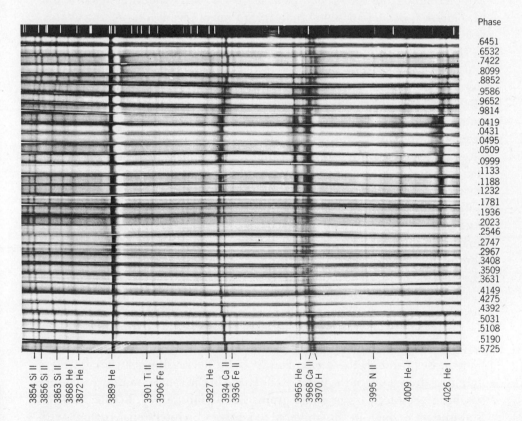

Phase

.6451
.6532
.7422
.8099
.8852
.9586
.9652
.9814
.0419
.0431
.0495
.0509
.0999
.1133
.1188
.1232
.1781
.1936
.2023
.2546
.2747
.2967
.3408
.3509
.3631
.4149
.4275
.4392
.5031
.5108
.5190
.5725

3854 Si II
3856 Si II
3863 Si II
3868 He I
3872 He I
3889 He I
3901 Ti II
3906 Fe II
3927 He I
3934 Ca II
3936 Fe II
3965 He I
3968 Ca II
3970 H
3995 N II
4009 He I
4026 He I

**Figure 15-5**
A sequence of spectra of Beta Lyrae that have been aligned by the interstellar line of Ca II 3934 Å. The wavy motions of the stellar lines result from the varying Doppler shift of the spectral lines as the primary component revolves in its orbit about the center of mass of the system. (Courtesy of the late O. Struve)

gases out into space. These gases do not take part in the periodic motion of the two stars.

In some stars, more than one interstellar line from a given transition in an element will appear, giving evidence that the interstellar gases are not uniformly distributed, but form numerous clouds with perhaps even more tenuous gas and dust between these clouds. Each cloud absorbs light independently of the others, and if each has a different velocity with respect to the Earth, it will produce an absorption line with a different Doppler shift from the others. In at least four stars the interstellar K line of calcium appears four times, indicating four different intervening clouds. One star has as many as seven components of the same interstellar line.

The density of interstellar gases is so low that by comparison even the emission nebulae seem almost congested. It has been estimated that in a cloud of interstellar gas there are on the average 200 atoms per cubic in., but between these clouds we may find only 20 atoms per cubic in. Here, then, is an opportunity for high-vacuum studies. Indeed, astronomical observations serve to supplement and extend the observations made in the physics laboratory, for in the celestial realm we find and study atoms at temperature and pressure extremes which we could not hope to obtain in terrestrial laboratories.

The composition of all interstellar gases, including diffuse nebulae, does not seem to differ much from what we find in the sun, in most of the stars, and, with the exception of hydrogen and helium, from what we find here on the Earth. The Earth as well as Mercury, Venus, the moon, and Mars do seem unusual in their comparative lack of hydrogen and helium.

There is evidence to support the idea that a good part of the interstellar dust is particles of carbon, perhaps like graphite. The main evidence is the manner in which interstellar dust reddens the light which passes through it. This reddening means that with decreasing wavelength less light passes through the dusty clouds; with decreasing wavelength more light is scattered out of the beam. Reddening of the same quantitative nature has been produced by graphite grains in the laboratory. Besides, carbon is abundant enough in the universe to supply the interstellar regions with the amount of dust we observe.

Thus our galaxy not only consists of billions of stars but also contains large amounts of gas and dust between them. This gas and dust is not distributed uniformly throughout the galaxy but appears in the form of individual clouds close to the galactic plane. If the density of a cloud is great enough, it will partially obscure the stars behind it and be called a dark nebula. If there are enough stars in the cloud, it will become a bright nebula. If those stars are B2 or later, the nebula will shine by reflected light and be called a reflection nebula. If the stars are B1 or earlier, the dust is forced away and the nebula will shine mainly by fluorescence, having absorbed ultraviolet radiation from the hot stars and emitting the visible light that makes it an emission nebula. If the gas and dust have a density too low to be readily noticeable as a dark nebula, they are considered as interstellar material whose presence is detected by the scattering of the blue light from the stars that shine through the nebula or by the interstellar absorption lines produced in the spectra of those stars.

## 15.4 Radio Astronomy

There is so much interstellar material that a major part of our galaxy remains concealed from visual observation. But the advent of radio astronomy has given us a tool for penetrating interstellar clouds because scattering decreases with increasing wavelength. Thus radio waves, with wavelengths a million or more times longer than that of light, travel through interstellar material without appreciable absorption or scattering.

Since radio waves are part of the electromagnetic spectrum, we can speak in terms of a continuous radio spectrum as well as of an emission or absorption radio spectrum. An emission radio line is equivalent to an emission line in the visible spectrum. There is also the possibility that a gas could selectively absorb radio radiation and produce the equivalent of an absorption line. Let us first consider the continuous spectrum in the radio region.

**a. The Radio Sky.** One of the radio astronomers' first tasks was to determine where in the sky the radio signals were coming from; in effect they asked, "What does the sky look like when 'seen' by radio waves?" The best way to answer such a question is to make a survey of the sky by recording the intensity of the radio radiation from every part of the celestial sphere. The results of such a survey look like a contour map. On a geographical contour map the contour lines connect points of equal elevations, whereas on a radio map of the sky the contour lines connect points of equal radio radiations intensity (Figure 15-6).

The results of many radio maps of the celestial sphere made at various wavelengths indicate that: (a) the stars are very weak emitters of radio energy, (b) the Milky Way, with all of its gas, is a major source of radio energy, and (c) there are many *discrete radio sources* in the sky.

The contour lines in the radio map of Figure 15-6 represents the amount of energy received by the radio telescope. The numbers of each contour line refer to particular units of energy. By comparing these numbers on the map it can be seen that in general the amount of energy increases toward the central plane of the Milky Way. In the direction of the galactic nucleus, the amount of energy received reaches a high of 150 units.

Very few stars are "seen" by radio telescopes. Stars simply do not radiate much energy in the radio or microwave region of the spectrum. The sun is a powerful radio source, but only because of its proximity to us.

**b. Free Electrons and a Continuous Spectrum.**  Energy from the gases between the stars originates in one of several ways. If the gas is largely ionized, it is composed mostly of free electrons and free protons. On occasions a free electron travels close to a free proton, but travels too fast to be captured. Nevertheless its line of travel is deviated by the electric attraction of the two particles. During the change in the direction of travel the electron emits energy; a change in direction of travel amounts to a transition from one "free orbit" to another "free orbit." Such a transition is called a *free–free transition*. Since these free orbits are not quantized, the free–free transitions may result in the emission of any amount of energy within certain limits. The limits are such that this energy is generally emitted in the radio region of the electromagnetic spectrum. And since the energy of the free orbits is not quantized, the radio energy emitted amounts to a continuous spectrum—all energies are emitted.

If the intensity of each wavelength of the continuous radio spectrum corresponds to Planck's law, then we surmise that the source is *thermal* in nature and the gas is largely ionized. The possibility exists that the continuous radio spectrum is synchrotron radiation, in which case it is polarized and its energy distribution is different from that of a thermal source (see p. 311 f).

**c. Bound Electrons and Radio Spectral Lines.**  As was pointed out earlier in this chapter, radio sources can also emit energy in a form equivalent to an emission line. One such radio emission line has a wavelength of 21 cm (about $8\frac{1}{3}$ in.). This line is caused by a transition of an electron within the hydrogen atom. This particular transition, however, is slightly different from those discussed so far.

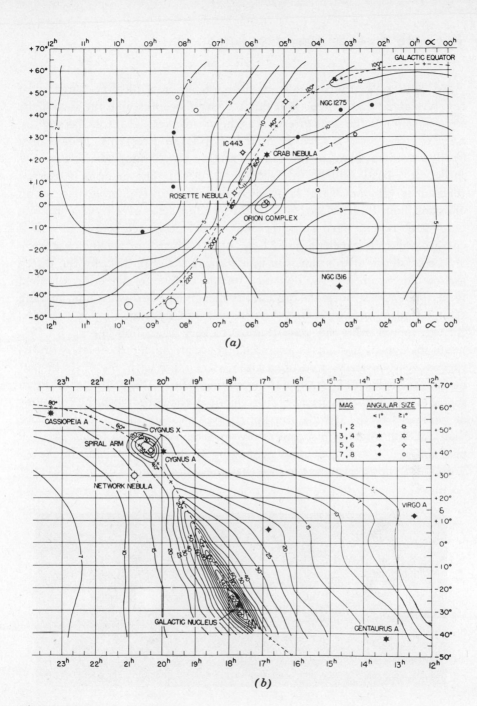

*(a)*

*(b)*

**Figure 15-6**

Radio map of the sky made at a wavelength of 120 cm with the Ohio State University radio telescope. (Radio Observatory, Ohio State University)

An orbital electron in an atom spins on an axis, like the Earth, as it revolves about the nucleus. But the nucleus also spins, like the sun. If the *electron spin* is in the same direction as the *nuclear spin,* the atom has slightly more energy than if the two spins are in opposite directions. Thus if they change from spinning in the same direction to spinning in the opposite direction, energy will be released (Figure 15-7). The amount of energy released is very small but is always the same and thus gives rise to an emission line. If the electron is in the ground state of a hydrogen atom when they reverse their spin, the amount of energy released or absorbed corresponds to a wavelength of 21 cm, or a frequency of 1,420 megacycles.

The existence of this emission line was predicted in 1945 by the Dutch astronomer Hendrik C. van de Hulst, but the chances of detecting it looked rather poor at first, since the probability of a given hydrogen atom's making such a transition is extremely low. Any given hydrogen atom, if undisturbed, should make this transition once every 11 million years!

But the hydrogen atoms in interstellar space are not left completely undisturbed. An interstellar hydrogen atom will collide with one of its distant neighbors (less than a centimeter away) once about every 300 years! Even though these collisions are very infrequent as compared with the collisions of molecules in the Earth's atmosphere (which occur at the rate of more than a billion times each second) they are frequent enough to increase the probability of a hydrogen atom's changing its electron spin. This increased probability, coupled with the fact that there are billions upon billions of

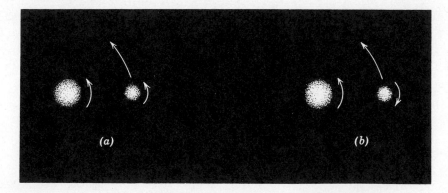

**Figure 15-7**

The 21-cm line in emission is caused by the electron spin flipping from (*a*) to (*b*).

hydrogen atoms in interstellar space, led radio astronomers to believe that the 21-cm line could be found.

The 21-cm line was finally detected in 1951. Since then it has assumed a major role in the study of our galaxy, for with it astronomers can detect clouds of neutral hydrogen from nearly the entire galaxy. The 21-cm line can only be emitted by a neutral hydrogen atom since it is a transition within that atom. The clouds of neutral hydrogen presumably contain ionized hydrogen as well, which emits a continuous spectrum of radio energy on which the 21-cm line is superimposed.

As we have seen, the interstellar clouds of gas and dust do not scatter the 21-cm line as they do visible light. Thus this radiation can travel through regions of space opaque to visible light. We should note, however, that the 21-cm line has also been detected in absorption which indicates that its energy can cause a hydrogen atom to change its electron spin back to the higher energy level where both electron and nucleus spin in the same direction.

The 21-cm line of hydrogen and other radio spectral lines offer an additional advantage not offered by the continuous radio emission: they can be used to detect radial velocities of clouds of gas by their Doppler shift. The intensive studies made with the 21-cm line, as well as with continuous radio emission, have given us a much better idea of the structure of our galaxy than we had at mid-century.

## 15.5 Interstellar Molecules

Downward transitions in atoms result in the emission of photons that are, for the most part, in the ultraviolet, visible, and infrared regions of the electromagnetic spectrum (the 21-cm line being a convenient exception). Molecules, on the other hand, have transition energies that range from the red part of the visible spectrum down to the radio region. Most of these transition energies, however, lie in the infrared, microwave, and radio regions. Since a good part of the infrared and microwave regions is blocked out by our filtering atmosphere (see Figure 2-18), a search for interstellar molecules has been and is now being conducted with the use of radio telescopes—and with startling success.

In addition to the molecules detected by interstellar lines in the visible part of the spectrum (CH and CN), the following molecules (with their radio spectral-line wavelengths) have been detected: $H_2O$, a spectral line with a wavelength of 1.35 cm; $NH_3$, a spectral line at 1.26 cm; $H_2CO$, a spectral line at 6.21 cm; and OH, four spectral lines between 17.4 and 18.6 cm. The formaldehyde molecule, $H_2CO$, is one of the more complex molecules yet found in interstellar clouds and indicates that chemical reactions have been more successful in interstellar spaces than previously thought possible. The formaldehyde molecule not only contains four atoms, but three different kinds of atoms.

The fact that each of these molecules has definite radio spectral lines permits us to use the Doppler principle to identify separate clouds containing the molecules if those clouds have different radial velocities. In the direction of a strong radio source in Sagittarius, for example, nine different components of $H_2CO$ spectral line have been detected, each with its own Doppler shift. It is interesting to note that the OH molecule is found to yield the same Doppler shifts in that same region of Sagittarius. These two molecules, OH and $H_2CO$, are often found in the same hydrogen clouds, with the OH molecule being the more abundant of the two.

It will be recalled from Section 13.4 that many of these sources of interstellar molecules have formed into small clouds of gas, only 2 A.U. or so in diameter. It may be that radio telescopes will give us our first definite observations of the formation of new planetary systems.

## BASIC VOCABULARY FOR SUBSEQUENT READING

Balmer continuum
Color excess
Diffuse nebula
Discrete radio source
Emission nebula

Fluorescence
Forbidden line
Metastable state
Reflection nebula

## QUESTIONS AND PROBLEMS

1. What is the difference in spectra between an emission nebula and a reflection nebula? Why does this difference exist?

2. Is it possible over the next million years or so for nebulae that are now dark to become bright? If so, what must happen?

3. Describe the process of fluorescence.

4. Explain why "forbidden" lines may be seen in emission nebulae.
5. Explain the process by which the 21-cm line is emitted. Why has this radio line achieved so much attention from the radio astronomer?
6. How can an astronomer detect the presence of clouds of neutral hydrogen; of ionized hydrogen?

## FOR FURTHER READING

Page, T., and L. W. Page, ed., *Stars and Clouds of the Milky Way,* The Macmillan Co., New York, 1968.

Steinberg, J. L., and J. Lequeux, *Radio Astronomy,* McGraw-Hill Book Co., New York, 1963.

Struve, O., and V. Zebergs, *Astronomy of the 20th Century*, Crowell, Collier and Macmillan, New York, 1962, Chapters XVII and XVIII.

Bowyer, C. S., "Galactic X-ray Astronomy," *Sky and Telescope,* p. 264 (Nov. 1965).

Clarke, R. W., "Locating Radio Sources with the Moon," *Scientific American,* p. 30 (June 1966).

Dieter, N., H. Weaver, and D. R. W. Williams, "The Interstellar Hydroxyl Radio Emission," *Sky and Telescope,* p. 132 (March 1966).

Friedman, H., "X-ray Astronomy," *Scientific American,* p. 36 (June 1964).

Robinson, B. J., "Hydroxyl Radicals in Space," *Scientific American*, p. 26 (July 1965).

Snyder, L. E., and David Buhl, "Molecules in the Interstellar Medium," *Sky and Telescope,* p. 267 (Nov. 1970); p. 345 (Dec. 1970).

"Graphite Grains in Interstellar Space," *Sky and Telescope*, p. 10 (July 1963).

"Infrared Nebulosity in Orion," *Sky and Telescope*, p. 217 (April 1968).

"Interstellar Formaldehyde Abundances," *Sky and Telescope*, p. 161 (March 1970).

"Interstellar Hydrogen Molecules Found," *Sky and Telescope*, p. 139 (Sept. 1970).

"Interstellar Water and Formaldehyde," *Sky and Telescope*, p. 271 (May 1969).

"Rocket Measurements of Interstellar Hydrogen," *Sky and Telescope*, p. 11 (July 1969).

# CHAPTER 16

CHAPTER · CHAPTER

# GALAXIES

Our intimate relation with our own galaxy is both an advantage and a disadvantage to the astronomer who wishes to learn more about its structure, rotation, and evolution. In Chapter 15 it is pointed out that since we are near the central plane of the galaxy and therefore in nearly the same plane as most of the interstellar material, much of the galaxy is hidden from visual observations. Radio observations, of course, tend to alleviate this disadvantage; nevertheless we would still like to *see* what our galaxy looks like. The problem is not unlike that of placing your head between the spokes of a large wagon wheel and then trying to figure out what the rest of the wheel looks like. Without previous knowledge of wagon wheels in general, the problem would be a difficult one. Correspondingly, a knowledge of galaxies in general is useful to the astronomer. Astronomers therefore study other galaxies in great detail, for not only do these studies help us interpret observations of our own galaxy, but they also add to our knowledge of how the universe is put together. Let us investigate the structure of galaxies in general before we consider our own in detail.

## 16.1 Hubble's Classification of Galaxies

Galaxies appear in every portion of the sky except along the Milky Way where the dust and gas of our own galaxy conceal the galaxies beyond. Hundreds of millions of galaxies are seen in the universe; the most distant ones visible in our large telescopes are estimated to be between 5 and 10 billion light years away. But galactic structure can be studied in only those that are much closer to us. The late Edwin P. Hubble, of the Mount Wilson Observatory, made an intensive study of galaxies and recognized three basic structures for the nearer galaxies: *elliptical*, *spiral*, and *irregular*. Each basic structure is further subdivided according to variations of shape within that group.

The elliptical galaxies (chapter opening photo of this chapter) seem to be similar in appearance to the globular clusters although they are much larger and contain many more stars. They vary in shape from spheres to ellipsoids whose major axis is about five times longer than the minor axis. Since rotation causes flattening, the faster a galaxy rotates the more ellipsoidal it is.

**Chapter Opening Photo**

Elliptical galaxy—a satellite of the Andromeda galaxy. (Photograph from the Hale Observatories)

Hubble's classification of elliptical galaxies is shown in Figure 16-1; the letter E (standing for elliptical) is followed by a number that indicates the degree of ellipticity. An E0 galaxy presents a circular disk. With increasing ellipticity the number following the E increases to an arbitrary maximum of 7. Since it is very difficult to determine the orientation in space of an elliptical galaxy, this classification applies to its shape only as it is seen from the Earth.

The spiral galaxies present a dynamic picture (Figure 16-2) that stands in contrast to the rather uniform elliptical galaxies. The first impression received upon looking at a spiral galaxy is that it is rotating about a central nucleus from which *spiral arms* curve out into space. Spiral galaxies were classified by Hubble into three subgroups: Sa, Sb, and Sc, according to the

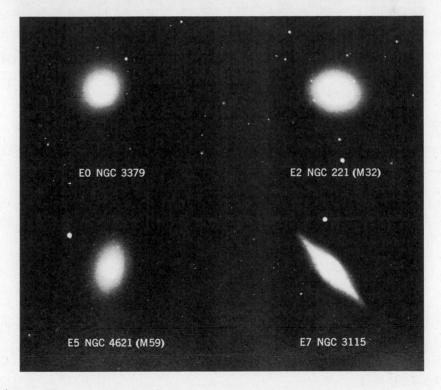

**Figure 16-1**

The shapes of elliptical galaxies vary from spherical to ellipsoidal. (Photographs from the Hale Observatories)

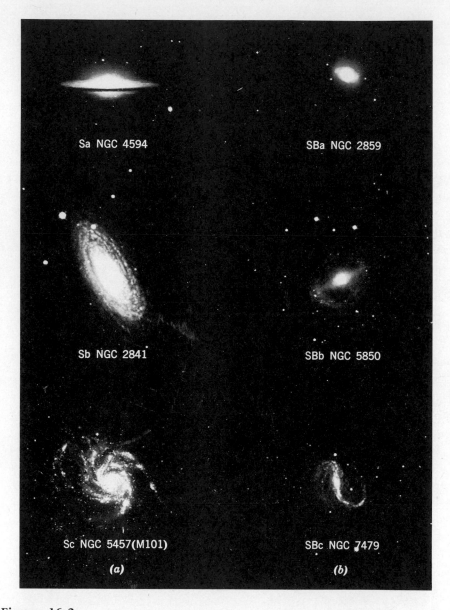

## Figure 16-2

Different types of (*a*) normal spiral galaxies and (*b*) barred spiral galaxies. (Photographs from the Hale Observatories)

amount of material in the arms relative to that of the nucleus, as well as to the degree of openness of the arms. Examples of this classification are seen in Figure 16-2a, under the heading of *normal spirals*.

In the same group with the normal spirals are the *barred spirals* (Figure 16-2b), so called because they have a "bar" of stars running through the nucleus. From the ends of this bar spiral arms may form, sometimes into a ring about the nucleus. Why some spiral galaxies assume the shape of normal and others of barred spirals is not known. The normal spirals, however, are more numerous: of the 600 galaxies studied by Hubble, 17% are elliptical; 50% are normal spirals; 30% are barred spirals; 3% are irregular.

The third type, the irregular galaxy, has no definite form. The two Magellanic Clouds are often considered to be members of this group, although some astronomers believe that the Large Magellanic Cloud should be classified as a barred spiral.

**a. Elliptical Galaxies.** It was not until the late 1940's that new photographic techniques made it possible to resolve one of the nearest elliptical galaxies, a companion of the Andromeda galaxy, into separate stars (opening photograph, Chapter 16). The distribution of stars in elliptical galaxies is such that the over-all shape of the galaxy does not change on photographs taken with different time exposures. A shorter exposure records only the more densely populated central regions, whereas a longer exposure records the outer regions of the galaxy. Thus it appears that the stars are placed rather symmetrically about the center but decrease in number from the center out.

The similarities of an elliptical galaxy, the nucleus of a spiral galaxy, and a globular cluster are striking. The brightest stars in each are red supergiants; the blue giants are conspicuous by their absence. It is not surprising, therefore, to learn that elliptical galaxies also contain RR Lyrae stars and are practically devoid of interstellar material. The elliptical galaxies are therefore classified as Population II. Since the lack of interstellar material makes them nearly transparent, more distant galaxies can sometimes be seen through their outer regions. It has become clear that the largest and most massive galaxies observed are elliptical galaxies.

**b. Normal Spiral Galaxies.** Spiral galaxies differ considerably from elliptical galaxies both in structure and in stellar content. The nucleus, of course, resembles an elliptical galaxy, but here the similarity ends. The spiral arms contain many blue-giant stars, open clusters, and gaseous nebulae. The Andromeda galaxy, M 31 (briefly discussed on p. 300 f), is typical of the normal spirals and is close enough to our own galaxy (2.2 million light years from the sun) to be faintly visible to the naked eye as a hazy patch in the sky. Moreover, in a photograph the details of the Andromeda galaxy can be clearly seen (Figure 16-3). The central portion is the nucleus composed of Population II stars. The spiral arms can be traced by following the lanes of emission nebulae, dark gases, galactic clusters and blue giants—all Population I objects.

Despite the brightness of the spiral arms, most of the light from the region outside the nucleus comes from a substratum of stars whose luminosity is equal to or less than that of giants. The arms are superimposed upon this substratum, which is nearly transparent, since more distant galaxies are visible through the parts of it that lie between the spiral arms. Observations indicate that the shape of the substratum is somewhat ellipsoidal. It extends above and below the plane of the galaxy, includes many globular clusters, is redder in color than are the spiral arms, and thus may be Population II.

Since we see the Andromeda galaxy at an oblique angle (the plane of the galaxy makes an angle of about 12° with the line of sight), we are able to determine by the Doppler shift the rate of rotation of those parts that are either approaching or receding from us. N. U. Mayall, of the Kitt Peak National Observatory, has studied the rotation by observing emission nebulae in the spiral arms. The emission nebulae give sharper spectral lines than the stars do and are therefore better suited for radial velocity measures. Mayall's results show that the inner portion of the galaxy rotates like a solid wheel; that is, the speed of the revolving stars increases with increasing distance from the center of the galaxy. The outer parts, however, rotate in a manner not unlike the motion of the planets about the sun; that is, the speed of the revolving stars decreases with increasing distance from the center.

If the vast majority of the mass of any system is at the center, we can expect the system to rotate in accordance with Kepler's third law. If, however, the mass of the galaxy is distributed with uniform density throughout the entire galaxy, we can expect it to rotate like a solid wheel. The Andromeda galaxy lies somewhere between these two extremes.

Since we know the distance of the Andromeda galaxy, we can determine its size from its angular diameter. The portion that appears in Plate 5 has a diameter of about 100,000 light years. Measurements made with a photocell, however, indicate that there is a fringe of stars extending beyond

**Figure 16-3**

Detail of the nucleus of the Andromeda galaxy. (Photograph from the Hale Observatories)

the obvious photographic boundaries. The diameter including this fringe is about 180,000 light years. From the size and rotation of the Andromeda galaxy, the mass of this system has been estimated to be $4 \times 10^{11}$ solar masses. Since the sun is roughly an average star in mass, we can assume that there are some 400 billion stars in the Andromeda galaxy!

**c. Barred Spiral Galaxies.** Barred spirals are similar to normal spirals in that their nuclei contain principally Population II objects and their arms are composed of Population I objects. Bright nebulosities as well as open clusters and lanes of dark obscuring matter can be seen in the spiral arms. The significant difference between normal and barred spirals can be seen by comparing Figure 16-4 with 16-5. The bar and the nucleus seem to be

**Figure 16-4**

The spiral galaxy in Triangulum, M 33. (Photograph from the Hale Observatories)

**Figure 16-5**

Barred spiral galaxy NGC 1300. (Photograph from the Hale Observatories)

rotating as a unit, like two spokes extending from the hub of a wheel. The spiral arms seem to lag behind the bar and the nucleus; thus they appear to rotate in a manner similar to that in which our planets revolve. At present there is no explanation for the structure of barred spirals. However, it is suspected that magnetic fields play a role in the structuring of galaxies including the barred spirals.

**d. Irregular Galaxies.** The irregular galaxies are quite different from both the elliptical and the spiral galaxies. They have no central nucleus, no spiral arms, and apparently no plane of symmetry. Common examples of this type of galaxy are the Magellanic Clouds (Figure 16-6).

The two *Magellanic Clouds* are located fairly close to our own galaxy. The Large Cloud is about 150,000 light years from us, and the Small Cloud

**Figure 16-6**

A photograph showing the Magellanic Clouds. Close to the Small Cloud is the globular cluster 47 Tucanae, which is a member of our galaxy. (Harvard College Observatory)

about 200,000 light years away. The diameter of the main body of the Large Cloud is roughly 25,000 light years; that for the small cloud is 8,500 light years. The mass of the Large Cloud is about 0.1 times that of our Milky Way Galaxy, that of the Small Cloud is about 0.02. The luminosities of the two clouds seem to be average for galaxies in general, our galaxy is much more luminous than most.

The Large Magellanic Cloud is a most interesting object for astronomers concerned with the evolution of stellar systems. It has an abundance of Population I objects: blue-giant stars, open clusters, obscuring gas and dust, and emission nebulae. In fact, the largest emission nebula known, the Tarantula nebula, lies in the Large Cloud. It is so big that were it to replace the Orion nebula, it would cover the entire constellation of Orion and be an extremely obvious feature in our night skies. But the most curious objects in the Large Cloud are the globular clusters, for many of these globular clusters have blue-giant stars! These globular clusters must be young and some may still be in the process of formation. These clusters are very regular and globular in shape but their color–luminosity diagrams look like those of the open clusters in our galaxy. These color–luminosity diagrams have a thin line of stars running right up the main sequence to the blue giants. Other globular clusters in the Large Cloud appear to be a billion years old, but still younger than any in our galaxy. It would appear as if star formation in the Magellanic Clouds has either started late or for some reason been retarded.

The Clouds have radial velocities of recession of 170 and 100 miles per second respectively when referred to the sun, which moves within our own galaxy. When their motions are referred to the center of our galaxy, however, we find that the Clouds have essentially zero radial velocity. This fact, coupled with their nearness to our galaxy, has led some astronomers to believe that they are revolving about our galaxy as satellites. Radio observations of the 21-cm line of neutral hydrogen indicate that the two Clouds are enclosed in a common envelope of hydrogen gas. But unlike some systems (Figure 16-7), there does not appear to be a bridge of stars connecting the Magellanic Clouds to our galaxy. The Andromeda galaxy also has two satellite galaxies, but both of these are elliptical galaxies.

**e. Dwarf Galaxies.**   Surprisingly enough the most common galaxy of all is the *dwarf galaxy* (Figure 16-8). These galaxies are difficult to detect, for they are not as obvious nor as luminous as the bigger galaxies, yet they exceed the bigger ones in number. There are more dwarf galaxies in the immediate neighborhood of our galaxy than all other kinds put together. Large numbers of dwarf galaxies have been detected in nearby clusters of galaxies. Some of these dwarf galaxies are regular in shape, and these tend to have Population II stars; the irregularly shaped dwarfs tend to have very luminous blue stars.

**Figure 16-7**
Spiral galaxy NGC 5194 with a bridge of stars connecting it to its satellite galaxy NGC 5195. (Photograph from the Hale Observatories)

**Figure 16-8**

The dwarf galaxy in Sextans. (Photograph from the Hale Observatories)

## 16.2  Very Active Galaxies

The discussion of galaxies so far has not introduced any new element into our concept of stellar systems. Population I objects appear in spiral arms, Population II objects in nuclei and elliptical galaxies. Galaxies rotate, but they move rather majestically. Even with an occasional supernova, galaxies seem to be rather placid; but this is not a safe conclusion. The galaxy M 82 (Figure 16-9*a*) was once classified as a peculiar galaxy; it certainly appears to be neither an elliptical nor a spiral galaxy. As optical identification of discrete radio sources proceeded, however, it became clear that M 82 is a very strong radio source. Subsequent observations have revealed that it is far from placid. A photograph taken with only the light of the red Hα line of hydrogen reveals a fine structure of filaments extending outward from the nucleus in a direction perpendicular to what appears to be the plane

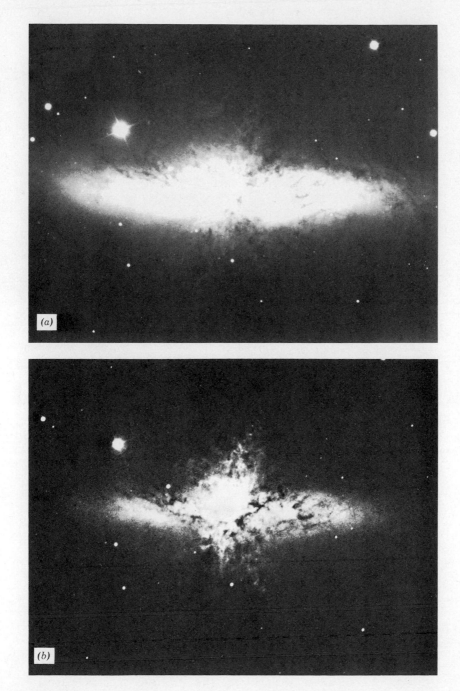

**Figure 16-9**

Exploding galaxy M 82 photographed in (*a*) white light, and (*b*) red light of hydrogen. (Photograph from the Hale Observatories)

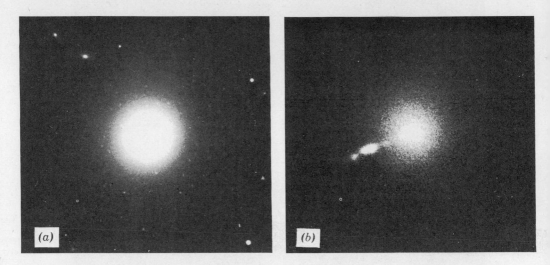

**Figure 16-10**

An elliptical galaxy, M 87 with (a) a long exposure, and (b) a short exposure revealing the jet-like source of radio energy. (Lick Observatory photograph)

of the galaxy (Figure 16-9b). Since the plane of the galaxy is inclined by about 8° from the line of sight, observations of radial velocity of these gases can be made. Their velocity indicates that the gases have been ejected by the nucleus. Furthermore, the velocity of the ejected gases increases in proportion to their distance from the center of the nucleus. The gases near the ends of the filaments are traveling outward at a velocity of 600 miles per second, and judging from their present distance—14,000 light years from the nucleus—they must have left the nucleus about 2 million years ago. The gases only half as far out have just one-half that velocity, so they, too, must have left the nucleus about 2 million years ago. In fact, it seems that all of the gas left at the same time—the galaxy suffered an explosion!

These ejected gases emit not only a bright-line spectrum but a continuous spectrum as well. The light of the continuous spectrum is polarized. This observation coupled with the fact that M 82 is a strong nonthermal radio source leaves little doubt that this galaxy emits synchrotron radiation. It would appear as if the magnetic field passes through the nucleus parallel to the axis of rotation of the galaxy, and the high-speed electrons are spiraling about the magnetic lines of force as they proceed outward from the nucleus.

(a)

(b)

**Figure 16-11**

A strong radio galaxy NGC 5128 in Centauris. (*a*) photograph, (*b*) regions marked *R* indicate where the radio energy originates. (Photograph from the Hale Observatories)

Another galaxy, M 87, when photographed on a long-time exposure appears as a normal elliptical galaxy (Figure 16-10a). But it, too, proves to be a source of strong radio energy, so it was investigated more thoroughly. A short-time exposure revealed a jet of gas extending outward from the nucleus (Figure 16-10b). This galaxy is also a source of X-ray radiation, indicating that the jet must contain very high-energy electrons indeed. The galaxy NGC 5128 (Figure 16-11a) is also a radio galaxy, but the radio energy does not all originate in the optical portion of the galaxy. A great deal of the radio energy comes from invisible gases on either side of and several diameters from the center of the galaxy (Figure 16-11b).

Details of these unusual galaxies are still being sought; the reasons why these galaxies emit such vast quantities of energy will become clear only after more details have been acquired, and discussed among the astronomers and then digested. Then perhaps a fresh idea will appear which may serve to describe these exploding (if they are) galaxies.

## 16.3  The Milky Way Galaxy

Which, if any, of these many and varied galaxies does our own Milky Way galaxy resemble?

Since our galaxy has a great deal of interstellar gas and dust and emission nebulae, it cannot be an elliptical galaxy. Since it has a very definite plane of symmetry, namely the galactic plane as defined by the Milky Way, it does not appear to be an irregular galaxy. We are left, therefore, with the tentative conclusion that our galaxy has a spiral structure. This conclusion is in harmony with a cursory glance at our galaxy, the shape of which can best be imagined by looking at the Milky Way, which forms a fairly narrow band around the celestial sphere. It must therefore be a fairly flat system, perhaps similar in cross section of the spiral galaxy in Figure 16-12 which is seen "edge on." The similarity is made more apparent by comparing this galaxy (NGC 4565) with a wide-angle photograph of the Milky Way (Figure 16-13). The difference between the two photographs, of course, is that one is viewed from the outside and the other is viewed from the inside.

Ours, then, is a spiral galaxy. We want to determine its size and structure —that is, locate the galactic nucleus (locate the position of the sun relative to the nucleus), observe evidence of spiral arms, and determine its mode of rotation.

a. The Galactic Nucleus. In finding the nucleus of our galaxy we automatically locate the relative position of the sun. This has been done by several methods, one of which is based on the location of globular clusters

**Figure 16-12**

Spiral galaxy NGC 4565 seen "edge on." (Photograph from the Hale Observatories)

**Figure 16-13**

Wide-angle photograph of the Milky Way. (Washburn Observatory)

(see p. 385). Of the 121 known globular clusters about 30 are concentrated in the constellation of Sagittarius. The remaining 90 are spread over a much larger part of the sky, the vast majority of them being in that half of the celestial sphere that is centered on Sagittarius. The distances of these globular clusters can be determined with a fair degree of accuracy, for not only do they contain RR Lyrae stars but many also inhabit regions of the sky that are relatively free from interstellar material.

Similar studies have been made of RR Lyrae stars not in globular clusters. These RR Lyrae stars form a system similar to that formed by the globular clusters but not as spherical; it is flattened slightly in the direction of the galactic plane, but again centered in Sagittarius. The nucleus is estimated to be 33,000 light years from the sun.

Studies using Cepheids, O- and B-type stars, and RR Lyrae stars led to the conclusion that the diameter of the galaxy is about 100,000 light years. This places the sun about two-thirds the way out from the center (the radius is 50,000 light years).

**b. The Galactic Plane.**   Having located the nucleus of our galaxy, we are left to find the central plane. A rough estimate of its location can be made by bisecting the Milky Way. It can be more precisely located by studying the distribution of certain objects over the celestial sphere. The objects that give the best results are the ones that can be seen for the greatest distance and therefore include the globular clusters, the O and B stars, the faint stars, and the Cepheids. The distribution of each of these is such that their number increases toward the galactic equator. Radio studies of the distribution of hydrogen gas, both ionized and neutral, have been invaluable in locating the central plane of our galaxy.

From observations that locate the central plane of our galaxy, it has become evident that the galaxy in the region of our sun is about 5,000 light years thick. The sun lies about 40 light years from the central plane and is oriented such that the south pole of the Earth is closer to the galactic plane.

**c. Galactic Coordinates.**   Having located both the nucleus of our galaxy and its central plane, it now becomes convenient to establish a system of galactic coordinates that can be used by astronomers when they discuss the location of objects in our sky but want to consider those objects relative to our galaxy, rather than the solar system.

The *galactic equator* has been established as the central plane of our galaxy. *Galactic latitude* is measured north and south from the galactic equator through 90° to the north galactic pole and the south galactic pole.

By definition, the north galactic pole lies in that hemisphere which includes the north celestial pole, although the two poles are 62° apart. The north galactic pole is between the constellations Bootes and Leo, in a region of very few stars.

*Galactic longitude* must be measured from some chosen point on the galactic equator. At a meeting of the International Astronomical Union in Moscow in 1958, the point from which galactic longitude is measured was selected as that point on the celestial sphere which is defined as the galactic center.* The location of this point was determined by observations of the 21-cm line and lies in the direction of Sagittarius. Galactic longitude is measured from this point eastward along the galactic equator through 360°.

There are, therefore, three coordinate systems on the celestial sphere. These are based on the three principal planes: the plane of the Earth's equator, the plane of the Earth's orbit about the sun, and the plane of the galaxy. The angle formed by the intersection of the galactic equator with the celestial equator is about 62°. Thus the plane of the Earth's equator makes an angle of 62° with the plane of the galaxy. The two planes intersect at two points in diametrically opposite parts of the sky. One of the intersections is in the constellation of Aquila, just south and west of the star Altair. The other intersection is just east of Orion, between Orion and the star Procyon.

**d. The Location of Spiral Arms.**  To locate the spiral arms, we can consider first the O- and B-type stars, since they are very luminous. The intrinsic luminosity of an O or B star can be determined by placing it on the H–R diagram, and its apparent brightness can be determined by means of photoelectric measurements. Its distance, however, cannot be determined directly by the inverse square law (as discussed on p. 277), for astronomers have accepted the fact that they can see these distant stars only through the scattering and absorbing interstellar material. This scattering and absorbing makes the stars appear fainter and thus more distant than they would be were there no interstellar material.

In order to determine the true distance as accurately as possible, some estimate must be made of the amount of light that has been lost in transit between the star and the Earth. One method is to determine the color excess of the star. The color excess, it will be recalled, is a measure of the amount of light scattered by interstellar material. The greater the color excess the

*Previous to this the origin of galactic longitude was in the constellation of Aquila at one of the two intersections of the galactic equator with the celestial equator.

more light lost in transit, and the farther away the star appears. The amount of light absorbed can be estimated from the amount scattered. By considering both of these factors, astronomers are able to make reasonable estimates of stellar distances by applying corrections to the inverse square law.

When the distances and directions of a large number of distant O and B stars have been determined we can see that they are not distributed randomly about the sun but fall into patterns that strongly suggest the existence of spiral arms. This indication of spiral structure has been verified by studies involving the space distribution of emission nebulae, which when coupled with the work on O and B stars and clouds of ionized hydrogen gas disclose the spiral features shown in Figure 16-14. The sun, it appears, is on the inner edge of the Orion arm.

The constellation Orion, with all of its activity of stellar formation has a galactic longitude of about 210°, so it would appear between the upper and the left-hand sides of Figure 16-14. The constellation Auriga has a galactic

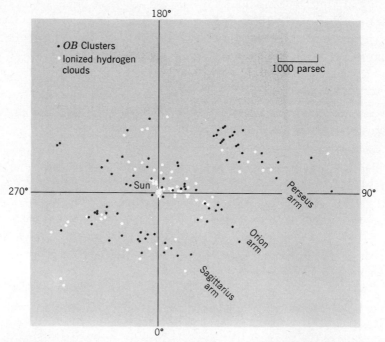

**Figure 16-14**

Spiral structure within about 3,000 parsecs of the sun as obtained from observations of early type stars and clouds of ionized hydrogen. (After W. Becker and R. Fenkart)

longitude of about 170°, Cassiopeia about 120°, and Cygnus about 60°. Sagittarius, of course, has a galactic longitude of 0°. It is in this direction that the great star clouds shown in Figure 15-3 appear.

## 16.4 Rotation of Our Galaxy

**a. Solid-Wheel Rotation.**   Having reached the conclusion that our galaxy is a normal spiral, we can now investigate the problem of rotation. If the entire galaxy rotates like a solid wheel, none of the stars will move with respect to one another, and we should notice no radial velocities or proper motions among the stars. Similarly, two people sitting on stationary horses on a merry-go-round do not move with respect to each other even though the merry-go-round is rotating.

**b. Keplerian Rotation.**   If, on the other hand, stars revolve about the center of our galaxy with motions comparable to the motions of the planets about the sun, we should definitely observe not only radial velocities but also proper motions.

Figure 16-15a shows three circular orbits concentric with the galactic center. If the stars revolve in these orbits in a manner that satisfies Kepler's third law, the inner star will travel faster than the other two stars. The middle star (our sun) will have an intermediate orbital velocity, and the outer star will be the slowest. The orbital velocity of each is represented by the length of the arrow.

With the sun in the position shown we would be able to detect a Doppler shift to the violet (negative radial velocity) for an inner star at *a*, as it is overtaking the sun because of its higher orbital velocity. We would observe the same Doppler shift for an outer star at *f*, since the sun is overtaking it. On the other hand, a Doppler shift to the red (positive radial velocity) would be observed for an inner star in position *c*, since it is leaving the sun behind. A similar shift would be observed for an outer star at *d*, for the sun is leaving it behind.

If inner and outer stars are at *b* and *e* respectively, their spectra would exhibit no Doppler shift. The orbital motion of each of these two stars, however, would be recognizable by their proper motions. The inner star would seem to travel in the direction indicated by the arrow because it is passing the sun, while the outer star would appear to travel in a direction opposite that of the arrow because it is falling behind the sun. A star at *b* has a galactic longitude of 0°, but since it is moving faster than the sun it will eventually exhibit a Doppler shift.

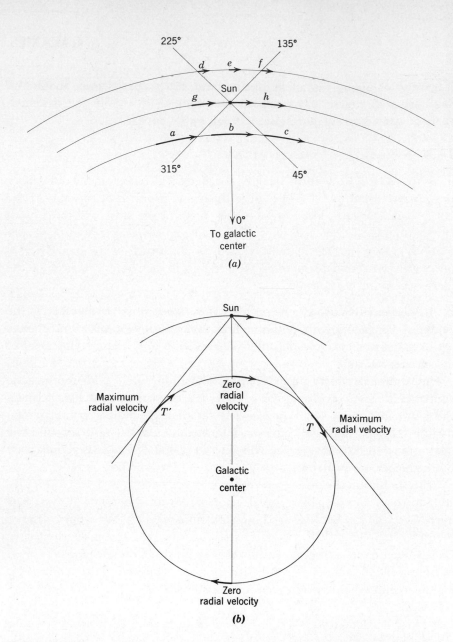

**Figure 16-15**

(*a*) The velocities of stars in the neighborhood of the sun showing Keplerian rotation. (*b*) The radial velocity of stars, resulting from rotation of the galaxy reaches a maximum at points *T* and *T'*. If the stars traveled in circular orbits their radial velocity would be zero at galactic longitude 0° and 180°.

Finally, if there were stars in the same orbit as the sun in either positions *g* or *h* they would exhibit no Doppler shift since they would be near the sun and traveling with the same speed. Thus, if the galaxy rotates according to Kepler's laws, a star's radial velocity will depend in part on its galactic longitude.

When the actual radial velocities of the distant O and B stars are examined, we find that in general they do indeed depend on the star's galactic longitude, although each star exhibits its own individual motion in addition to this over-all galactic rotation. The O and B stars at galactic longitude 45° (see Figure 16-15a), in general, exhibit a positive radial velocity indicating recession. At 135° galactic longitude they exhibit a negative radial velocity, at 225° a positive radial velocity, and at 315° a negative radial velocity. This evidence further substantiates the belief that our galaxy is a spiral, for the rotation of its outer regions is Keplerian (that is, it is similar to the planetary motion about the sun) and thus resembles the rotation of other spiral galaxies.

The amount of a star's radial velocity resulting from galactic rotation depends not only on its galactic longitude but also on its distance from the sun. Radial velocities increase for stars ever more distant from the sun until the position *T* (Figure 16-15b) is reached. At this point a given star's radial velocity is at a maximum, since it is moving directly away from the sun (and at position *T′*, directly toward the sun). Radial velocities should then begin to decrease until they reach zero for those stars in our galaxy that are diametrically opposite the sun.

c. **Rotation and Galactic Structure.**   We can see O and B stars along the galactic equator for a distance of only about 15,000 light years and since our sun is 33,000 light years from the center, the optically observed O and B stars do not include a very large portion of our galaxy. Radio observations, however, have extended this limit of observation to nearly the entire galaxy, and since radial velocity resulting from galactic rotation depends on the location of an object in our galaxy, radio astronomers find themselves in the unique position of being able to determine the spiral structure of most of our galaxy. Their observations all hinge on the fact that the 21-cm line, like every spectral line, is subject to the Doppler principle.

Figure 16-16a indicates why the radio astronomer has been so successful. If clouds of neutral hydrogen emitting radio radiation at 21 cm are at *a* and *b*, each will have a different radial velocity and thus display a different

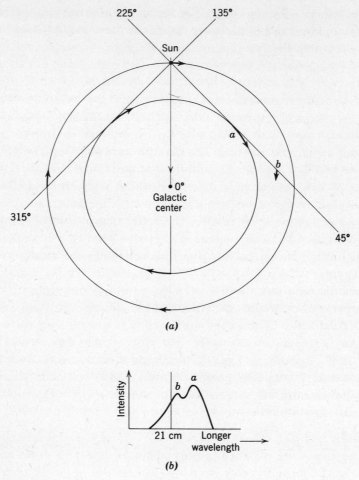

**Figure 16-16**

(*a*) The location of clouds of neutral hydrogen emitting the 21-cm line determines their radial velocity relative to the sun. (*b*) The 21-cm line may show two or more humps indicating two or more clouds of neutral hydrogen with different radial velocities.

Doppler shift. A radio telescope pointed in the direction of *a* will record the 21-cm line as it is shown in Figure 16-16*b*. The larger hump on the curve results from radio emission of the nearer cloud, which displays a larger Doppler shift to longer wavelengths than the more distant cloud at *b*, because *a*'s motion is carrying it *directly* away from the Earth. In some cases more than two humps are seen on the 21-cm line, indicating that there are more than two hydrogen clouds in a given direction, each at different distances from the sun.

By employing this technique for other parts of the galaxy, radio astronomers have been able to locate the positions of clouds of neutral hydrogen in almost the entire galaxy. Since neutral hydrogen is found mostly in spiral arms, the cloud positions also locate the spiral arms of our galaxy. The results are shown in Figure 16-17. The conclusion drawn from the work of radio and optical astronomers is that our galaxy is a normal spiral. There is increasing evidence to indicate that perhaps it is a fairly open Sc type spiral; it may resemble the spiral galaxy shown in Figure 16-18.

From our knowledge of the size and rotation of our galaxy, it is estimated that it has a mass of $2 \times 10^{11}$ suns! There are roughly 200 billion stars in our galaxy!

Although the general features of the structure and rotation of our galaxy are understood, there are many aspects that are not clear. A few examples will emphasize the uncertainties of our knowledge.

The stars do not actually revolve about the galactic center in neat concentric circular orbits. Their orbits are slightly elliptical; thus the distance of each star from the galactic center varies. If we observe stars at 0° and 180° galactic longitude, we do detect Doppler shifts, indicating motion away from and toward the galactic center. Thus each star has its own intrinsic motion superimposed upon the rotation of the galaxy. It is significant that this intrinsic motion is much larger for Population II stars than for Population I stars.

Recent studies of B stars and clouds of neutral hydrogen indicate that the central regions of our galaxy appear to be rotating much like a solid wheel, and the outer parts move more like the planets about the sun. This manner of galactic rotation, of course, is consistent with the rotation of the Andromeda galaxy.

Galactic rotation can be shown the most clearly if we plot the orbital velocity against the distance from the galactic center (Figure 16-19). If our entire galaxy were to rotate like a solid wheel, the orbital velocity of stars would increase in direct proportion with increasing distance from the center of the galaxy, as shown on the graph. If the rotation is Keplerian, the orbital velocities of the stars should decrease with increasing distance from the galactic center in accordance with the curve on the graph. The orbital velocity derived from the B stars is shown. In the region near the sun, this rotation is similar to "planetary" rotation, but it becomes more like "solid wheel" rotation closer to the center of the galaxy. The exact shape and slope of this curve is one of the uncertainties in galactic rotation.

**Figure 16-17**

A drawing of the Milky Way galaxy as observed by the 21-cm line. The circle marks the center of the galactic nucleus, and the dot in the small circle marks the position of the sun. This drawing was prepared by Westerhout of the Leiden Observatory, but the portion of the Milky Way visible only in the southern hemisphere was observed by Kerr in Australia. The portion omitted lies behind the galactic nucleus. (Leiden Observatory)

**Figure 16-18**

NGC 5457, an open spiral galaxy in the constellation of Ursa Major. (Photograph from the Hale Observatories)

It appears as if the hydrogen clouds, as observed by means of the 21-cm line, rotate in a manner different from the B stars. This seems to correspond to studies of star formation in nebulae—either the stars acquire a velocity different from that of the nebula, or the velocity of the nebula changes after the stars have formed.

The fact that the apparent motion of a star (or group of stars) in the galaxy depends on its distance from the sun was used to make the first determination of luminosity for the Cepheids and RR Lyrae stars (see p. 297). Essentially, the method is to determine the average motion for a group of stars, let us say RR Lyrae stars, with respect to the sun. From this

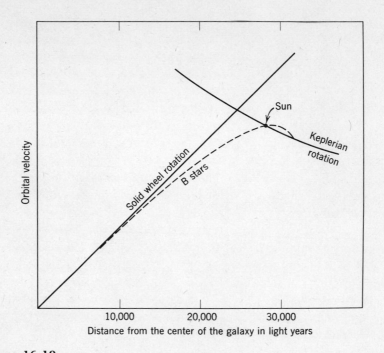

**Figure 16-19**

A graph plotting rotational velocity of our galaxy as a solid wheel, as Keplerian rotation, and by observations of B-type stars.

motion we can determine their average distance, and from their distance and their average brightness we can find their average luminosity. Since this method uses relative motions of groups of stars with respect to the sun, it has been called the method of *mean parallaxes*. Granted, it gives us only an average and would therefore be unworkable if many types of stars were considered together. It works, however, if stars of only one class are considered together; for example, we can determine the average luminosity of a group of A-type stars, or RR Lyrae stars, or Cepheids of a given period and population.

The sun's motion about the center of the galaxy is of considerable interest to us, for to calculate the radial velocity of any object with respect to the galaxy we must subtract the sun's motion from the observed motion of the object. From studies of the motions of globular clusters, RR Lyrae stars, and near-by galaxies, it appears that the sun is moving around the center of the galaxy with an orbital velocity of about 200 miles per second. Its direction of motion is toward galactic longitude 90°, that is, at right angles to a line joining the sun with the center of the galaxy. The sun appears to travel in a circular orbit whose radius is 33,000 light years. At the observed velocity it will take the sun about 200 million years to complete one trip around its orbit. Most of the stars in the sun's neighborhood share this motion; the Population II stars, however, do not. Since most of the stars in the sun's neighborhood do share the sun's motion, their velocity with respect to the sun is small—roughly about 20 miles per second. The Population II stars, on the other hand, have velocities with respect to the sun of as much as 100 miles per second. This high relative velocity gave them the name *high-velocity stars*, a name which increased knowledge has made quite inappropriate.

The RR Lyrae stars are fairly representative of high-velocity stars and can be used to illustrate their motion. The average motion of the RR Lyrae stars is about 80 miles per second *with respect to the sun*, but in a direction opposite to the sun's motion. This means that they are moving about the galaxy more slowly than the sun is. If we subtract the sun's motion from their motion we find that they have orbital velocities of about 120 miles per second (200 − 80 = 120).

## 16.5 Spiral Arms

One of the many unsolved problems of the structure of spiral galaxies is the very existence of the spiral arms. If galaxies rotate as they are observed to do, then each galaxy must have made many revolutions since its formation. Our sun, for example, is 4.6 billion years old and it makes 1 revolution of the galaxy in 200 million years; it has, therefore, made a total of about 23 revolutions. The question now arises, why aren't the spiral arms twisted about the galaxy so tightly as to be unrecognizable? How do Sc galaxies maintain their open shape? What mechanism produces the stability which maintains the barred spiral arms? A different and yet related question concerns the existence of vast amounts of gas and very young stars in the spiral arms. Why hasn't all of this gas condensed into stars long ago? Why are stars still forming in a galaxy that is perhaps 10 billion years old?

Observations by radio telescopes of the 21-cm line indicate rather conclusively that hydrogen gas is being ejected by the nucleus of our galaxy (and the nuclei of other spirals as well). The amount of gas ejected is estimated to be about 1 solar mass per year, and it leaves the nucleus with a velocity of about 30 miles per second. Perhaps this is enough to replenish the gas which has formed the stars now visible, but by itself it does not account for the structure and stability of the spiral arms.

One of the suggestions to account for the existence of the spiral arms considers the speed of revolution of the material about the galactic center. It has been proposed that because of irregularities in the gravitational field of the galaxy the speed of revolution of the stars and clouds changes slightly. Presumably, their speed decreases in certain regions. This decrease in speed results in an accumulation of material.

By way of analogy, a reduction in the speed of cars on a freeway will cause an accumulation of automobiles called a traffic jam. The cars don't have to stop. An accident along side the freeway may cause the traffic to slow from an average speed of 65 to 40 mph. The resulting accumulation has been experienced by many exasperated commuters. Even after the accident has been cleared up, the accumulation of cars persists. So a decrease in the speed of the material revolving about the center of the galaxy may cause an accumulation that we see as the spiral arms. That is, the spiral arms may be celestial traffic jams!

Measurements of motions of stars and gases within a spiral arm indicate that the two classes of objects move differently. It would seem as if the stars form in the gases only to acquire a velocity different from the gas which nurtured them; the stars must leave the gases and populate the regions between the spiral arms. And between the spiral arms there has indeed been observed a substratum of red-dwarf stars. These stars must have left the spiral arms billions of years ago. The gas which forms these stars is then replenished by the ejection of gases from the nucleus. This leaves the question of how long can a nucleus supply gas to support spiral arms? Do galactic nuclei shrink with age? Do spiral arms then lose whatever strength they have?

## 16.6 Galactic Fossils

It is very striking indeed that there seems to be a significant relationship between the orbital motions and locations of stars, and their ages:

1. The globular clusters are the oldest members of our galaxy and they form a spherical system concentric with the galactic nucleus. Their

orbital velocities are low, and their orbits are eccentric and inclined at large angles from the galactic plane.

2. The O and B stars are the youngest members and they form the flattest system, being nearly coincident with the galactic plane. Their orbits are very nearly circular and their orbital velocities are higher than those of any other stars.

3. The other stars seem to form a continuum between these two extremes.

At one extreme are the O and B stars, which are very young Population I stars. The late main-sequence stars (K and M) are usually older than the O and B stars. They form a system that is fairly flat, but their orbits are more likely to carry them farther from the galactic plane than do the orbits of the O and B stars. Their orbits are also less circular than those of the youngest stars, and their orbital velocities a little lower. The RR Lyrae stars are Population II stars but are younger than the globular clusters. They form a system that, although it extends far from the galactic plane, is slightly flattened and thus does not extend as far as the globular clusters. Their orbits are also eccentric and inclined from the galactic plane but not as much as the orbits of the globular clusters. Their orbital velocities are low but not as low as those of globular clusters.

The ages of stars are also related to their location in the galaxy in that the older stars are more likely to congregate near the galactic nucleus. Thus their orbits are more likely to be smaller than those of younger stars.

All these facts assembled are wonderful material for speculative thought; it appears that our galaxy, like a tree, has left rings that tell of its past. When our galaxy was at the age during which its globular clusters were forming from gaseous clouds of hydrogen, it must have had a shape quite similar to that now defined by the system of globular clusters—a nearly spherical shape. Furthermore, it must have rotated with a velocity nearly equal to that of the globular clusters—a low velocity. We speculate further that apparently the galaxy then began to contract, but the globular clusters, having formed, maintained their orbits. As the galaxy contracted it rotated more rapidly to conserve its angular momentum. In rotating more rapidly it began to flatten out and after roughly 10 billion years it has become the flat system that we now observe. The youngest stars and the remaining clouds of hydrogen are all part of this flat system; the older stars are not. With increasing age, the older stars form ever more spherical systems and on the average have smaller orbits and lower orbital velocities.

Their orbits, however, can be more eccentric and more inclined from the galactic plane than those of younger stars. Three stages of galactic evolution are depicted in Figure 16-20.

This is not meant to imply that there are no Population II objects in the galactic plane, for there are. Some may even have orbits that lie in the galactic plane. Those Population II objects in orbits inclined from the galactic plane must still cross it twice in one revolution. During this crossing they are temporarily (thousands of years) close to the galactic plane.

That so many globular clusters are concentrated near the galactic nucleus indicates that during the time when the globular clusters were forming, the shape of our galaxy may have resembled that of a small elliptical galaxy of type E0, with a high concentration of stars near the center, that gradually thins out with increasing distance from the center.

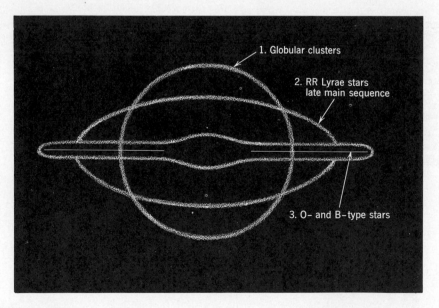

**Figure 16-20**

Speculation leads to these three stages of galactic evolution, because a star's location in our galaxy as well as its orbital inclination from the galactic plane and its speed of revolution about the galactic center all depend to a certain degree on the age of the star.

This comparison of our galaxy at this stage of development with elliptical galaxies is interesting; unfortunately, it is misleading. Although our galaxy in the early stages of its development may have had a shape similar to that of elliptical galaxies, it must have been different because elliptical galaxies contain little apparent interstellar gas and dust. Thus star formation must be nearly at an end in these Population II objects.

The terms Population I and Population II now take on added significance. They represent the extremes in age. There are stars of intermediate age and thus of intermediate population class. Since we assume that star formation has been going on continually, the gradation from Population I to Population II must be continuous. The sun appears to be an old Population I star.

## 16.7 Evolution of Galaxies

Galaxies have a very strong tendency to form in clumps and clusters. Very often galaxies will form together in a very tight assemblage (see Figure 17-3). It is presumed that galaxies so close together are under the gravitational influence of one another. If this is true, then the relative masses of each galaxy can be estimated from their motions; the more massive galaxies will move the slowest, the least massive will move the fastest. So reasoned Thornton Page of Van Vleck Observatory, Wesleyan University, who then set about to determine the relative masses of many galaxies from a number of such close assemblages. Surprisingly enough he found that the E0 galaxies, the very spherical ones, are consistently nearly 30 times as massive as the open Sc spirals. Furthermore he found a gradation in mass from the E0 galaxies to the Sc galaxies. The dwarf galaxies, however, are not part of this gradation, their masses are even less than those of the Sc galaxies.

It is well known that elliptical galaxies do not rotate as fast as the spiral galaxies. However, the measure of rotation should include the amount of matter rotating, so we employ angular momentum as this measure. As in the formation of the solar system, the angular momentum of a galaxy should remain constant since no appreciable outside forces act to reduce it. But even with its greater mass the angular momentum of an elliptical galaxy is less than that of a spiral galaxy.

H. C. Arp, of the Hale Observatories, plotted the mass of the galaxies against their angular momentum per unit mass and found a very interesting relationship (Figure 16-21). It is significant that the most massive galaxies have the least amount of angular momentum per unit mass; and that there should be a relatively uniform gradation between the E0 and the Sc galaxies. Arp suggests that this gradation results from the conditions present during the formation of the galaxies.

If the galaxies form from blobs of gas of nearly equal mass, then all galaxies should have equal mass; but if the gas from which each galaxy forms has a different amount of angular momentum, then the resulting galaxies might be different. Arp suggests that during the contraction of a large nonrotating blob of gas the density of the gas remained fairly uniform so that star formation proceeded at a fairly uniform rate and for only a relatively short period of time. All of the gas would have condensed into

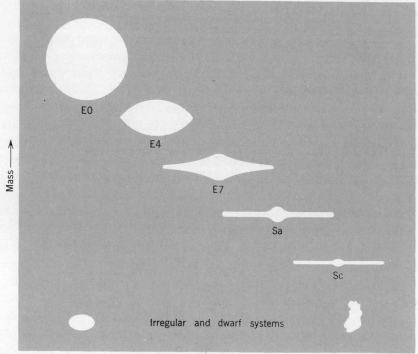

**Figure 16-21**

As the angular momentum per unit mass of galaxies increases, their total mass decreases.

stars and all of the stars would have nearly the same age. In other words, an elliptical galaxy would form from a nonrotating blob of gas. Those blue-giant stars formed during this early age would long since have evolved into white dwarfs. There being no more gas to form new stars, no successive generations of blue giants would form.

On the other hand, if the initial blob of gas were rotating—if it had an appreciable amount of angular momentum—then as the gas contracted it would have rotated still faster to conserve angular momentum. As it rotated faster, gases would have been thrown out to form an equatorial bulge, and later a nebular disk reminiscent of the nebular disk about the proto-sun. The formation of the nebular disk would have decreased the density of the gas and consequently decreased the rate of star production. With the increased rotation much of the gas may have been thrown free of the forming galaxy resulting in a decrease in mass. Evidence for this idea is supported in the stages of evolution of our own galaxy.

Consequently, it is suggested that the amount of angular momentum of the initial blob of gas is the main factor determining whether the gas will become an E0, E1, E2, . . . , Sa, Sb, or Sc galaxy.

It is generally accepted that during the life-time of a spiral galaxy the production of supernovae and other gas-ejecting stars replenishes some of the interstellar material used up in star formation. If that initial inter-stellar material is nearly pure hydrogen, however, these gas-ejecting stars "contaminate" the interstellar gases. During the lifetime of a star hydrogen is converted into helium, which is in turn converted into carbon, oxygen, and the heavier elements. By ejecting gases back into space stars spew forth some of these heavier elements as well. Consequently, the first stars to form in a spiral galaxy should have a low percentage of metals; stars formed later in the life of a galaxy should have a higher percentage of metals. The most recently formed stars should have the highest per-centage of metals of all the stars. It is well established that Population I stars have a higher percentage of metals than do the Population II stars, so the idea of an increasing metallic content of the interstellar gases is well supported. Presumably, all of the stars of the elliptical galaxies have the same, or nearly the same, chemical composition.

This suggestion of galactic evolution is one of the first to really bring some of the various shapes and forms of galaxies into a unified principle of formation. There are certainly many galaxies which do not fall into the straight sequence E0 to Sc, such as the barred spirals, the dwarf systems, the irregular systems, the exploding galaxies, etc. But more progress is made when astronomers have a framework—a hypothesis—upon which to work. If that hypothesis is later shown to be incorrect, astronomers have still learned and benefited from it. This suggestion of galactic evolution will serve astronomers very well, even if it cannot answer all of their questions now. The details will be forthcoming with additional studies, both observational and theoretical. This is the manner in which ideas develop into good descriptions of our universe.

## BASIC VOCABULARY FOR SUBSEQUENT READING

Barred spiral galaxy

Dwarf galaxy

Elliptical galaxy

Galactic equator

Galactic latitude

Galactic longitude

Irregular galaxy

Normal spiral galaxy

Population I objects

Population II objects

## QUESTIONS AND PROBLEMS

1. Explain the coordinate system that is most convenient for studies of our galaxy.
2. What observations have led to our understanding of the general shape of the Milky Way galaxy and to a knowledge of the location of the sun with respect to the center?
3. Which objects can best be used to trace the spiral arms of our galaxy? How can each of these objects best be observed?
4. Describe the general shape of the spiral arms of our galaxy which are near our sun.
5. Discuss the "galactic fossils" of our Milky Way galaxy.

## FOR FURTHER READING

Baade, W., *Evolution of Stars and Galaxies*, Harvard University Press, Cambridge, Mass., 1963.

Bok, B. J., and P. Bok, *The Milky Way*, Harvard University Press, Cambridge, Mass., 1957.

Hodge, P. W., *Galaxies and Cosmology*, McGraw-Hill paperback, New York, 1966.

Page, T., and L. W. Page, ed., *Beyond the Milky Way*, The Macmillan Co., New York, 1969.

Page, T., and L. W. Page, ed., *Stars and Clouds of the Milky Way*, The Macmillan Co., New York, 1968.

Page, T., ed., *Stars and Galaxies*, Prentice-Hall paperback, Englewood Cliffs, N.J., 1962.

Struve, O., and V. Zebergs, *Astronomy of the 20th Century*, Crowell, Collier and Macmillan, New York, 1962, Chapters XIX and XX.

Arp, H., "On the Origin of Arms in Spiral Galaxies," *Sky and Telescope*, p. 385 (Dec. 1969).

Berge, G. L., and G. A. Seielstad, "The Magnetic Field of the Galaxy," *Scientific American*, p. 46 (June 1965).

Bok, B. J., "An International Discussion of Milky Way Research," *Sky and Telescope*, p. 4 (July 1963).

Bok, B. J., "The Large Cloud of Magellan," *Scientific American*, p. 32 (Jan. 1964).

Bok, B. J., "The Spiral Structure of Our Galaxy," *Sky and Telescope*, p. 392 (Dec. 1969); p. 21 (Jan. 1970).

Hodge, P. W., "Dwarf Galaxies," *Scientific American*, p. 78 (May 1964).

Hodge, P. W., "Globular Clusters in the Magellanic Clouds," *Sky and Telescope*, p. 72 (Feb. 1961).

Hodge, P. W., "The Sculptor and Fornax Dwarf Galaxies," *Sky and Telescope*, p. 336 (Dec. 1964).

Iben, I., "Globular-Cluster Stars," *Scientific American*, p. 26 (July 1970).

Page, T., "Galaxies and Quasars at Prague," *Sky and Telescope*, p. 372 (Dec. 1967); p. 16 (Jan. 1968); p. 90 (Feb. 1968).

Page, T., "The Evolution of Galaxies," *Sky and Telescope*, part I, p. 4 (Jan. 1965); part II, p. 81 (Feb. 1965).

Westerlund, B. E., "Report on the Magellanic Clouds," *Sky and Telescope*, p. 23 (July 1969).

Weymann, R. J., "Seyfert Galaxies," *Scientific American*, p. 28 (Jan. 1969).

"Age of Our Galaxy," *Sky and Telescope*, p. 287 (May 1964).

"Distribution and Motions of Supergiant Stars," *Sky and Telescope*, p. 162 (March 1970).

"Galaxies with Bright Infrared Cores," *Sky and Telescope*, p. 357 (June 1970).

"Gould's Belt: An Expanding Group of Stars," *Sky and Telescope*, p. 93 (Feb. 1968).

"The Heart of the Milky Way," *Sky and Telescope*, p. 282 (May 1965).

"Helium and the Galaxy's Age," *Sky and Telescope*, p. 219 (April 1965).

"Movie of the Milky Way's Hydrogen Clouds," *Sky and Telescope*, p. 92 (Feb. 1970).

"Our Galaxy as Mapped with RR Lyrae Stars," *Sky and Telescope*, p. 95 (Feb. 1968).

"X-ray Galaxies," *Scientific American*, p. 50 (April 1966).

CHAPTER 17 CHAPTER

# THE UNIVERSE AND RELATIVITY

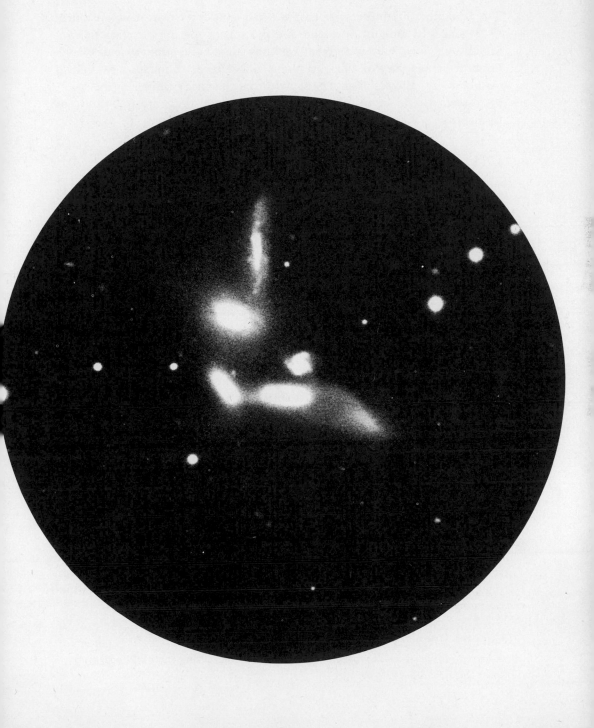

We define the universe as that which includes all matter in existence. Since the amount of matter and the size of the universe are not yet known, we must start with what we know about the *observable* universe. From their analyses of the observable universe astronomers are able to formulate hypotheses about the structure of the entire universe (the study of *cosmology*) and its origin, evolution, and even its future (the study of *cosmogony*).

## 17.1 The Distribution of Galaxies

The basic building block in the universe is the galaxy, just as the basic building block in a galaxy is the star. Astronomers observe all the visible galaxies projected onto the celestial sphere, and from this projection conclude that galaxies are not distributed at random throughout the universe.

C. D. Shane, of the Lick Observatory, has presented conclusive evidence than the galaxies are clumped together into clusters, Figure 17-1*a*. The survey plates for the proper motion study at the Lick Observatory (see p. 264) contain thousands of galaxies as well as the foreground stars in our own galaxy. Shane and his co-workers counted the number of galaxies in each square degree of the sky and plotted these counts on a large "celestial" map. They then drew contour lines joining those regions of the sky that had the same number of galaxies. A small portion of their "galactic contour map" is shown in Figure 17-1b). Each of the areas of closed contour lines represents a cluster of galaxies, and this is only a small portion of the entire celestial sphere! Although there also appear to be stray galaxies that are not contained in any cluster, they are the exception and not the rule.

Fritz Zwicky, of the Mount Wilson and Palomar Observatories, has concluded that many of these clusters have spherical symmetry; that the space density of the galaxies increases toward the center of the cluster; and that the very bright elliptical galaxies are concentrated toward the center, with the spiral and irregular galaxies more prevalent in the outer regions. Some of the clusters may contain as many as 100,000 galaxies but these are exceptionally rich.

### Chapter Opening Photo

A group of five galaxies with clouds of material connecting them. (Photograph from the Hale Observatories)

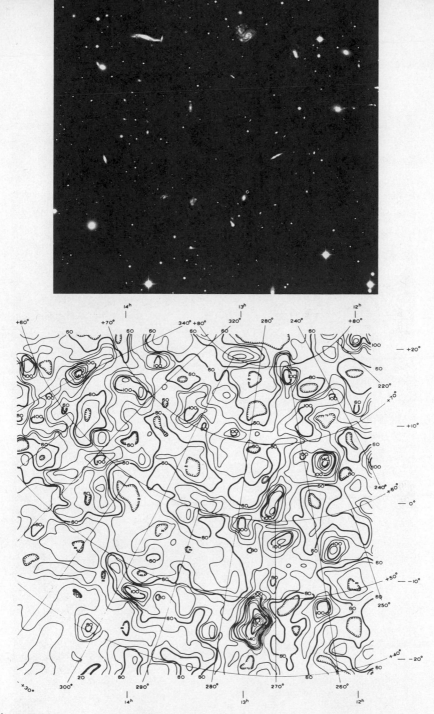

**Figure 17-1**

(*a*) A cluster of galaxies in the constellation Hercules (*top*). (Photograph from the Hale Observatories.) (*b*) The distribution of galaxies in a portion of the celestial sphere as drawn by C. D. Shane indicates that galaxies tend to form in clusters (*bottom*). (Lick Observatory photograph)

(c)

**Figure 17-2**

Photographs of clusters of galaxies all enlarged the same amount to show the decrease in apparent size and brightness of the galaxies with increasing distance. (*a*) Part of the Coma Berenices cluster estimated to be 220 million light years away. (*b*) Corona Borealis cluster estimated to be 740 million light years away. (*c*) Hydra cluster 2,100 million light years away. (Photographs of the Hale Observatories)

Our own galaxy is one of about 17 (there is some doubt of membership for a few) forming a rather small cluster called the *Local Group*. Our Local Group also contains the Magellanic Clouds, the Andromeda galaxy with its two satellites, M33 (the spiral in Triangulum, Figure 16-4), and others. If the Large Magellanic Cloud is classified as a barred spiral, the Local Group contains all types of galaxies. The Andromeda galaxy and our galaxy are by far the largest members of this group; M33 is a poor third, and the Magellanic Clouds are fourth and fifth in size. All these galaxies are included in a volume of space about 3 million light years in diameter.

Other well-known clusters of galaxies include: the Coma Berenices cluster with about 9000 members; the Corona Borealis cluster with at least 400 members; the Hydra cluster with over 200 members (Figure 17-2).

Each cluster of galaxies is a group bound within itself, held together by the mutual gravitational attraction of all the members. But this gravitational field does not restrict the motion of the member galaxies within the cluster; galaxies will have intrinsic velocities within the cluster of as much as 600 miles per second.

Zwicky has commented that each cluster appears to be an isolated unit and that no clusters form in pairs, groups, or clusters of clusters, which fact may have broad implications for Newton's law of gravitation applied to such large distances. Zwicky suggests that because of the large distances separating clusters of galaxies, attraction ceases and repulsion sets in, for the clusters of galaxies seem to avoid one another.

This certainly cannot be said of the galaxies inside the clusters, for there is overwhelming evidence that galaxies tend to associate closely with one another. There are many galaxies that exist as pairs, others as three and more (Figure 17-3). The distance between galaxies in a cluster is not as great in comparison to their size as is the distance between the stars in comparison to their size. On the average, galaxies have a diameter of about $5 \times 10^4$ light years and in a cluster they have an average separation of roughly $30 \times 10^4$ light years. Thus they are separated, on the average, by a distance roughly six times their diameters, whereas stars in the region of the sun are separated by a distance of about 30 million solar diameters.

Zwicky amply shows that intergalactic gas and dust does exist. He points out that the number of faint clusters of galaxies is much reduced in regions of the sky covered by near-by rich clusters of galaxies. For example, there is an average of only 8 faint clusters seen through the central region of the Coma Berenices cluster in an area of the celestial sphere of 36 square degrees, but in regions where no near-by rich clusters exist there is an average of 70 clusters for the same area.

## 17.2 Distances of the Galaxies

Although we can learn a great deal from the study of the distribution of galaxies on the two-dimensional celestial sphere, we must consider the universe as being three-dimensional and study the distances of the galaxies.

Let us review the ever-present problem of distance determination. For only the nearest stars does the method of heliocentric parallax do us any good. Beyond about 50 light years, heliocentric parallax begins to

**Figure 17-3**
A group of five galaxies, called Stephan's Quintet, forming a compact system. (Photograph from the Hale Observatories)

fail and other less direct methods must be used. An astronomer can measure the apparent brightness of a star whose intrinsic luminosity is known, such as a Cepheid or RR Lyrae star. From the brightness and luminosity the distance can be determined by the inverse square law. The astronomer may have to account for the loss of light from scattering by interstellar matter, in which case his results are likely to be less accurate.

**a. RR Lyrae Stars and Cepheids.** The distances of the nearest galaxies, the Magellanic Clouds, have been determined by using RR Lyrae stars, for their period–luminosity relation is more reliable than that of the

Cepheids. The Cepheids must be used for more distant galaxies such as the Andromeda galaxy and M33, since RR Lyrae stars cannot be seen in them. For still more distant galaxies not even the Cepheids can be seen and the astronomer must resort to other methods.

**b. Ten Brightest Stars.**   One method that has been used in the past is to estimate the luminosity of the brightest stars in a given galaxy. This method is based on the assumption that the brightest stars in each galaxy are of nearly the same luminosity. This luminosity can be estimated from studies of the stars in our galaxy as well as of the brightest stars in near-by galaxies of known distance. From this luminosity and the measured brightness of the stars the distance of the galaxy can then be estimated. But even this method fails with galaxies at still larger distances, so we must resort to other procedures.

**c. Brightest Globular Clusters.**   Many galaxies are surrounded by a halo of globular clusters similar to ours which, because each cluster contains thousands of stars, can be seen at greater distances than can the brightest stars in any galaxy. The brightest globular clusters can be used to estimate distances by the same method used for the brightest stars. Still more distant galaxies, however, are too far away for even the globular clusters to be seen, much less individual stars, except for an occasional supernova.

**d. Supernovae.**   But an occasional supernova is of great help to the astronomer. As indicated on p. 309 supernovae can be used as a tool to estimate distance, since supernovae of either type I or II have fairly consistent luminosities within these types. Supernovae of type II reach luminosities of about 200 million suns and may thus be used to determine the distances of galaxies too far away for globular clusters to appear (Figure 17-4). Although supernovae, when observed, prove very valuable for the measurement of distances, they are sporadic and thus not readily available. Therefore other distance indicators are more frequently used.

**e. Ten Biggest Galaxies.**   The realization that galaxies form in clusters had led astronomers to try to use the galaxies themselves as distance indicators. The biggest galaxies in each cluster are of about the same size and luminosity. If this luminosity and size can be determined, they could be used to estimate the distances of clusters of galaxies. Figure 17-2 indicates how this method works. The clusters shown in each plate are at different

**Figure 17-4**

A supernova appeared in the galaxy NGC 7331 in 1959. The upper photograph was taken before the supernova erupted. (Lick Observatory photograph)

distances from us; thus the galaxies in the Coma Berenices cluster, which is the closest of the three, appear larger than those in the other two clusters. The galaxies in the Hydra cluster appear the smallest, since it is the most distant.

As astronomers seek to measure greater and greater distances they, of necessity, rely on those methods used to determine lesser distances. Although most distance measurements ultimately rely on the astronomical unit, some do not and consequently serve as a check. The distance of the Crab nebula (p. 309), for example, was determined by its rate of expansion. But all distances outside our galaxy depend on the period—luminosity relationship of the Cepheids. When it became clear that there are at least two types of Cepheids (p. 298), the distance determined by means of Cepheids and consequently all intergalactic distances had to be increased by a factor of two, and the universe became twice as large.

In answer to the inevitable question, "Are the distances now accepted correct?" we can say only that these distances are the best values available and that they appear to be more nearly correct than the previously accepted ones. Before the distance sale was changed it was thought that the Andromeda galaxy was much closer and thus it (along with all the other galaxies) gave the impression of being smaller than it now appears to be. Astronomers felt uncomfortable about this, for it placed us in the unique position of belonging to the largest galaxy in the universe. But with the increase in the distance scale by a factor of 2, all the other galaxies have doubled their size. Andromeda is now larger than our own galaxy, and all the other galaxies are more nearly comparable in size to ours. Hence astronomers are more confident with the present distance scale than they were with the old one. As for our own galaxy, its size was determined by RR Lyrae stars, not by Cepheids, and thus it was unaffected by the change in the distance scale. Now it is recognized that the E0 galaxies are the biggest of all.

**f. The Red-Shift.**   There is yet another method of estimating distances of galaxies. In 1929 Hubble announced the results of studies which clearly indicated that with the exception of our Local Group *all* the galaxies have positive radial velocities, that is, a Doppler shift to the red. Furthermore, the farther the galaxy is from us the larger the red-shift. The spectral lines normally used to measure the red-shift are the H and K lines of calcium, and the extent of their shift can be seen in Figure 17-5. The galaxy in the top spectrogram has a velocity of *only* 750 miles per second, and the lines

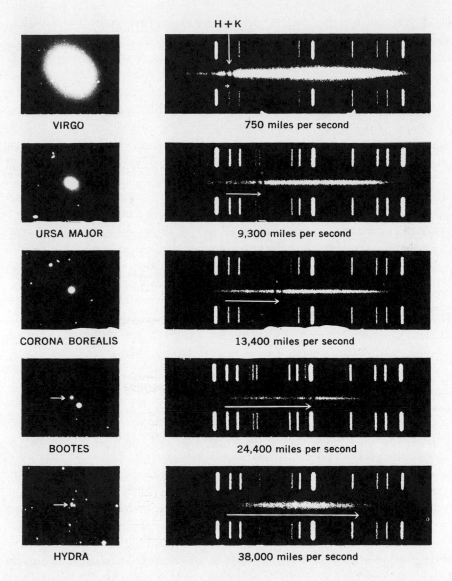

H + K

VIRGO

750 miles per second

URSA MAJOR

9,300 miles per second

CORONA BOREALIS

13,400 miles per second

BOOTES

24,400 miles per second

HYDRA

38,000 miles per second

## Figure 17-5

The actual spectrograms of five galaxies showing the extent of the shift of the H and K lines of calcium to the longer wavelength. The apparent diameter of the galaxies decreases with increasing distance and they appear ever fainter. (Photographs from the Hale Observatories)

have been shifted the distance shown by the arrow. In each succeeding spectrogram the H and K lines are shifted ever farther to the red until in the last plate they are shifted almost to the limit of sensitivity of the photographic emulsion used.

The problem of distance determination then became one of establishing the relationship between radial velocity and distance. This relationship can be shown by plotting radial velocities against distances for galaxies whose distances have been determined by other methods. The resulting graph (Figure 17-6) is a straight line. Thus if the radial velocity of a galaxy is measured, its distance can be estimated by consulting the graph. The steepness (or slope) of the graph can also be used to determine the distance directly. The slope of the curve, called the *Hubble constant*, indicates that the radial velocity will increase by 18 mi/sec for every million light years distance. A galaxy at 200 million light years distance should have a radial velocity of 3,600 mi/sec. A galaxy with a radial velocity of 5,400 mi/sec should be 300 million light years away from us.

Accordingly, the clusters whose spectra appear in Figure 17-5 are at the following distances: Corona cluster, 740 million light years; Bootes cluster, 1,360 million light years; Hydra cluster, 2,100 million light years. These distances are so great that they are incomprehensible.

It appears that the red-shift, discovered by Hubble, is a fundamental characteristic that may help man in his endeavor to understand the universe: (1) With the exception of our Local Group, *all* the galaxies have their spectra shifted to the red. If this is interpreted as a Doppler shift we are left with the conclusion that all these galaxies are receding from us! (2) The farther the galaxy is from us the larger the red-shift. Thus the velocity of recession increases with increasing distance. (3) These velocities of recession reach values interestingly close to the velocity of light.

Although it may be possible to interpret the red-shift as resulting from causes other than radial velocity, astronomers generally agree that the radial velocity interpretation is the most plausible. The natural consequence of this interpretation is that the universe is expanding. Thus the term red-shift has become synonymous with expansion of the universe.

The fact that the red-shift increases with increasing distance from our galaxy places a maximum limit to the size of the observable universe. If galaxies exist at such distances that their radial velocities equal or exceed that of light, we will never be able to observe them because the light will never reach us.

Since all the galaxies are receding from us, we could infer that our galaxy is at the center of the universe, were it not that such an inference would place us only a step away from Plato, who put the Earth at the center of the universe. Astronomers are unwilling to accept any interpretation that places us in such a unique position. In general they prefer to base their thinking on the *cosmological principle: the appearance of the universe does not depend upon the observer's position*. That is, with the exception of local and small differences, the universe should appear the same to all observers no matter where they may be located—in our galaxy or in a very distant galaxy.

To explain the observed expansion on these grounds, the structure of the universe must be considered. Discussions of the structure of the universe, however, involve the theory of relativity expounded by Albert

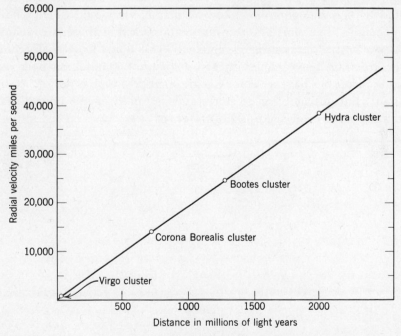

**Figure 17-6**

The velocity of recession of galaxies increases with increasing distance from us. It must be remembered, however, that these distances represent the location of these clusters hundreds of millions of years ago.

Einstein. Therefore it becomes necessary to consider certain aspects of that theory before proceeding with the structure of the universe.

## 17.3  The Principle of Special Relativity

At the turn of this century there were many observations conflicting with the physics derived since the time of Galileo. Yet that physics—now called *classical physics*—has proved extremely successful. For example, astronomers used Newtonian physics to predict the existence and location of an unseen planet, Neptune. Scientists still use it to place satellites in the orbit of their choosing, and to aim lunar probes at a particular location on the moon. It works. But it has its limitations.

In 1888 the famous Michelson–Morley experiment was performed which showed that the measured velocity of light is independent of the motion of the observer. For example, if we were to measure the velocities of light emitted by star *a* (Figure 17-7) and by star *b*, we would find that those two velocities are equal! The measurement of each of those two velocities is independent of the motion of the Earth. This is not true of cars moving along a freeway; two cars moving at a velocity of 60 miles per hour but in opposite directions have a velocity with respect to each other of 120 miles per hour. But automobiles do not travel at speeds approaching that of light, and that is what makes the difference.

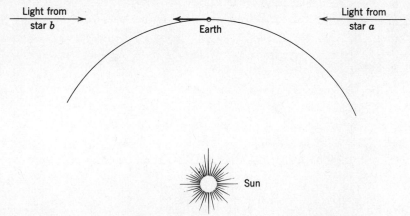

**Figure 17-7**

The Michelson–Morley experiment showed that the velocity of light as measured from the Earth of either star *a* or *b* is the same despite the motion of the Earth.

Einstein derived his *principle of special relativity* to describe the actions of systems moving relative to each other at *constant* velocities. The consequences of this principle are many, the concepts are profound and in conflict with our common sense, that is, the everyday garden variety of common sense.

The two basic postulates of special relativity are: (1) the velocity of light in free space is finite and constant for all observers, and (2) the laws of physics should be valid for all observers even if those observers move relative to one another. By adopting these two postulates Einstein showed that the measured values mass, length, and time change when transferred from one system to another if those two systems are moving at relativistic speeds (speeds close to that of light) with respect to each other.

If people outfit two spacecraft with essentially identical items and embark on a very rapid space flight such that the two spacecraft A and B travel at a relativistic speed with respect to each other, they will observe that objects on the *other* spacecraft appear both more massive and yet smaller—but smaller in only one direction, the direction of relative motion—than those once identical objects on their own spacecraft. They will find that the clocks on the *other* spacecraft will run slowly compared to the clocks on their craft. Passengers in both craft A and B will notice the same effects; the measurements of mass, length, and time depend upon the motion of the observer relative to that which he observes.

Ridiculous! says the skeptic. A kilogram is a kilogram, and a second of time is a second of time! But it has been proved beyond the shadow of a doubt, claims the physicist. Electrons have been accelerated to speeds close to that of light and their mass does indeed increase; that is, a greater force is required to obtain the same acceleration to still greater speeds. Atomic particles called pions which serve as good clocks, because they decay into other particles in a measurable duration of time, have been accelerated to relativistic speeds and they decay more slowly!

In proposing these fundamental ideas Einstein did not give any reason for the apparent change in mass, length, and time when the observer moves relative to that which he observes. He simply showed conclusively that *these observations are the natural consequence of our living in a universe in which the velocity of light is finite and constant for all observers regardless of their motion*. The principles of special relativity have been proved valid and are the foundation for much of modern physics and astrophysics.

One of the many consequences of these fundamental concepts is the equivalence of energy and mass: $E = mc^2$. Recall that it is the conversion of mass to energy which permits astronomers to account successfully for the tremendous outpouring of energy from stars for billions of years. Another consequence of the postulates of special relativity is the fact that in free space no material object can travel with a velocity equal to or greater than that of light. This applies to electrons—this has been verified—and to galaxies.

One of the nonrelativistic implications of the finite velocity of light is that astronomers really are cosmic historians; today they *observe* the past. Astronomers can only say what happened on the sun 8 minutes ago; they give the date of the supernova that resulted in the Crab nebula as A.D. 1054, even though the light from that explosion had already been traveling in space for about 4,000 years before it reached the Earth. When they observe galaxies estimated to be 2 billion light years from the Earth, they see those galaxies as they were 2 billion years ago, not as they are "now." Therefore what use is it to say a certain galaxy is 2 billion light years away from us? None whatsoever, for all that an astronomer can say is that the galaxy's position 2 billion years ago is 2 billion light years away from our position at the present time. He cannot and never will be able to measure how far apart we are at any one given instant of time. There is no "now" in the universe because light has a finite velocity, and measurements of such vast distances become misleading.

This condition results from the fact that we cannot really separate time and distance as two separate and independent entities. If we wish to identify a certain event we must not only say where it occurred but when. If two people arrange to meet in the northwest corner of the 50th floor of the Empire State building, they have specified the three space coordinates: one along the north-south line, the other along the east-west line and the third along the up-down line. But they must also state at what time they will meet. Thus it takes four dimensions to identify a particular event, three of space and one of time, and these four dimensions are inextricably bound together.

## 17.4 The Theory of General Relativity

Einstein also considered systems which move relative to each other with an accelerated motion. In doing this he evoked the *principle of equivalence: no instrument can determine any difference between the effects of gravity and the effects of acceleration.* As a thought experiment the reader should consider two spacecraft in "free space," that is, space free of any gravitational field. Let the first spacecraft *not accelerate*; its rocket motors are turned off. Without a gravitational field how will the occupants feel? Will they fall? No, of course not. The second spacecraft should *accelerate*; its rocket motors are turned on. How will the passengers of the accelerating spacecraft feel? Will they accelerate with the spacecraft? What force will act on them to accelerate them? After reasonable consideration the student should come to the conclusion that if the second spacecraft accelerates at a rate of 32 ft/sec$^2$, the passengers would feel the same as if they were here on Earth. They would weigh the same; things would fall "down" the same as they do here on Earth. Down is toward that part of the spacecraft which exerts a force on and accelerates them. But, asked Einstein, what would happen to a beam of light shining at right angles to the direction of travel? He answered that question by showing that the beam of light would appear to bend! It would bend "downward" relative to the ship. So if a gravitational field is equivalent to an acceleration, a beam of light should bend downward in a gravitational field—a strong gravitational field.

Einstein predicted that light from a star which travels close to the surface of the sun will have its path deflected by the gravitational field of the sun (Figure 17-8). The apparent position of star $a'$ would be radially displaced from its true position $a$ in a direction away from the sun. The amount of this displacement increases with star light which passes ever closer to the edge of the sun. It has been observed in many total solar eclipses, for it is only at this time that stars so close to the sun's apparent disk can be seen. But this displacement is not a good quantitative verifica-

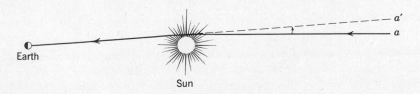

**Figure 17-8**

The light from a more distant star that passes close to the sun is affected by the sun's gravitational field in that the light bends toward the sun assuming a new direction of travel.

tion of general relativity; the observed deflection is about 28% greater than the predictions of general relativity.

Continued interest in the problems proposed by the theory of general relativity has brought forth a competing theory, the *scalar–tensor theory*, by Robert Dicke of Princeton University. As frequently happens in science, two theories will be considered by both the theoretical astronomer and the observational astronomer. That theory which best accounts for the ensuing observations will be the one accepted. Unless, of course, a third and better theory is proposed in the mean time!

## 17.5 The Structure of the Universe

In discussing the gravitational effect on light Einstein brought an entirely new concept to the attention of astronomers and cosmologists. If light travels unhampered by matter, it will travel in a straight line. Sometimes, however, a straight line is not a straight line as we commonly know it. If we define a straight line as the shortest distance between two points in space we have a definition that is completely general, but not so obvious.

**a. A Straight Line and the Shape of Space.**    To illustrate, let us consider the shortest distance *along the surface* of the Earth between San Francisco and London. This shortest distance is an arc of a *great circle*\* on the Earth. An airplane pilot following this course must travel farther north than either San Francisco or London (see Figure 17-9).

But, you may argue, this course is not a straight line, it is an arc of a circle—which it is. Can you, however, distinguish between a straight line and the path of light through space? What constitutes a straight line for the beam of light as it passes close to the sun? That beam of light when viewed end-on appears as a point—the image of a star! Doesn't a straight line appear as a point when viewed end-on? Doesn't that beam of light continue to travel in a straight line even when near the sun?

Now we are confounded by geometry—but after all, geometry is a study of space, and space is a basic ingredient of the universe. If we define a straight line as the shortest distance between two points, we are no better off. It has been shown that light follows the shortest distance between two points—when it does not pass from one medium to another (and the deflection of a star's light by the sun does *not* depend upon the

---

\*A great circle is a circle whose center is coincident with the center of the sphere on which it is drawn.

solar corona). Einstein suggested that the universe may not be describable with Euclid's geometry. In fact, Einstein suggested that *a gravitational field curves space!* In curved space a straight line is not Euclidean, it becomes a *geodesic: the shortest distance between two points in space, curved or Euclidean.*

It is difficult enough to imagine "space" even without defining it, and an attempt to conceive of "curved space" seems at first quite hopeless. Space presumably is that which is between the atoms and dust particles in the interstellar regions. Space, then, being "emptiness," must be continuous; emptiness cannot be broken up into discrete particles with nothing between. So how can we possibly imagine space to be curved?

We can best imagine it if we reiterate what has been said—namely, that light travels along a geodesic in space, and if its path deviates from a

**Figure 17-9**

The great circle route from San Francisco to London.

straight line as defined by Euclidean geometry, the space must deviate in some way from Euclidean space.

Two intersecting straight lines in Euclidean geometry define a plane. A triangle, which in Euclidean geometry also defines a plane, has three angles the sum of which must add up to exactly 180°. But two geodesics in curved space do not define a Euclidean plane, and thus the three angles of a triangle in curved space do not add up to 180°. The angles of a *spherical triangle* must exceed 180° but be *less than* 540° (Figure 17-10).

If the mass of the sun is enough to curve space in its immediate vicinity, perhaps all the mass in the universe is enough to curve all the space in the universe. Maybe a geodesic in the universe is not a Euclidean straight line. Perhaps Euclidean geometry is so fundamental to us only because our everyday straight lines and planes are so extremely small when compared to the universe. A ship crossing the Pacific Ocean seems to be continually traveling in a straight line, yet we know that it is traveling on a spherical Earth and thus it is actually traveling along a Euclidean circle. The world was once thought to be flat; it has been proved to be nearly spherical. Our universe can be thought of as being a Euclidean universe, but it is possible that it is not.

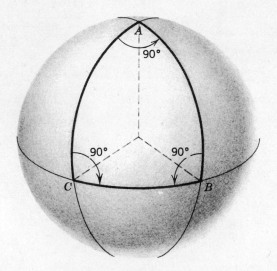

**Figure 17-10**
The three angles of this spherical triangle add up to 270°.

**b. Non-Euclidean Space.** To illustrate a non-Euclidean universe, we must use our imagination. Let us imagine a bug that has only one dimension —a creature that can move and conceive of motion only along a straight line, or only along the circumference of a circle (Figure 17-11a). If this bug resides in a straight line, he can travel forward or backward for an infinite distance and not retrace his steps, but he cannot move to the right or left. If he lives on the circumference of a circle he can still travel forward or backward an infinite distance. In order to travel an infinite distance he must first complete one trip around the circle and then retrace his steps for an infinite number of revolutions. He cannot travel along a radius of the circle; in fact he cannot even imagine a radius for he can conceive of only one dimension—the forward and backward directions.

Now let us imagine a two-dimensional creature—a beetle (Figure 17-11b). He can reside in a Euclidean plane and move either forward or backward an infinite distance, or to the right or left an infinite distance without retracing his steps. On the other hand, if he lives on the surface of a sphere, he can travel either forward or backward, or to the right or left an infinite distance, but he will eventually return to his starting place and retrace his

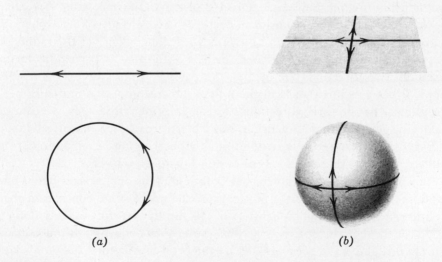

(a)                              (b)

**Figure 17-11**

(a) A one-dimensional bug can travel along a line or along a circumference of a circle. (b) A two-dimensional beetle can travel along a plane or along the surface of a sphere.

steps. He cannot travel or even conceive of motion along the radius of the sphere.

A creature living in a straight line or in a plane would live in a Euclidean universe of one and two dimensions respectively. A creature living on the circumference of a circle or on the surface of a sphere would live in a non-Euclidean universe of one and two dimensions respectively. Geodesics in these non-Euclidean universes are not Euclidean straight lines.

Do we, as three-dimensional creatures, live in a universe that is Euclidean and thus one step up from the line-plane sequence? Or do we live in a non-Euclidean universe that is one step up from the circle–sphere sequence? If our universe is Euclidean it could be infinite in extent; we could imagine ourselves traveling in space along a straight line for an infinite distance and never retracing our steps. If we live in a non-Euclidean universe which is the third in the sequence circle–sphere–universe, we could travel along a geodesic for an infinite distance but would eventually return to our starting place and begin to retrace our steps.

A Euclidean universe is easier to imagine, for it would be three-dimensional and infinite in extent. A straight line would be a Euclidean straight line. A non-Euclidean universe of three dimensions is *impossible* to imagine. The two-dimensional beetle in the surface of his spherical universe cannot possibly conceive of a third dimension, yet the radius of his sphere is in that third dimension. It is a line that is perpendicular to the only two directions the beetle can imagine. If our universe is of this type, it has a radius that is in the fourth dimension and is mutually perpendicular to the three directions we can imagine. Obviously we cannot conceive of this fourth dimension, but we can treat it mathematically; we can study its properties by extending our observations from a circle to a sphere and from there to this type of non-Euclidean universe.

Einstein treated the universe as a four-dimensional space–time continuum. The Earth travels in a region of the space–time continuum that is affected by the gravitational field of the sun. It therefore travels along a geodesic through this space–time continuum. Such concepts, although difficult to imagine, have been verified by the orbit of Mercury. Newton's law of gravitation is not able to account for the orbit of Mercury in that the major axis of its orbit rotates more rapidly than gravitational theory predicts. Mercury moves in a manner shown in the exaggerated drawing in Figure 17-12. Einstein's concept of the universe is mathematically able to account for this peculiar motion, which Mercury displays more than any other planet because it is so close to the sun.

In his scalar–tensor theory, however, Robert Dicke maintains that the structure and rotation are not what they appear to be. Careful observations indicate that the sun is less spherical than had previously been thought, and this lack of spherical structure alters the application of Einstein's theory of general relativity to the motion of Mercury. This new application is not in good agreement with observation. So the question remains open.

**c. Counts of Galaxies and the Shape of Space.**   Since we have no proof that the universe is Euclidean, it is of interest to consider the possibility that it is not. How can we determine the shape of our universe? To answer this question let us again resort to the two-dimensional analogy. We can place dots on a plane in a random fashion but with an overall uniform density and let these dots represent clusters of galaxies in a Euclidean two-dimensional universe (Figure 17-13). Assume that dot A represents the cluster of galaxies in which our two-dimensional creature exists. If we count the clusters of galaxies in ever-increasing circles about him we find that their number increases as the square of the distance away from him. There are 3 dots in the first circle whose radius is 1 unit, 12 dots $(3 \times 2^2)$ in the circle of radius 2 units, 27 $(3 \times 3^2)$ in the circle of radius 3 units, etc.

Now let us represent a non-Euclidean two-dimensional universe by placing dots on a sphere (such as ink dots on a balloon) and label one of them A (Figure 17-14a). If our beetle now counts these dots about him he will find that their number will increase less rapidly than the square of the distance from him. There may be 3 dots in the first circle but only 11 in the second and 25 in the third. The reason for this is that there is less

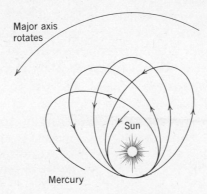

Major axis
rotates

Sun

Mercury

**Figure 17-12**

The major axis of Mercury's orbit rotates according to Einstein's theory of relativity and not according to Newton's law of gravitation.

area on the surface of a sphere of a given radius than there is in a plane within a circle of the same radius, as may be seen by taking one half of a grapefruit peel and flattening it out on a table. The grapefruit peel will split in several places. This shows that there is less area on the surface of the half grapefruit peel than in a circle on a plane with a radius equal to one-fourth the circumference of the grapefruit (Figure 17-14*b*).

It is also conceivable that non-Euclidean two-dimensional space could exist in which the number of dots in ever-increasing circles would increase more rapidly than the square of the distance out. Such space would take the shape of a western saddle (Figure 17-15*a*). It is not difficult to imagine a flimsy saddle that could be flattened out on the floor of the saddle room and in so doing the saddle would have to fold in several places in order to lie flat on the floor (Figure 17-15*b*). Thus there is more area on a saddle within a circle of given radius than on a plane within a circle of the same radius.

Space that follows a form similar to the Euclidean plane is called *zero-curvature* space; that which appears on the surface of a sphere is called

**Figure 17-13**

In Euclidean space, the number of randomly spread dots included within ever larger circles increases as the square of the radii for two-dimensional space and as the cube of the radii of spheres for three-dimensional space.

*positively curved* space; and the space on a saddle is *negatively curved* space.

To determine the shape of our universe we must count clusters of galaxies in ever-increasing *volumes* of space, since our space is three-dimensional. Therefore if the number of clusters increases as the *cube* of the distance from us, our universe is Euclidean. If the number of clusters increases less rapidly than the cube of the distance from us, our universe has a positive curvature (circle–sphere). If the number of clusters of galaxies increases more rapidly than the cube of the distance from us, our universe has a negative curvature (saddle). Such counts have been made, but uncertainties in distance and thus in time have so far prevented them from becoming valid tests.

## 17.6 Expansion of the Universe

Expansion can be explained in both a Euclidean and a non-Euclidean universe. If the universe is Euclidean, we must imagine that the space between the clusters of galaxies becomes larger as the universe expands, just as the space between the raisins in bread becomes larger as the bread is baked. The clusters themselves do not become larger but the space between them increases. The larger the distance between any two clusters the more rapidly the clusters recede from one another. Thus the rate of expansion increases with increasing distance between clusters of galaxies. The effect would appear the same from anywhere in the universe.

If the universe is non-Euclidean, we can still account for the observed effects of expansion. Let us recall the two-dimensional beetles on the surface of a balloon on which dots represent the clusters of galaxies. If the balloon is further inflated the distance between any two dots will increase. Dots farther apart recede from each other more rapidly than those close together, for there is more rubber between these dots to stretch. Again the rate of expansion increases with increasing distance and the effect is the same no matter from which dot the observations are made. This analogy fails in one respect—the dots on an inflating balloon will become larger whereas the clusters of galaxies remain the same size.

## 17.7 The Origin of the Universe

The shape of our universe, whatever it may be, has not yet been determined, but it must be related in some way to the origin of the universe. Did our universe have a beginning? If so, how did it begin? And when?

(a)

**Figure 17-14 (a)**

In positively curved space, the number of randomly spread dots included within ever larger circles increases less rapidly than the square of the radii for two-dimensional space and less rapidly than the cubes of the radii of spheres for three-dimensional space.

There are currently two main theories concerning the evolution of the universe. One is called the *big bang theory*, for it assumes a definite beginning of rather dramatic proportions. The other, the *steady-state theory*, maintains that the universe has an infinite age and thus no beginning or ending.

**a. The Big Bang Theory.** According to the big bang theory all the matter in the universe was once contained in a *primeval nucleus*, the density of which was perhaps equal to the density of the nucleus of an atom. This primeval nucleus exploded. Shortly after it exploded, it had a temperature of 250 million degrees Kelvin. As the universe has continued to expand it has continued to cool to its present overall temperature of about 50°K (−370°F).

By determining the rate of expansion and the size of the present universe it is possible to retrace the process to find the date of the explosion. The date considered at present is roughly 10 billion years ago. Presumably the rate of expansion decreased as the universe expanded just as the velocity of a rock thrown in the air decreases. The rate of decrease depends on the size of the universe and its mass, for as the universe expands the kinetic energy of the galaxies is converted into potential energy.

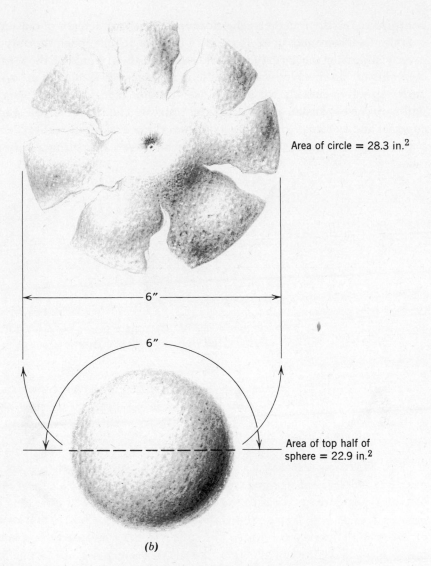

Area of circle = 28.3 in.$^2$

6"

6"

Area of top half of sphere = 22.9 in.$^2$

(b)

## Figure 17-14 (b)

There is more area included within a circle of given radius if that circle is on a plane than if the circle is on a sphere with the radius along the circumference of that sphere.

If Newton's law of gravitation holds in such a universe, there are basically two possible modes of expansion. If the universe is expanding fast enough, the clusters of galaxies exceed the escape velocity of the universe and will continue to recede from each other forever. The space density of clusters of galaxies then continues to decrease until in some distant epoch the only galaxies visible from the Earth (if it should still exist then) would be those in our own cluster. If, however, the universe is not expanding at such a "breakneck" speed, eventually the mutual gravitational forces of all the galaxies will stop the expansion and reverse the process. The universe will start to contract and eventually return to its primeval nucleus. Presumably it could expand again, repeating the process over and over, becoming a *pulsating universe.*

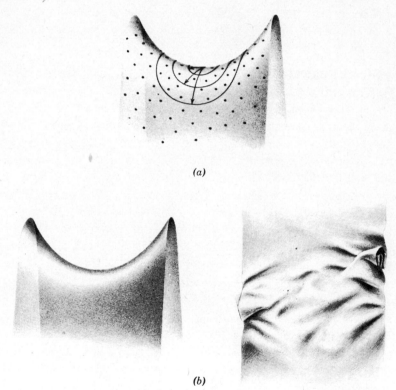

(a)

(b)

**Figure 17-15**

(a) In negatively curved space the number of randomly spread dots included within ever larger circles increases more rapidly than the square of the radii for two-dimensional space and more rapidly than the cubes of the radii of spheres for three-dimensional space. (b) There is less area included within a circle of given radius if that circle is on a plane than if the circle is on a saddle with the radius following the curvature of that saddle.

The theory of a pulsating universe implies that the universe has many beginnings, for it would start anew every time it expanded from the primeval nucleus. But the pulsations could have been going on for an infinite time, and we could never determine this, for each time it contracts into the primeval nucleus all traces of its past life must be completely erased forever. The best we can do would be to determine how long it has been since the most recent primeval nucleus.

There is also the possibility that the universe started as an infinitely thin gas from which it contracted into the primeval nucleus. It then expanded, but on this theory the rate of expansion would be too great to allow it to contract again and it would continue to expand for an infinite length of time.

**b. The Steady-State Theory.** The idea of the steady-state universe results from an extension of the cosmological principle. Proponents of this theory make the point that Einstein showed that the universe is a four-dimensional space–time continuum. Therefore, the cosmological principle should be restated: *the appearance of the universe does not depend upon the observer's position in either space or time.* That is, with the exception of local and small differences, the universe should appear the same to all observers no matter where they may be located or when—at what astronomical epoch—they should observe it. The universe should appear the same, in its overall aspects, if viewed 5 billion years ago or 5 billion years from now. Surely some stars will have been born and died out; individual galaxies will have changed, but the overall aspects of the universe should remain constant throughout all space and time.

The steady-state theory assumes that the universe never began, or to be more precise, that it is in a process of continual creation. As the universe expands, its overall density decreases, but there is a universal lower limit to this density, below which it cannot exist. As the density approaches this lower limit, more matter is created to increase the density once more. Thus as the universe continues to expand, new matter is created to fill the void. The newly formed matter is hydrogen that eventually forms clusters of galaxies. Each new cluster of galaxies lives out its life as the universe continues to expand, while new clusters of galaxies form. New galaxies form, and old ones die, but the universe is always at the same density and there are always galaxies of different ages. Thus the universe will be the same if examined at any epoch. Although individual galaxies and clusters will have changed, the overall picture will not have changed. This is what is meant by a "steady-state" universe that is not only infinite in age but infinite in extent.

**c. Possible Observational Tests.** Which theory, if either, is correct is difficult to tell. Future observations should reveal the answer. If we can find very old galaxies and very young galaxies, the steady-state theory will gain favor. However, the number of old galaxies in any one part of the universe should be rather small, since they are diluted by newer galaxies as the universe continues to expand.

Another observation that might help to reveal the answer lies in the rate of expansion of the galaxies as their distance from us increases. If the steady-state theory is correct the rate of expansion should increase steadily with increasing distance, for the universe, by and large, is the same throughout all space and time. If, however, the rate of expansion should increase more rapidly, it would support the big bang theory, for as we see ever more distant galaxies we also see back in time and thus observe the rate of expansion of the universe as it existed a billion years ago or more. Let us recall that according to the big bang theory the rate of expansion should decrease with increasing time; as we look *back* in time, then, this rate should increase.

Still another critical test is the observation of the density of galaxies in the universe which, according to the steady-state theory, is constant over all time and space. The proponent of the big bang theory, on the other hand, says that the density of galaxies in space is continually decreasing with time; thus as we look *back* in time by observing ever more distant galaxies, the density should increase. The most distant observable galaxies are perhaps 5 to 10 billion light years away; according to the big bang theory these should be about 5 to 10 billion years younger than those closest to us and consequently in an earlier stage of development. The steady-state advocate, on the other hand, says that since the universe is the same over all time and space, the very distant galaxies should be similar to those closest to us.

## 17.8 Observations of the Universe

To settle any scientific question, we have learned to rely on observations. A statement should be questioned unless it can be supported with observations, and the more observations the better. Both of these theories of the universe are being questioned, and observations, some unexpected, continue to keep the question exciting.

**a. Quasars.** The correlation of discrete radio sources with images on photographic plates has led astronomers down many new paths. Exploding galaxies was one such path. Another path is the very puzzling objects, discovered in 1960, which have been dubbed *quasars*, short for quasi-stellar sources. (When an object is discovered it must be named for purposes of classification and conversation, but its name may have nothing to do with its actual structure—since its structure may not be known when the name is applied. For example, quasi-stellar source means essentially the same thing as asteroid! But there the similarity ends.)

Quasars were observed first as discrete radio sources and so individuals are named according to their number in the *Third Cambridge* catalog: 3C 48, 3C 273, etc. They were later identified as bluish "stars" (Figure 17-16), with unrecognizable spectral lines. Martin Schmidt of the Mount Wilson and Palomar Observatories, however, showed that these strange spectral lines are really lines of the far ultraviolet displaced into the near ultraviolet and visible part of the spectrum. Quasars suffer a severe case of red-shift.

The red-shift of 3C 9 is so large that the change in wavelength exceeds the unshifted wavelength by a factor of 2! This means that the Lyman-$\alpha$ line of hydrogen with an unshifted wavelength of 1216 Å has an apparent wavelength of 3648 Å ($1216 + 2 \times 1216$). The Balmer $H\alpha$ line with an unshifted wavelength of 6563 Å appears in the infrared region at 19,689 Å. To translate this into a radial velocity using the classical relationship

$$\frac{\Delta\lambda}{\lambda_0} = \frac{v}{c}$$

would yield a velocity of recession twice that of light! How can we see it, if the Earth and that quasar are receding from each other at twice the speed of light?

Clearly there is a discrepancy! This discrepancy, however, is easily resolved. All we have to do is to realize that the speeds involved are too great to use Newtonian physics; we must employ the principle of special relativity. It, too, produces an algebraic expression for the Doppler shift:

$$\frac{\Delta\lambda}{\lambda_0} = \sqrt{\frac{c + v}{c - v}} - 1$$

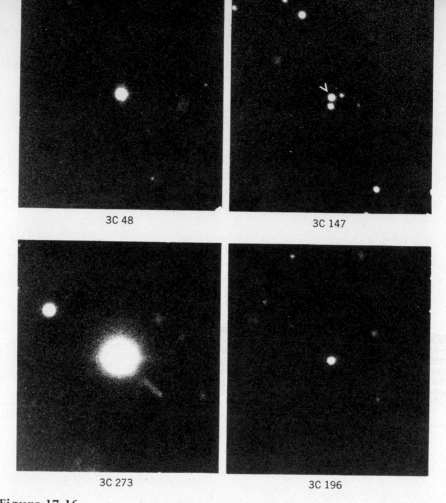

3C 48

3C 147

3C 273

3C 196

**Figure 17-16**

Four different quasars. The jet can be seen extending from 3C 273. (Photographs from the Hale Observatories)

where $v$ is the relative velocity of the source and $c$ the velocity of light. By using this equation from special relativity, we determine the radial velocity of 3C 9 to be $0.8c$, or 80% the speed of light!

It is easier, however, to discuss the Doppler shift of quasars in terms of the fraction $\Delta\lambda/\lambda_0$, called simply the *red-shift*. If the red-shift is 1, the spectral lines are shifted an amount equal to their laboratory wavelength $\lambda_0$. The Lyman-alpha line, for example, would have an observed wavelength of 1216 Å + 1216 Å = 2432 Å. The observed wavelength of the Lyman-alpha line of a quasar with a red-shift of 2 would be 1216 Å + 2432 Å = 3648 Å.

Continued observations of quasars indicate that their density in space increases with increasing red-shift until a red-shift of 2.5 is reached. Then their numbers seem to decline even though their brightnesses have not yet reached the limit of our telescopes. Apparently the decline in numbers is real.

If the red-shift results from the expansion of the universe, then as we consider ever greater red-shifts, we are considering ever more distant quasars. But as we observe ever more distant quasars, we are also looking back in time. If their numbers begin to decrease after a red-shift of 2.5, then their density in space—and thus in time—must actually decrease. But why?

Recognizing uncertainties in estimating their distances and thus their age, quasars with a red-shift of 2.5 are roughly 8.5 billion light years away from us. According to the big bang theory, the age of the universe is roughly 10 billion years. It would therefore appear that quasars formed after the universe began. It seems that most of them formed when the universe was about 1.5 billion years old. Then as the universe aged, their numbers declined, so their activity must not be long lasting. We see no quasars nearby, so they must not be forming this late in the age of the universe.

The observations of quasars clearly tip the scales toward the big bang theory. Quasars seem to have been a passing phase, early in the life of the universe, which because of the expansion and great distances are only now becoming visible to us. Astronomers are cosmic historians.

If quasars are so distant from us, however, why are they so easily visible? They simply must radiate fantastic amounts of energy, perhaps 10 or 100 times the amount of energy radiated by the Milky Way galaxy! Just how they manage this vast expenditure of energy is as yet an unsolved mystery.

The mystery is further complicated by the observations that their brightnesses and thus their luminosities vary significantly in a matter of days. How such an apparently large, massive, and luminous object can change its luminosity so suddenly is not at all understood.

The mystery of the universe is still very much with us, even if its nature is changing.

**b. The Primeval Fireball.** The quasars, even though they remain unexplained, support the big bang theory of the universe. Another unexpected discovery, this one in 1965, also gives definite evidence in favor of the big bang theory.

Two radio astronomers at Bell Telephone Laboratories, A. A. Penzias and R. W. Wilson, detected what was for them a disturbing background noise in their radio telescope. Since the radio signals come uniformly from all directions in the sky, it was proven not to be of terrestrial origin. Thorough investigation showed that it was blackbody radiation from a source whose temperature is about $3°K$ (see Section 9–2, p. 223 f).

At the same time this investigation was proceeding, Robert Dicke and some of his friends at Princeton University (only 30 miles away) were independently investigating theoretically what would happen to the energy of the *primeval fireball*, the big bang, that started our universe expanding. They had come to the conclusion that although the initial temperature of the explosion may have been $10^{10}$ degrees Kelvin, and the wavelength of the radiation correspondingly very short, the temperature would have cooled because of expansion. The temperature they predicted from their theoretical studies was about $5°K$, nearly the same as that causing the radiation found by Penzias and Wilson at Bell Telephone Labs.

Continued studies have indicated that the observed radiation may indeed be left over from the initial explosion of the universe. The radiation has remained in the universe and has continually lost energy, until it now represents not a body at $10^{10}$ degrees Kelvin, but one of only $3°K$. Radiation of this nature is strong evidence in favor of the big bang theory, for it plays no role whatsoever in the steady-state theory.

Having learned something of the changes that observations have forced man to make in his thinking of the universe we should be ready to expect an explanation of quasars which stems not from the present ideas of physics and astronomy, but perhaps from some fundamental mechanism not yet imagined by man. Startling observations demand bold explanations; new ideas stimulate new observations. This is the way science progresses.

## BASIC VOCABULARY FOR SUBSEQUENT READING

| | |
|---|---|
| Cosmogony | Positively curved space |
| Cosmology | Quasars |
| Geodesic | The red-shift |
| Negatively curved space | Zero curvature space |

## QUESTIONS AND PROBLEMS

1. List in order the methods of distance determination used by astronomers, beginning with the method that can be used only for the nearest stars and ending with the method used for the most distant galaxies.
2. Discuss three possible shapes our universe might take.
3. Outline the steady-state theory and the big bang theory of the universe. Compare the predictions made by each that lend themselves to observations, and that might be used to support one theory over the other.
4. If the number of quasars increases with increasing distance from our Milky Way galaxy, which of the two main theories of the universe would this observation support and why?
5. If a quasar has a red-shift of 1.5, what is the observed wavelength of the Lyman-alpha line?

## FOR FURTHER READING

Barnett, L., *The Universe and Dr. Einstein*, William Morrow and Co., New York, 1957.

Hodge, P. W., *Concepts of the Universe*, McGraw-Hill paperback, New York, 1969.

Hodge, P. W., *Galaxies and Cosmology*, McGraw-Hill paperback, New York, 1966.

Munitz, M. K., *Theories of the Universe*, The Free Press, New York, 1957.

Page, T., and L. W. Page, ed., *Beyond the Milky Way*, The Macmillan Co., New York, 1969.

Bunner, A. N., "High-Energy Cosmic Rays," *Sky and Telescope*, p. 204 (Oct. 1967).

Burbidge, E. M., and C. R. Lynds, "The Absorption Lines of Quasi-stellar Objects," *Scientific American*, p. 22 (Dec. 1970).

Fichtel, C. E., "Gamma-Ray Astronomy," *Sky and Telescope*, p. 85 (Feb. 1968).

Ginzburg, V. L., "The Astrophysics of Cosmic Rays," *Scientific American*, p. 51 (Feb. 1969).

Green, L. C., "The Fourth 'Texas' Symposium" (The Primeval Fireball), *Sky and Telescope*, p. 153 (Sept. 1967).

Green, L. C., "Observational Aspects of Cosmology," *Sky and Telescope*, p. 199 (April 1966).

Green, L. C., "Quasars Six Years Later," *Sky and Telescope*, p. 290 (May 1969).

Green, L. C., "Relativistic Astrophysics," *Sky and Telescope*, part I, p. 145 (March 1965); part II, p. 226 (April 1965).

Robinson, I., et al., "Relativistic Astrophysics—A Report on the Second Texas Symposium," *Physics Today*, p. 17 (July 1965).

Schmeidler, F., "The Einstein Shift—An Unsettled Problem," *Sky and Telescope*, p. 217 (April 1964).

Schmidt, M., and F. Bello, "The Evolution of Quasars," *Scientific American*, p. 54 (May 1971).

"Black Holes," *Scientific American*, p. 54 (Oct. 1970).

"Dead Galaxies," *Scientific American*, p. 50 (April 1969).

"The Extragalactic Distance Scale," *Sky and Telescope*, p. 93 (Feb. 1970).

"Light Variations in Quasars," *Sky and Telescope*, p. 18 (July 1969).

"The Local Group of Galaxies," *Sky and Telescope*, p. 21 (Jan. 1964).

"New Measurements of the Cosmic Microwave Background," *Sky and Telescope*, p. 10 (Jan. 1968).

"Quasar Hypotheses," *Scientific American*, part I, p. 38 (Aug. 1964); part II, p. 50 (Feb. 1966).

"A Radio Test of Relativity," *Sky and Telescope*, p. 138 (Sept. 1970).

"Structural Changes in a Quasar," *Sky and Telescope*, p. 158 (March 1968).

"Where Did Cosmic Helium Originate?" *Sky and Telescope*, p. 88 (Feb. 1965).

"White Dwarfs and the Einstein Red Shift," *Sky and Telescope*, p. 78 (Feb. 1968).

"X-Rays from Outside the Galaxy," *Sky and Telescope*, p. 145 (March 1968).

# INDEX

**483**

# DATE DUE

| 3/12/19 | | | |
|---------|---|---|---|
| 3/5/19 | | | |
| 4/22/19 | | | |
| | | | |
| | | | |
| | | | |
| | | | |
| | | | |
| | | | |
| | | | |
| | | | |
| | | | |
| | | | |
| | | | |
| | | | |
| | | | |
| | | | |